Foundations of Quantitative Finance

Chapman & Hall/CRC Financial Mathematics Series

Aims and scope:

The field of financial mathematics forms an ever-expanding slice of the financial sector. This series aims to capture new developments and summarize what is known over the whole spectrum of this field. It will include a broad range of textbooks, reference works and handbooks that are meant to appeal to both academics and practitioners. The inclusion of numerical code and concrete real-world examples is highly encouraged.

Series Editors:
M.A.H. Dempster
Centre for Financial Research
Department of Pure Mathematics
and Statistics
University of Cambridge, UK

Rama Cont
Department of Mathematics
Imperial College, UK

Dilip B. Madan
Robert H. Smith School of Business
University of Maryland, USA

Robert A. Jarrow
Lynch Professor of Investment Management
Johnson Graduate School of Management
Cornell University, USA

Commodities: Fundamental Theory of Futures, Forwards, and Derivatives Pricing, Second Edition
M.A.H. Dempster, Ke Tang

Foundations of Quantitative Finance
Book I: Measure Spaces and Measurable Functions
Robert R. Reitano

Introducing Financial Mathematics: Theory, Binomial Models, and Applications
Mladen Victor Wickerhauser

Foundations of Quantitative Finance
Book II: Probability Spaces and Random Variables
Robert R. Reitano

Financial Mathematics: From Discrete to Continuous Time
Kevin J. Hastings

Financial Mathematics : A Comprehensive Treatment in Discrete Time
Giuseppe Campolieti and Roman N. Makarov

For more information about this series please visit: https://www.crcpress.com/Chapman-HallCRC-Financial-Mathematics-Series/book-series/CHFINANCMTH

Foundations of Quantitative Finance

Book II: Probability Spaces and Random Variables

Robert R. Reitano

Brandeis International Business School
Waltham, MA

CRC Press
Taylor & Francis Group
Boca Raton London New York

CRC Press is an imprint of the
Taylor & Francis Group, an **informa** business

A CHAPMAN & HALL BOOK

First edition published 2023
by CRC Press
6000 Broken Sound Parkway NW, Suite 300, Boca Raton, FL 33487-2742

and by CRC Press
4 Park Square, Milton Park, Abingdon, Oxon, OX14 4RN

CRC Press is an imprint of Taylor & Francis Group, LLC

Library of Congress Cataloging-in-Publication Data

Names: Reitano, Robert R., 1950- author.
Title: Foundations of quantitative finance. Book II, Probability spaces and random variables / Robert R. Reitano.
Other titles: Probability spaces and random variables
Description: First edition. | Boca Raton : CRC Press, 2023. | Includes bibliographical references and index.
Identifiers: LCCN 2022025709 | ISBN 9781032197180 (hardback) | ISBN 9781032197173 (paperback) | ISBN 9781003260547 (ebook)
Subjects: LCSH: Finance--Mathematical models. | Probabilities. | Random variables.
Classification: LCC HG106 .R448 2023 | DDC 332.01/5195--dc23/eng/20220601
LC record available at https://lccn.loc.gov/2022025709

ISBN: 978-1-032-19718-0 (hbk)
ISBN: 978-1-032-19717-3 (pbk)
ISBN: 978-1-003-26054-7 (ebk)

DOI: 10.1201/9781003260547

Typeset in CMR10
by KnowledgeWorks Global Ltd.

Publisher's note: This book has been prepared from camera-ready copy provided by the authors.

to Dorothy and Domenic

Contents

Preface   xi

Author   xiii

Introduction   xv

1 Probability Spaces   **1**
1.1 Probability Theory: A Very Brief History 1
1.2 A Finite Measure Space with a "Story" 2
    1.2.1 Bond Loss Example 7
1.3 Some Probability Measures on $\mathbb{R}$ 11
    1.3.1 Measures from Discrete Probability Theory 11
    1.3.2 Measures from Continuous Probability Theory 16
    1.3.3 More General Probability Measures on $\mathbb{R}$ 20
1.4 Independent Events 20
    1.4.1 Independent Classes and Associated Sigma Algebras 24
1.5 Conditional Probability Measures 27
    1.5.1 Law of Total Probability 28
    1.5.2 Bayes' Theorem 32

2 Limit Theorems on Measurable Sets   **35**
2.1 Introduction to Limit Sets 35
2.2 The Borel-Cantelli Lemma 38
2.3 Kolmogorov's Zero-One Law 43

3 Random Variables and Distribution Functions   **47**
3.1 Introduction and Definitions 47
    3.1.1 Bond Loss Example (Continued) 50
3.2 "Inverse" of a Distribution Function 52
    3.2.1 Properties of F^* 54
    3.2.2 The Function F^{**} 60
3.3 Random Vectors and Joint Distribution Functions 62
    3.3.1 Marginal Distribution Functions 65
    3.3.2 Conditional Distribution Functions 67
3.4 Independent Random Variables 69
    3.4.1 Sigma Algebras Generated by R.V.s 70
    3.4.2 Independent Random Variables and Vectors 71
    3.4.3 Distribution Functions of Independent R.V.s 74
    3.4.4 Independence and Transformations 75

4   Probability Spaces and i.i.d. RVs   **77**
  4.1   Probability Space $(\mathcal{S}', \mathcal{E}', \mu')$ and i.i.d. $\{X_j\}_{j=1}^{N}$ 78
     4.1.1   First Construction: $(\mathcal{S}'_F, \mathcal{E}'_F, \mu'_F)$ 79
  4.2   Simulation of Random Variables - Theory   81
     4.2.1   Distributional Results   81
     4.2.2   Independence Results   86
     4.2.3   Second Construction: $(\mathcal{S}'_U, \mathcal{E}'_U, \mu'_U)$ 88
  4.3   An Alternate Construction for Discrete Random Variables   91
     4.3.1   Third Construction: $(\mathcal{S}'_p, \mathcal{E}'_p, \mu'_p)$ 93

5   Limit Theorems for RV Sequences   **99**
  5.1   Two Limit Theorems for Binomial Sequences   99
     5.1.1   The Weak Law of Large Numbers   100
     5.1.2   The Strong Law of Large Numbers   103
     5.1.3   Strong Laws versus Weak Laws   108
  5.2   Convergence of Random Variables 1   108
     5.2.1   Notions of Convergence   109
     5.2.2   Convergence Relationships   111
     5.2.3   Slutsky's Theorem   115
     5.2.4   Kolmogorov's Zero-One Law   118

6   Distribution Functions and Borel Measures   **123**
  6.1   Distribution Functions on $\mathbb{R}$ 125
     6.1.1   Probability Measures from Distribution Functions   126
     6.1.2   Random Variables from Distribution Functions   129
  6.2   Distribution Functions on $\mathbb{R}^n$ 130
     6.2.1   Probability Measures from Distribution Functions   131
     6.2.2   Random Vectors from Distribution Functions   135
     6.2.3   Marginal and Conditional Distribution Functions   136

7   Copulas and Sklar's Theorem   **137**
  7.1   Fréchet Classes   137
  7.2   Copulas and Sklar's Theorem   140
     7.2.1   Identifying Copulas   144
  7.3   Partial Results on Sklar's Theorem   145
  7.4   Examples of Copulas   149
     7.4.1   Archimedean Copulas   150
     7.4.2   Extreme Value Copulas   154
  7.5   General Result on Sklar's Theorem   158
     7.5.1   The Distributional Transform   160
     7.5.2   Sklar's Theorem - The General Case   164
  7.6   Tail Dependence and Copulas   165
     7.6.1   Bivariate Tail Dependence   165
     7.6.2   Multivariate Tail Dependence and Copulas   170
     7.6.3   Survival Functions and Copulas   173

8   Weak Convergence   **179**
  8.1   Definitions of Weak Convergence   180
  8.2   Properties of Weak Convergence   184
  8.3   Weak Convergence and Left Continuous Inverses   189
  8.4   Skorokhod's Representation Theorem   191

8.4.1   Mapping Theorem on $\mathbb{R}$ 192
8.5   Convergence of Random Variables 2 194
    8.5.1   Mann-Wald Theorem on $\mathbb{R}$ 194
    8.5.2   The Delta-Method 195

9   Estimating Tail Events 1   201
9.1   Large Deviation Theory 1 202
9.2   Extreme Value Theory 1   206
    9.2.1   Introduction and Examples 206
    9.2.2   Extreme Value Distributions 210
    9.2.3   The Fisher-Tippett-Gnedenko Theorem 212
9.3   The Pickands-Balkema-de Haan Theorem 223
    9.3.1   Quantile Estimation 223
    9.3.2   Tail Probability Estimation 224
9.4 γ in Theory: von Mises' Condition 229
9.5   Independence vs. Tail Independence 234
9.6   Multivariate Extreme Value Theory   235
    9.6.1   Multivariate Fisher-Tippett-Gnedenko Theorem 236
    9.6.2   The Extreme Value Distribution G 238
    9.6.3   The Extreme Value Copula C_G 241

References   249

Index   253

Preface

The idea for a reference book on the mathematical foundations of quantitative finance has been with me throughout my professional and academic careers in this field, but the commitment to finally write it didn't materialize until completing my first "introductory" book in 2010.

My original academic studies were in "pure" mathematics in a subfield of mathematical analysis, and neither applications generally nor finance in particular were even on my mind. But on completion of my degree, I decided to temporarily investigate a career in applied math, becoming an actuary, and in short order became enamored with mathematical applications in finance.

One of my first inquiries was into better understanding yield curve risk management, ultimately introducing the notion of partial durations and related immunization strategies. This experience led me to recognize the power of greater precision in the mathematical specification and solution of even an age-old problem. From there my commitment to mathematical finance was complete, and my temporary investigation into this field became permanent.

In my personal studies, I found that there were a great many books in finance that focused on markets, instruments, models and strategies, and which typically provided an informal acknowledgment of the background mathematics. There were also many books in mathematical finance focusing on more advanced mathematical models and methods, and typically written at a level of mathematical sophistication requiring a reader to have significant formal training and the time and motivation to derive omitted details.

The challenge of acquiring expertise is compounded by the fact that the field of quantitative finance utilizes advanced mathematical theories and models from a number of fields. While there are many good references on any of these topics, most are again written at a level beyond many students, practitioners and even researchers of quantitative finance. Such books develop materials with an eye to comprehensiveness in the given subject matter, rather than with an eye toward efficiently curating and developing the theories needed for applications in quantitative finance.

Thus the overriding goal I have for this collection of books is to provide a complete and detailed development of the many foundational mathematical theories and results one finds referenced in popular resources in finance and quantitative finance. The included topics have been curated from a vast mathematics and finance literature for the express purpose of supporting applications in quantitative finance.

I originally budgeted 700 pages per book, in two volumes. It soon became obvious this was too limiting, and two volumes ultimately turned into ten. In the end, each book was dedicated to a specific area of mathematics or probability theory, with a variety of applications to finance that are relevant to the needs of financial mathematicians.

My target readers are students, practitioners and researchers in finance who are quantitatively literate, and recognize the need for the materials and formal developments presented. My hope is that the approach taken in these books will motivate readers to navigate these details and master these materials.

Most importantly for a reference work, all ten volumes are extensively self-referenced. The reader can enter the collection at any point of interest, and then using the references cited, work backwards to prior books to fill in needed details. This approach also works for a course on a given volume's subject matter, with earlier books used for reference, and for both course-based and self-study approaches to sequential studies.

The reader will find that the developments herein are at a much greater level of detail than most advanced quantitative finance books. Such developments are of necessity typically longer, more meticulously reasoned, and therefore can be more demanding on the reader. Thus before committing to a detailed line-by-line study of a given result, it is always more efficient to first scan the derivation once or twice to better understand the overall logic flow.

I hope the additional details presented will support your journey to better understanding.

I am grateful for the support of my family: Lisa, Michael, David, and Jeffrey, as well as the support of friends and colleagues at Brandeis International Business School.

Robert R. Reitano

Brandeis International Business School

Author

Robert R. Reitano is Professor of the Practice of Finance at the Brandeis International Business School where he specializes in risk management and quantitative finance. He previously served as MSF Program Director and Senior Academic Director. He has a PhD in mathematics from MIT, is a fellow of the Society of Actuaries, and is a Chartered Enterprise Risk Analyst. Dr. Reitano consults in investment strategy and asset/liability risk management, and previously had a 29-year career at John Hancock/Manulife in investment strategy and asset/liability management, advancing to Executive Vice President and Chief Investment Strategist. His research papers have appeared in a number of journals and have won both the Annual Prize of the Society of Actuaries as well as two F.M. Redington Prizes of the Investment Section of the Society of the Actuaries. Dr. Reitano serves on various not-for-profit boards and investment committees.

Introduction

Foundations of Quantitative Finance is structured as follows:

Book I: *Measure Spaces and Measurable Functions*

Book II: *Probability Spaces and Random Variables*

Book III: *The Integrals of Riemann, Lebesgue and (Riemann-)Stieltjes*

Book IV: *Distribution Functions and Expectations*

Book V: *General Measure and Integration Theory*

Book VI: *Densities, Transformed Distributions, and Limit Theorems*

Book VII: *Brownian Motion and Other Stochastic Processes*

Book VIII: *Itô Integration and Stochastic Calculus 1*

Book IX: *Stochastic Calculus 2 and Stochastic Differential Equations*

Book X: *Classical Models and Applications in Finance*

The series is logically sequential. Books I, III, and V develop foundational mathematical results needed for the probability theory and finance applications of Books II, IV, and VI, respectively. Then Books VII, VIII, and IX develop results in the theory of stochastic processes. While these latter three books introduce ideas from finance as appropriate, the final realization of the applications of these stochastic models to finance is deferred to Book X.

This Book II, *Probability Spaces and Random Variables*, both applies the foundational materials on measure spaces developed in Book I, and sets the stage for probability theory and finance applications in future books. Probability theory is often thought of in terms of its discrete or continuous models. It was Andrey Kolmogorov (1903–1987) who first recognized that measure theory provided the foundational basis for probability theory, and that these other models where merely special cases of this general framework.

After providing a brief history of the development of probability theory, Chapter 1 introduces this general probability space framework, noting that a probability space $(\mathcal{S}, \mathcal{E}, \mu)$ is "simply" a finite measure space with $\mu(\mathcal{S}) = 1$. The development both echoes and extends the measure space ideas of Book I to include ideas unique to probability theory, such as independent events and conditional probability measures. Popular examples from discrete and continuous probability theory are presented, and these theories are seen to be special cases within the general measure theoretic framework.

Chapter 2 then develops what is arguably the most foundational limit theorem on the limit superior of a sequence of measurable sets, the remarkably useful Borel-Cantelli lemma. This result provides tangible criteria for determining when such a limiting set has the critical probability of 0 or 1. In applications of this result throughout later books, there will often be no other way to address such questions.

Also derived is Kolmogorov's important result on limit sets, known as the Kolmogorov 0-1 law. This law applies to tail events, which are members of the tail sigma algebra, and states that such events must have probability 0 or 1. It generalizes the Borel 0-1 law, a corollary to the above lemma, since the limit superior is seen to be a tail event.

Random variables, which are "simply" measurable functions defined on probability spaces, provide the unifying theme of Chapters 3 to 5. Chapter 3 introduces random variables and random vectors and the associated distribution functions including, in the vector case, the variously defined marginal and conditional distribution functions.

The "left continuous" inverse of a distribution function is studied, and will be seen to be critical in generating samples of random variables in Chapter 4, in the development for copulas in Chapter 7, and in the study of extreme value theory in Chapter 9. Also included is a study of independent random variables and sigma algebras, and associated results.

Applying the infinite dimensional product measure space construction of Chapter 9 of Book I, Chapter 4 develops theoretical frameworks for collections of independent, identically distributed (i.i.d.) random variables using three constructions. The first construction implements the Book I approach directly, using the Borel measure induced by the distribution function of the given random variable X.

After developing ideas underlying stochastic simulations, the second construction for i.i.d. random variables obtains an infinite dimensional space using the Borel measure induced by the uniform distribution. The associated independent random variables distributed as X are then defined with the left continuous inverse of the distribution function of X. The third construction is a modification of the first in the case where X is discrete, and thus the original real probability space can be greatly simplified. Both the second and third constructions modify the Book I development by reducing the original space with essential subsets.

Chapter 5 then turns to limit theorems for random variable sequences, a topic that will be extended later in this book, as well as in other books as more tools are introduced. Initially focusing on binomial random variable sequences, the associated weak law of large numbers, known as Bernoulli's theorem, and the strong law of large numbers, known as Borel's theorem, are derived.

The chapter then initiates the more general study of modes of convergence of random variables, investigating convergence in probability, convergence with probability 1, and convergence in distribution. The traditional version of Slutsky's theorem, to be generalized in Book VI, and Kolmogorov's zero-one law on tail events, are derived.

The focus of Chapter 6 is to investigate and rationalize the many notions of "distribution function." In Book I, such functions are seen to be characterized by certain properties, and are both induced by Borel measures and can be used to create Borel measures. In earlier chapters of the current book, distribution functions are defined by random variables and can also induce random variables.

Starting with distribution functions on $\mathbb{R}$, this chapter rationalizes these perspectives, connecting functions with given properties, functions induced by random variables, and functions induced by Borel measures, in various ways. This investigation is then generalized to distribution functions on $\mathbb{R}^n$.

Chapter 7 then develops the theory of copulas, which are "simply" joint distribution functions with uniform marginal distribution functions. This theory has its mathematical roots in Book I's Chapter 8 on general Borel measures on $\mathbb{R}^n$. The goal of the chapter is the derivation of Sklar's theorem. It states that given any joint distribution function, there exists a copula, which when valued on that distribution's marginal distributions, reproduces the original joint distribution.

Put another way, Sklar's result states that given only the marginal distributions, all the complexity of the original joint distribution can be reproduced with the right copula. Since marginal distributions are relatively easy to estimate, it is no surprise that this theory has found many applications in finance and elsewhere. Many examples of copulas are illustrated, and the theory of extreme value copulas is introduced. These results are then extended to results on tail dependence, where survival copulas are introduced to simplify the analysis.

Critical to the final chapter's study of extreme value theory, Chapter 8 introduces and develops some fundamental results on weak convergence of distribution functions, and more generally increasing functions. Properties of weak convergence are developed, including Helly's selection theorem and Prokhorov's theorem, and then results are obtained between weak convergence of distribution functions and weak convergence of the left continuous inverses.

These results provide the tools for Skorokhod's representation theorem, and the mapping theorem on $\mathbb{R}$. Turning back to the random variable point of view, these results apply to prove the Mann-Wald theorem, which addresses when various modes of random variable convergence are preserved under measurable transformations, and a proof of the delta method.

Finally, Chapter 9 returns to the study of tail events. The first section is dedicated to large deviation theory, which studies tail probabilities related to the average of n independent random variables. It investigates estimates of the type developed for the Chapter 5 study of the binomial laws of large numbers.

The second and larger investigation is on extreme value theory, which studies the limiting distribution of the maximum of n independent random variables. The entire investigation is facilitated by the use of the associated left continuous inverse functions, and the crowning result is the Fisher-Tippett-Gnedenko theorem, which characterizes all such limiting distributions. This result is recast and applied in the form of another key result, the Pickands-Balkema-de Haan theorem, which provides more explicit results on tail probabilities.

The final section develops multivariate extreme value theory, utilizing the Chapter 7 results on extreme value copulas and the Fisher-Tippett-Gnedenko theorem for the marginal distributions. This investigation will be continued in Book IV with the tools of Book III.

I hope this book and the other books in the collection serve you well.

Notation 0.1 (Referencing within FQF Series) *To simplify the referencing of results from other books in this series, we will use the following convention.*

A reference such as "Proposition I.3.33" is a reference to Proposition 3.33 of Book I, while Chapter III.4 is a reference to Chapter 4 of Book III, and so forth.

1

Probability Spaces

1.1 Probability Theory: A Very Brief History

In this section we provide a hundred thousand foot view of the historical development of this subject. The development of these theories required the contributions of numerous mathematicians, and in the light touch of this section we can neither identify all of the key contributors to these developments, nor identify all the key contributions of those identified.

Historically, **probability theory** seems to have begun with investigations into "games of chance," the traditional framework for many models and applications in what is now called "discrete" probability theory. One early study was done by **Gerolamo Cardano** (1501–1576), a mathematician and gambler who wrote one of the first books on the subject, *Liber de ludo aleae* ("Book on Games of Chance"). Though written in the early 1500s, it was not published until almost 100 years after his death.

The approach to "probabilities" at that time has come to be known as the "classical interpretation." The probability of an outcome with a given property, or "event," is defined as the ratio of the number of possible outcomes which possess the given property, to the total number of all outcomes possible. This definition was deemed applicable when the totality of all outcomes is finite, and all outcomes can be argued to be "equally likely" to occur. For example, the probability of 2 heads and 1 tail in 3 flips of a "fair" coin is $3/8$, since there are 8 possible outcomes in the sequence of 3 flips, and there are 3 outcomes with the desired property. Because it is a fair coin, each triplet of H/Ts could be argued to be equally likely.

This model well fits many games of chance: coins, dice, playing cards, etc., where the total number of outcomes is finite, and one can generally defend the "equally likely" assumption. But this interpretation does not work when it is apparent that not all outcomes are equally likely. For example, in the coin toss model above, if the coin is weighted so that H appears twice as often as T, the counting of outcomes does not provide a correct answer. Moreover, there is some circularity of logic in this approach, in that we are defining the probability of an event in cases where we deem the "likelihood" of all outcomes to be equal. Indeed, what can "likelihood" mean if not "probability."

This classical interpretation evolved in the 1700s into the "frequentist interpretation" of probability theory. In this interpretation, the probability of an event is defined in terms of a limit of ratios or proportions, as the number of possible outcomes of an experiment increases without limit. This ratio is defined as in the classical theory, in terms of the number of outcomes with the given property as a proportion of the total number of outcomes. This approach avoids the circularity of logic of the classical approach, since it is entirely observational in concept, and no assumption is needed on the relative likelihoods of outcomes. It also provides a definition without limiting the totality of outcomes to be finite.

For example, to say that the probability of a head on a flip of a fair coin is $1/2$ means that as the number of flips increases without bound, the proportion of heads will "converge" to $1/2$, being informal for now about what is meant by such convergence. That this should occur is intuitively appealing but requires a more formal statement and proof which addresses both the calculation of flip sequence probabilities, and the meaning of convergence. Certainly not all such infinite sequences will have this property, and indeed there will be one with all Hs, and some with 90% Hs, etc.

Notable contributors in this evolution were **Jacob Bernoulli** (1655–1705) with his *Ars Conjectandi* ("The Art of Conjecturing"), published posthumously in 1713; and **Abraham de Moivre** (1667–1754) with *The Doctrine of Chances* in 1718. Bernoulli proved an early version of the law of large numbers, while De Moivre derived a special case of the central limit theorem, applying it to approximating binomial probabilities. In other work he also applied this probability framework to the development of actuarial tables which could be used to estimate the costs of providing life insurance and other life-contingent benefits.

Another key contributor was **Pierre-Simon Laplace** (1749–1827) with *Théorie analytique des probabilités* ("Analytical Theory of Probability") in 1812, introducing the moment generating function and the method of least squares, and applying calculus to probability theory.

The next major break-through required a measure theory, the study of which began with the work of **Henri Lebesgue** (1875–1941) and the development of the Lebesgue approach to integration. This approach was introduced in Book I and will be further developed in Book III. Lebesgue's work followed the formal development by **Bernhard Riemann** (1826–1866) of the Riemann approach to integration. Using his newly developed theory, Lebesgue proved that a bounded function $f(x)$ is Riemann integrable on an interval $[a, b]$ if and only if it is continuous "almost everywhere," which is to say, except on a set of Lebesgue measure 0. Also critical for such investigations was the formalization of the notion of "limit" by **Augustin-Louis Cauchy** (1789–1857).

The convergence of probability theory and measure theory occurred first in the work of **Andrey Kolmogorov** (1903–1987) in his *Foundations of the Theory of Probability* in 1933. This work introduced an axiomatic framework for probability theory that both merged and generalized the discrete and continuous models of the theory. Kolmogorov's approach is central to all modern developments of this theory.

1.2   A Finite Measure Space with a "Story"

A sometimes confounding reality to the aspiring student of mathematical finance is that while probability theory can be seen as an application of measure theory, or, measure theory seen as a generalization of probability theory, in practice these theories can look quite distinct because of the notational conventions that have become standardized in the literature. We begin with a formal definition of a **probability space**. It will look familiar to the reader of Book I.

Definition 1.1 (Probability space and measure) *A **probability space** is a triplet, $(\mathcal{S}, \mathcal{E}, \mu)$, which identifies a **sample space** $\mathcal{S}$, a collection of subsets called **events** which form a **sigma algebra** $\mathcal{E}$, and a **set function**:*

$$\mu : \mathcal{E} \to [0, 1],$$

*which is a **probability measure**.*

That $\mathcal{E}$ is a sigma algebra means, by Definition I.2.5, that:

1. $\emptyset, \mathcal{S} \in \mathcal{E}$;

2. *If $A \in \mathcal{E}$, then $\widetilde{A} \in \mathcal{E}$, where $\widetilde{A}$ denotes the **complement** of A:*

$$\widetilde{A} \equiv \{s \in \mathcal{S} | s \notin A\};$$

3. *If $A_j \in \mathcal{E}$ for $j = 1, 2, 3...$, then $\bigcup_j A_j \in \mathcal{E}$.*

That μ is a probability measure means that:

4. $\mu(\mathcal{S}) = 1$;

5. *If $A \in \mathcal{E}$, then $\mu(A) \geq 0$ and $\mu(\widetilde{A}) = 1 - \mu(A)$.*

6. *If $A_j \in \mathcal{E}$ for $j = 1, 2, 3...$ are **mutually exclusive events**, also called **disjoint events** or **pairwise disjoint events**, that is:*

$$A_j \bigcap A_k = \emptyset \text{ for } j \neq k,$$

*then μ is **countably additive**:*

$$\mu\left(\bigcup_j A_j\right) = \sum_j \mu(A_j). \tag{1.1}$$

*An event $A \in \mathcal{E}$ is called a **null event under** μ if $\mu(A) = 0$. If A is a null event and every $A' \subset A$ satisfies $A' \in \mathcal{E}$, then the triplet, $(\mathcal{S}, \mathcal{E}, \mu)$, is called a **complete probability space**.*

Remark 1.2

1. *In the definition of a complete probability space, note that when $A' \in \mathcal{E}$, of necessity $\mu(A') = \mu(A - A') = 0$. This is because $A' \subset A$ implies by 1.1 that:*

$$\mu(A) = \mu(A') + \mu(A - A'),$$

and the result follows since μ is non-negative valued.

2. *The notation used for the triplet that defines a probability space is not standardized, and different texts will invariably use different conventions for sample spaces, sigma algebras, events and probability measures. Inevitably, this will also happen in this series of books.*

Thus a probability space is simply a **measure space** by Definition I.2.23, which is **finite** with $\mu(\mathcal{S}) = 1$. Similarly, a complete probability space is simply a **complete finite measure space** by Definition I.2.48. But these measure spaces are "special" in the sense that they are typically mathematical representations of some real-world phenomenon or "experiment" for which the outcome is uncertain. The collections of these outcomes, the sigma algebra of events, are assumed to occur with given likelihoods or "probabilities," based on observation and/or a reasonable mathematical model. In this sense, every probability space has a "story" associated with it that identifies the events related to this real-world phenomenon.

The collection of events $\mathcal{E}$ represents the collection of subsets of the sample space on which the probability measure μ is defined and on which it specifies probabilities. To be useful, this collection must satisfy the properties of a sigma algebra because it logically must contain the "empty event" $\emptyset$ and the "certain event" $\mathcal{S}$, as well as any events defined as the complement of other events, or defined as finite or countable unions of other events.

By **De Morgan's laws** of Exercise I.2.2, and named for **Augustus De Morgan** (1806–1871), it is then also the case that the intersection of finitely or countably many events is an event. That is, $\bigcap_j A_j \in \mathcal{E}$ since:

$$\bigcap_j A_j = \left(\widetilde{\bigcup_j \tilde{A}_j}\right).$$

Similarly, if $A, B \in \mathcal{E}$, then the **set difference** $A - B \in \mathcal{E}$:

$$A - B \equiv A \bigcap \tilde{B}.$$

A probability measure is simply a measure by Definition I.2.23, that specifies probabilities of events, and logically must satisfy the properties that define a measure. For example, the empty event should logically have probability 0, the certain event probability 1, and otherwise events are assigned probabilities in a way that is consistent with sigma algebra and measure manipulations. For example, since for any $A \in \mathcal{E}$ it is the case that $A \bigcup \tilde{A} = \mathcal{S}$ and these sets are disjoint, item 5 is logically compelled by finite additivity. One similarly justifies the usefulness and consistency with applications of requiring probability measures to be countably additive over disjoint events.

Remark 1.3 (On probability spaces) *In "discrete" sample spaces, which contain finitely or countably many sample points, $\mathcal{E}$ almost always contains each of the sample points and hence all subsets of $\mathcal{S}$. Consequently discrete probability spaces are always complete. In a general sample space that is uncountably infinite, the collection of events will virtually always be a proper subset of the collection of all subsets in the same way that the sigma algebra $\mathcal{M}_L$ of Lebesgue measurable sets on $\mathbb{R}$ is a proper subset of the power set $\sigma(P(\mathbb{R}))$. See Section I.2.4.*

*Given the generality of the sigma algebra structure of the collection of events, it is clear that on any given sample space, any number of sigma algebras can be defined. The same statement applies to probability measures. For example, there might be two sigma algebras $\mathcal{E}$ and $\mathcal{E}'$ with $\mathcal{E} \subset \mathcal{E}'$, so every event in $\mathcal{E}$ is an event in $\mathcal{E}'$. In that sense, $\mathcal{E}'$ is a "**finer**" sigma algebra because it contains more events, and $\mathcal{E}$ a "**coarser**" sigma algebra because it contains fewer events. It can also be the case that there are two sigma algebras where neither $\mathcal{E} \subset \mathcal{E}'$ nor $\mathcal{E}' \subset \mathcal{E}$ is true.*

Some questions that arise in probability spaces include the following, where we note the answers already developed from Book I.

1. ***Define $\mathcal{E}$ generated by a collection of sets $\mathcal{D}$:** If $\mathcal{D}$ is a given collection of subsets of $\mathcal{S}$, is there a sigma algebra $\mathcal{E}$ that contains the sets in $\mathcal{D}$, meaning $\mathcal{D} \subset \mathcal{E}$? As one example, if $\mathcal{E}$ and $\mathcal{E}'$ are two sigma algebras on $\mathcal{S}$, is there a sigma algebra, $\mathcal{E}''$ so that $\mathcal{E} \cup \mathcal{E}' \subset \mathcal{E}''$? For another example, if $f : \mathcal{S} \to \mathbb{R}$ is a given function, is there a sigma algebra that contains all sets of the form, $f^{-1}[(a, b)]$ for all open intervals $(a, b) \subset \mathbb{R}$?*

   The answer to the existence question is "yes" as there is always the power sigma algebra $\sigma(P(\mathcal{S}))$. But this is not a useful answer because in addition to a sigma algebra of events, a given application will also require a probability measure on this sigma algebra. Consequently, given the difficulty and sometimes impossibility of defining a measure on the power sigma algebra, one is typically less ambitious and may instead ask, is there a smallest sigma algebra which includes the collection $\mathcal{D}$? Because intersections of sigma algebras create sigma algebras, the smallest such sigma algebra is simply the intersection of all sigma algebras which contain the sets of $\mathcal{D}$. See Example I.2.7 and Proposition I.2.8.

However, before defining something as the "intersection of all sigma algebras which possess a given property," it is critical that it be verified that this collection of sigma algebras is not empty. If this collection is empty this would render this construction meaningless. But in all cases the collection of sigma algebras is always non-empty since it includes the power sigma algebra, and so the existence of a smallest sigma algebra $\mathcal{E}$ implied by this construction is always valid.

2. **Define μ induced by a set function on a collection $\mathcal{D}$:** *If $\mathcal{D}$ is a given collection of subsets of $\mathcal{S}$, and μ a set function defined on $\mathcal{D}$ which satisfies properties of a measure, can the definition of μ be extended to a sigma algebra $\mathcal{E}$ for which $\mathcal{D} \subset \mathcal{E}$? If so, is such an extension unique?*

*This is a more difficult question than 1, but was also addressed in Book I which developed constructions of this type in the context of Lebesgue and Borel measures and then generalized this process. In the Lebesgue case of Chapter I.2, $\mathcal{D}$ was the collection of open intervals and a set function $|A|$ was defined on these sets to equal the interval length of A. In the Borel case of Chapter I.5, $\mathcal{D}$ was a semi-algebra of right semi-closed intervals and a generalized interval length set function $|A|_F$ used. So $|(a,b)| \equiv b-a$ for Lebesgue measure and $|(a,b]|_F \equiv F(b)-F(a)$ for Borel measures for appropriately restricted functions F. These set functions were used as the basis for defining outer measures applicable to all sets in $\sigma(P(\mathbb{R}))$, respectively m^* and $\mu_{\mathcal{A}}^*$, where $\mathcal{A}$ denoted the algebra generated by the semi-algebra. Sigma algebras were then selected based on a criterion called **Carathéodory measurability** in Definitions I.2.33 and I.5.18, respectively.*

In the case of Lebesgue outer measure m^*, the Carathéodory measurable sets were seen to form a complete sigma algebra $\mathcal{M}_L$, and m^* was then seen to be a measure on this sigma algebra. This sigma algebra contained the Borel sigma algebra $\mathcal{B}(\mathbb{R})$, which is the smallest sigma algebra that contains the open intervals. The Borel measure development was analogous in that Carathéodory measurable sets were again seen to form a complete sigma algebra $\mathcal{M}_{\mu_F}(\mathbb{R})$, which contained $\mathcal{B}(\mathbb{R})$.

*The general extension of an initial set function or pre-measure on a semi-algebra to a measure on a complete sigma algebra that contains this semi-algebra was developed in Chapter I.6. These results were summarized in the **Carathéodory Extension theorems** of Propositions I.6.2 and I.6.13 and named for **Constantin Carathéodory (1873–1950)**, as well as the **Hahn–Kolmogorov theorem** of Proposition I.6.4 and named for **Hans Hahn (1879–1934)** and **Andrey Kolmogorov (1903–1987)**.*

These extension theorems are key in measure theory, and specifically in probability theory because it is usually easy to prescribe probabilities of simple sets like intervals. These theorems then identify conditions under which such definitions can be extended to a complete sigma algebra and thereby produce a probability space with the desired probability structure. Further, this extension was proved to be unique on the smallest sigma algebra which contains the original semi-algebra (Proposition I.6.14), and to the completion of this sigma algebra (Proposition I.6.24).

3. **Create the completion of a probability space:** *If $(\mathcal{S}, \mathcal{E}, \mu)$ is a probability space which is not complete, can $\mathcal{E}$ be expanded to a sigma algebra $\mathcal{E}'$ that is complete and hence contains all subsets of $\mathcal{E}$-null sets? And if so, can the definition of μ be expanded to $\mathcal{E}'$ without changing its values on $\mathcal{E}$?*

The answer to this question is "yes" as was demonstrated in Section I.6.5. In the construction of Lebesgue measurable sets $\mathcal{M}_L$, this issue did not arise because $\mathcal{M}_L$ already contained all sets of Lebesgue outer measure 0 and hence was complete. The *same was true for the Borel measure spaces $\mathcal{M}_{\mu_F}(\mathbb{R})$ of Chapter I.5 as well as all constructions under the extension theorems of Chapter I.6 which utilize an outer measure approach.*

However, the Borel sigma algebra $\mathcal{B}(\mathbb{R})$ is not complete under either Lebesgue or Borel measures. The completion theorem of Proposition I.6.20 allows one to always assume that a sigma algebra is complete, or has been replaced by its "completion," knowing that the measures of the original sets will not be changed in the process.

The significance of ensuring that a sigma algebra is complete is that one does not then need to worry about the effect of sets of measure 0 on the measurability of functions. As proved in Proposition I.3.16 (by way of Exercise I.3.15), if $f : (\mathcal{S}, \mathcal{E}, \mu) \to \mathbb{R}$ is a measurable function with $\mathcal{E}$ complete, and $g(x) = f(x)$ except on a set of measure 0, then $g(x)$ is also measurable.

Example 1.4 (Product sample space) *Given a probability space $(\mathcal{S}, \mathcal{E}, \mu)$, define the n-trial sample space, denoted $\mathcal{S}^n$, by:*

$$\mathcal{S}^n = \{(s_1, s_2, ..., s_n) | s_j \in \mathcal{S}\}.$$

Intuitively, each point of $\mathcal{S}^n$ represents an "n-sample" of points from the sample space $\mathcal{S}$, though formally this is an example of the construction in Chapter I.7. There we constructed a complete sigma algebra of events $\mathcal{E}^n$ and probability measure μ_n, as follows.

For $\mathcal{E}^n$ we began with a collection of sets $\mathcal{A}'$ in $\mathcal{S}^n$ of the form $\prod_{j=1}^{n} A_j$, where $A_j \in \mathcal{E}$, and on such sets defined a set function μ_0 by:

$$\mu_0 \left(\prod_{j=1}^{n} A_j \right) = \prod_{j=1}^{n} \mu(A_j). \tag{1}$$

This collection of sets is a semi-algebra by Proposition I.7.2, and applying the extension process of Chapter I.6, the measure space $(\mathcal{S}^n, \mathcal{E}^n, \mu_n)$ is finally constructed in Proposition I.7.20. The punchline is that $\mathcal{A}' \subset \mathcal{E}^n$, and on $\mathcal{A}'$:

$$\mu_n = \mu_0. \tag{2}$$

This identity is also valid on the algebra $\mathcal{A}$ generated by $\mathcal{A}'$.

Now define a collection of sets $\{B_k\}_{k=1}^{n} \subset \mathcal{A}'$ by:

$$B_k = \prod_{j=1}^{n} A_{jk},$$

*where $A_{jk} \in \mathcal{E}$ for all j, k, and $A_{jk} = \mathcal{S}$ when $j \neq k$. This is a collection of **independent events** by Definition 1.15. This means as in (1.26) that for every subset $K \subset \{1, 2, ..., n\}$ with $K \equiv \{k_1, k_2, ... k_m\}$:*

$$\mu_n \left(\bigcap_{i=1}^{m} B_{k_i} \right) = \prod_{i=1}^{m} \mu_n (B_{k_i}). \tag{3}$$

To see this, note that:

$$\bigcap_{i=1}^{m} B_{k_i} = \prod_{k=1}^{n} A'_k,$$

where $A'_k = A_{kk}$ for $k \in K$, and $A'_k = \mathcal{S}$ otherwise. Thus by (1) and (2):

$$\mu_n \left(\bigcap_{i=1}^{m} B_{k_i} \right) = \prod_{i=1}^{m} \mu(A_{k_i k_i}) = \prod_{i=1}^{m} \mu_n (B_{k_i}).$$

Now let X be a random variable on $(\mathcal{S}, \mathcal{E}, \mu)$, which will simply mean that $X : \mathcal{S} \to \mathbb{R}$ is measurable in the sense that $X^{-1}(A) \in \mathcal{E}$ for every Borel $A \in \mathcal{B}(\mathbb{R})$. Then random variables

$\{X_j\}_{j=1}^n$ *can be defined on* $(\mathcal{S}^n, \mathcal{E}^n, \mu_n)$, *each with the same distribution function as* X, *and such that these random variables are* **independent** *by Definition 3.47.*

All of the above comments will remain true when $n = \infty$, *by replacing the construction of Chapter I.7 with that of Chapter I.9.*

In Chapter 4, we formalize the notion of generating "random" samples of a random variable.

1.2.1   Bond Loss Example

We develop a finance example below to demonstrate that the development of a formal probability space can be subtle.

Example 1.5 (Probability space of bond defaults) *Let:*

$$\mathcal{P} \equiv \{B_1, B_2, ..., B_N\},$$

denote a bond or loan portfolio where each bond B_j *has a loan amount* f_j, *and a risk class which might be defined in terms of third-party credit ratings for bonds, or internal risk assessment criteria for other types of loans. For simplicity we refer to these as bonds, and for each, denote the probability of default in one year by* $q(B_j)$. *The probability of receiving scheduled repayments during the year is then* $1 - q(B_j)$, *and sometimes denoted* $p(B_j)$. *Our goal is to produce a probability space of all possible outcomes of portfolio defaults in one year.*

1. *(**A First Attempt and Failure**) We begin with a natural model of defining the sample space* $\mathcal{S} = \mathcal{P}$ *as the collection of bonds, and the sigma algebra* $\mathcal{E}$ *as the power sigma algebra* $\mathcal{E} = \sigma(P(\mathcal{S}))$. *That is,* $\mathcal{E}$ *contains every subset of* $\mathcal{P}$ *and the empty set* $\emptyset$. *This choice reflects the observation that any subset of* $\mathcal{S}$ *has the potential of defaulting, including the empty set, and hence every subset is potentially an "event."*

   The probability measure μ *is defined on* $\mathcal{E}$ *for any subset of* $\mathcal{S}$ *as the probability of that subset defaulting in one year, and all other bonds making scheduled payments. So if* $J \subset \{1, 2, ..., N\}$ *is a subset and the event* A *is given as:*

   $$A = \{B_j | j \in J\},$$

   define:

   $$\mu(A) \equiv \prod_{j \in J} q(B_j) \prod_{j \in J'} (1 - q(B_j)). \tag{1}$$

   Here J' *denotes all indexes not contained in* J. *Here and below we explicitly allow* $J = \emptyset$.

   The reader may note that the formula for μ *reflects the assumption that bond defaults are "independent." We will develop this notion more formally later, but for now intuition suffices. If we had* N *coins, each with probability of heads* $q(B_j)$, *then* $\mu(A)$ *is precisely the probability of obtaining heads only on the jth flips where* $j \in J$.

   While this formula assigns a probability to each event in $\mathcal{E}$, *the resulting structure* $(\mathcal{S}, \mathcal{E}, \mu)$ *is not a probability space. The problem with this construction is not* $\mathcal{E}$, *which is obviously a sigma algebra, but with* μ, *which is easily seen to not be a probability measure on* $\mathcal{E}$:

   (a) *If* $J = \emptyset$, *then* $A = \emptyset$ *but* $\mu(A) = \prod_{j=1}^{N}(1 - q(B_j)) \neq 0$ *unless at least one* $q(B_j) = 1$. *Of course we can attempt to fix this by simply declaring the* above μ formula to only apply to $J \neq \emptyset$, and then for $J = \emptyset$ and $A = \emptyset$ we **define** $\mu(A) = 0$. Though not an aesthetically pleasing approach, this problem could be fixable with a definitional override, though definitional consistency questions then emerge.

(b) If $J = \{1, 2, ..., N\}$ then $A = \mathcal{S}$ and $\mu(A) = \prod_{j=1}^{N} q(B_j) \neq 1$ unless all $q(B_j) = 1$. Again, we could resort to a definitional override.

(c) Parts a and b disguise a bigger problem. That is, for A defined above, if:

$$\widetilde{A} \equiv \{B_j | j \in J'\},$$

then as can be verified:

$$\mu(\widetilde{A}) \equiv \prod_{j \in J'} q(B_j) \prod_{j \in J} (1 - q(B_j)) \neq 1 - \mu(A).$$

(d) Finally, generalizing part c, μ is not finitely additive on $\mathcal{E}$. For example, despite $\{B_1\}$ and $\{B_2\}$ being disjoint sets in $\sigma(P(\mathcal{S}))$, and $\{B_1\} \cup \{B_2\} = \{B_1, B_2\}$:

$$\mu(\{B_1, B_2\}) \neq \mu(\{B_1\}) + \mu(\{B_2\}).$$

2. **(Discussion)** What is needed is a more thoughtfully defined sample space, sigma algebra of events, and probability structure. To be sure, the probability assignments are correct in terms of quantifying the probability of a given collection of bonds defaulting. The question is, how can the sample space $\mathcal{S}$ and event space $\mathcal{E}$ be defined so that:

(a) the measure of $\mathcal{S}$ is 1;

(b) complementary events in $\mathcal{S}$ will be assigned complementary probabilities, thereby resolving the problem in 1.c above, and by 2.a also solving 1.a;

(c) disjoint events in $\mathcal{S}$ union to an event with the appropriate additive probability.

3. **(A Second Attempt and Success)** Given the bond portfolio $\mathcal{P}$, define the sample space $\mathcal{S}$ as the collection of all outcomes of default in one year, specified as N-tuples of bonds defined by all J and J'. Specifically,

$$\mathcal{S} = \{(B_J, B_{J'}) | J \subset \{1, 2, ..., N\}\},$$

where B_J denotes the defaulting subset, and $B_{J'}$ the complementary non-defaulting subset.

Thus $\mathcal{S}$ is the collection of all 2^N default outcomes for the portfolio. Define $\mathcal{E} = \sigma(P(\mathcal{S}))$, the power sigma algebra of subsets of $\mathcal{S}$, and on $\mathcal{E}$ define μ additively. First, if $A_i \in \mathcal{S}$, then define $\mu(A_i)$ by the formula in (1) above. If $A \in \mathcal{E}$, then by definition $A = \bigcup_i A_i$ for disjoint $\{A_i\} \subset \mathcal{S}$, and define:

$$\mu(A) = \sum_i \mu(A_i). \tag{2}$$

Then $\mu(\emptyset) = 0$ by this definition since the empty set produces a vacuous summation. Further:

$$\mu(\mathcal{S}) = \sum_{A \in \mathcal{S}} \mu(A) = 1,$$

since this is equivalent to:

$$\sum_J \left[\prod_{j \in J} q(B_j) \prod_{j \in J'} (1 - q(B_j)) \right] = 1,$$

which in turn follows from an expansion of

$$\prod_{j=1}^{N} (q(B_j) + [1 - q(B_j)]) = 1.$$

It is then also the case that for $A \in \mathcal{E}$ defined this way, $\widetilde{A}$ is the collection of default outcomes not in A, and thus $\mu(\widetilde{A}) = 1 - \mu(A)$ follows from $\mu(\mathcal{S}) = 1$. That μ is finitely additive over disjoint elements of $\mathcal{E}$ follows by the same argument.

With this construction, $(\mathcal{S}, \mathcal{E}, \mu)$ is now a probability space. Each event in $\mathcal{E}$ represents a subset of the various default outcomes that can occur in one year, while $\mu(A)$ for $A \in \mathcal{E}$ is the probability that one of these default outcomes is realized in the next year.

Remark 1.6 (On Example 1.5) *It can be argued that a lot of work went into defining a legitimate probability space in part 3, but that having done this, we did not gain any additional insights into the default probability model that was contemplated in part 1. Even without the formal measure-theoretical framework, most finance practitioners would have little problem specifying probabilities of events $A = \bigcup_i A_i$ for disjoint $\{A_i\} \subset \mathcal{S}$ defined in part 3, even without the formality offered there.*

If there is a lesson in the above example, it is in realizing that a given finance or other model may well be framed in the units of part 1, whereby we view a bond portfolio as the sample space and the possible default outcomes in one year as the sigma algebra. While workable intuitively, and also rigorously in defining probabilities, this framework is technically not equivalent to a probability space model. What was needed above was a transition from this "natural" model, to the "probability space" model of a collection of possible default outcomes as sample points with sets of outcomes as the sigma algebra.

In practice, the details of this construction are sometimes avoided and the basic model and probabilities specified as in part 1. However, with a little more effort it will always be true that there exists a probability space $(\mathcal{S}, \mathcal{E}, \mu)$ that is consistent with the specified probabilities.

Example 1.7 (Probability space of bond default losses) *As above, let $\mathcal{P} \equiv \{B_1, B_2, ..., B_N\}$ denote a bond or loan portfolio where each bond B_j has a loan amount f_j and probability of default $q(B_j)$ in one year. In this continuation of the above example we seek to recognize the various levels of dollar loss upon default, and so the goal is to produce a probability space which characterizes all possible loss outcomes due to portfolio defaults in one year.*

*The loss recognized by the lender on default is often called the **loss given default** (LGD) or **loss ratio** for each loan, denoted l_j, and specifies the proportion of the loan amount f_j lost by the lender given a default on loan B_j. Sometimes the **loan recovery rate** r_j is specified instead, representing the relative amount of f_j recovered by the lender given a default of loan B_j. In general, loss plus recovery need not equal 1.0, meaning $l_j + r_j = 1.0$.*

This is because the lender will accrue interest for a period after default, and therefore the amount owed will be $(1 + i)f_j$. For example, if $f_j = 100$ and $i = 0.02$ and the lender recovers 60, the recovery rate is $r_j = 0.6$, but the loss ratio is $l_j = 0.42$.

If it is hypothesized that loan B_j has a loss ratio of l_j on default, and that this LGD variable is constant, then the above probability space in part 3 is adequate. We simply define $\mathcal{S}' = \{(L_J, L_{J'}) | J \subset \{1, 2, ..., N\}\}$, where $L_J = \{l_j f_j\}_{j \in J}$, $L_{J'} = \{0\}_{j \in J'}$ denote the set of fixed dollar loss outcomes associated with a given default outcomes. The sigma algebra $\mathcal{E}'$ and μ are then defined as above.

If the loss ratio is not constant for any given bond B_j, then the sigma algebra defined by default outcomes is too course to identify all possible events. For example, the event defined by $J \equiv \{1\}$ in the constant LGD space is effectively the event that bond B_1 defaults. When LGD is not constant, this "event" is the union of events defined by the different values that LGD can achieve. Thus what is needed is a refinement of $\mathcal{S}$ and $\mathcal{E}$ which will in effect split each of the default events into finer events which identify the loss levels potentially incurred by the defaulted bonds which each event identifies.

As a simple example of a somewhat more general model, assume that each l_j can assume 3 values: $l_j = 0.25, 0.50, 1.00$, on default of B_j. Define a new sample space $\mathcal{S}'$ in terms of the original default sample space $\mathcal{S}$ of part 3 above. Specifically, if $(B_J, B_{J'}) \in \mathcal{S}$ with index set $J = (j_1, j_2, ..., j_M)$, this sample point will become 3^M sample points in $\mathcal{S}'$, each one representing one of the possible loss outcomes for these M bonds. In detail, the 3^M points in $\mathcal{S}'$ will be of the form:

$$(L_J^K, L_{J'}) \equiv (l_{j_1}^{k_1} f_{j_1}, l_{j_2}^{k_2} f_{j_2}, ..., l_{j_M}^{k_M} f_{j_M}, 0, ...0).$$

Here $K \equiv K_J = (k_1, k_2, ..., k_M)$ denotes one of the 3^M M-tuples of points where each k_i equals $1, 2, 3$, with respective $l_{j_i}^{k_i}$-values equal to $0.25, 0.50, 1.00$.

So while the original sample space $\mathcal{S}$ contained 2^N points, equaling the number of subsets of the collection of N bonds, $\mathcal{S}'$ will contain 4^N points. To see this, note that there are $\binom{N}{j}$ sample points of $\mathcal{S}$ with j defaulted bonds (see (1.9) below), and hence there are $\binom{N}{j} 3^j$ sample points of $\mathcal{S}'$ with j losses. Then by (1.10):

$$\sum_{j=0}^{N} \binom{N}{j} 3^j = \sum_{j=0}^{N} \binom{N}{j} 3^j 1^{N-j} = 4^N.$$

This generalizes, in that if each l_j can assume M values, $\mathcal{S}'$ will contain $(M+1)^N$ points.

Defining $\mathcal{E}' \equiv \sigma(P(\mathcal{S}'))$, the probability measure on $\mathcal{E}$ must also be refined. If $A \in \mathcal{S}'$ contains the bonds defined by a non-empty index subset $J = (j_1, j_2, ..., j_M)$, and loss ratios specified by $K_J = (k_1, k_2, ..., k_M)$, then define:

$$\mu'(A) = \prod_{j \in J} q(B_j) \Pr(k_j) \prod_{j \in J'} (1 - q(B_j)). \tag{1}$$

Here given a default of bond B_j, $\Pr(k_j)$ denotes the probability that the loss ratio is given by k_j as above.

Now extend μ' to $\mathcal{E}'$ additively as in part 3 of Example 1.5.

To prove that:

$$\mu'(\mathcal{S}') = 1,$$

note that by (1) this is equivalent to:

$$\sum_J \left[\sum_{K_J} \prod_{j \in J} q(B_j) \Pr(k_j) \prod_{j \in J'} (1 - q(B_j)) \right] = 1. \tag{2}$$

Letting $p_{j,a} \equiv \Pr(k_j = a)$, and so $p_{j,1} + p_{j,2} + p_{j,3} = 1$, we have as in Example 1.5:

$$\begin{aligned}
1 &= \sum_J \left[\prod_{j \in J} q(B_j) \prod_{j \in J'} (1 - q(B_j)) \right] \\
&= \sum_J \left[\prod_{j \in J} q(B_j) (p_{j,1} + p_{j,2} + p_{j,3}) \prod_{j \in J'} (1 - q(B_j)) \right].
\end{aligned}$$

Fixing J:

$$\prod_{j \in J} q(B_j) (p_{j,1} + p_{j,2} + p_{j,3}) = \sum_{K_J} \prod_{j \in J} q(B_j) \Pr(k_j),$$

and (2) is proved.

It then follows that $(\mathcal{S}', \mathcal{E}', \mu')$ is a probability space of bond default losses.

1.3   Some Probability Measures on $\mathbb{R}$

As will be seen below in Chapter 3 on random variables, probability measures on $\mathbb{R}$ are commonly induced by more general probability spaces $(\mathcal{S}, \mathcal{E}, \mu)$. But such measures can also arise in a specific application for which $\mathbb{R}$ is the appropriate sample space.

1.3.1   Measures from Discrete Probability Theory

There are infinitely many discrete probability measures possible on $\mathbb{R}$, where by "discrete" is meant that there is a finite or countable subset $\{x_i\} \subset \mathbb{R}$ and associated probabilities $\mu(\{x_i\}) = p_i$ with $0 \leq p_i \leq 1$ and $\sum_i p_i = 1$. Such a specification always produces a measure μ on $\mathbb{R}$, and specifically on the power sigma algebra $\sigma(P(\mathbb{R}))$ of all sets, since for any $A \in \sigma(P(\mathbb{R}))$:

$$\mu(A) \equiv \sum\nolimits_{x_i \in A} p_i. \tag{1.2}$$

Because the sum of a convergent series of positive terms is well defined and independent of the order of the summation, the definition of μ in (1.2) is seen to be countably additive and produces a probability space $(\mathbb{R}, \sigma(P(\mathbb{R})), \mu)$.

Definition 1.8 (Defining functions of discrete probability theory) *In discrete probability theory, the measure μ is interpreted as a **probability function (p.f.)** or **probability density function (p.d.f.)**, and typically denoted $f(x)$. So:*

$$f(x_i) \equiv \mu(\{x_i\}),$$

and thus:

$$f(x) \equiv \begin{cases} p_i, & x = x_i, \\ 0, & otherwise. \end{cases} \tag{1.3}$$

*This function has an associated **distribution function (d.f.)** or sometimes **cumulative distribution function (c.d.f.)**, denoted $F(x)$, and defined:*

$$F(x) = \sum\nolimits_{y \leq x} f(y), \tag{1.4}$$

noting that this is at most a countable sum.

Example 1.9 (Formal construction of a discrete probability space)
With the representation in (1.4), note that $F(x)$ is increasing and right continuous. Thus by the development in Chapter I.5 and culminating in Proposition I.5.23, $F(x)$ gives rise to a Borel measure μ_F defined on a complete sigma algebra of Carathéodory measurable sets, there denoted $\mathcal{M}_{\mu_F}(\mathbb{R})$. Since $F(x)$ has limits of 0 and 1 at $-\infty$ and ∞, respectively, it follows from that development that μ_F is a probability measure. We demonstrate that with μ defined in (1.2), $\mathcal{M}_{\mu_F}(\mathbb{R}) = \sigma(P(\mathbb{R}))$ and:

$$(\mathbb{R}, \sigma(P(\mathbb{R})), \mu) = (\mathbb{R}, \mathcal{M}_{\mu_F}(\mathbb{R}), \mu_F). \tag{1}$$

To this end, recall that the Chapter I.5 extension process began with defining a set function $|\cdot|_F$ on the semi-algebra $\mathcal{A}'$ of right semi-closed intervals by:

$$|(a, b]|_F \equiv F(b) - F(a).$$

But then by (1.2):

$$F(b) - F(a) = \sum\nolimits_{x_i \in (a,b]} p_i = \mu\,[(a, b]],$$

and so:

$$\mu = |\cdot|_F, \ \ on \ \mathcal{A}'. \tag{2}$$

The set function $|\cdot|_F$ *extends by addition to the algebra* $\mathcal{A}$ *of finite disjoint unions of such intervals, and is a measure* $\mu_{\mathcal{A}}$ *on* $\mathcal{A}$ *by Proposition I.5.13, and so by (2):*

$$\mu = \mu_{\mathcal{A}}, \ \ on \ \mathcal{A}.$$

Let μ_F *denote the Proposition I.5.23 extension of the measure* $\mu_{\mathcal{A}}$ *on* $\mathcal{A}$ *to a measure on* $\mathcal{M}_{\mu_F}(\mathbb{R})$. *Since* μ *is also well defined on this sigma algebra, Proposition I.6.14 assures that* $\mu = \mu_F$ *on* $\sigma(\mathcal{A}) \subset \mathcal{M}_{\mu_F}(\mathbb{R})$, *the smallest sigma algebra that contains* $\mathcal{A}$. *Thus* $\mu = \mu_F$ *on the Borel sigma algebra* $\mathcal{B}(\mathbb{R})$ *since* $\sigma(\mathcal{A}) = \mathcal{B}(\mathbb{R})$ *by Proposition I.8.1. Finally by Proposition I.6.24,* $\mu = \mu_F$ *on* $\sigma^{C_{\mu_F}}(\mathcal{A})$, *the Proposition I.6.20 completion of* $\sigma(\mathcal{A})$ *with respect to* μ_F, *and also* $\sigma^{C_{\mu_F}}(\mathcal{A}) = \mathcal{M}_{\mu_F}(\mathbb{R})$. *Combining obtains:*

$$\mu = \mu_F, \ \ on \ \mathcal{M}_{\mu_F}(\mathbb{R}). \tag{3}$$

We claim that:

$$\mathcal{M}_{\mu_F}(\mathbb{R}) = \sigma(P(\mathbb{R})), \tag{4}$$

and this with (3) then proves (1).

To this end, recall the outer measure $\mu_{\mathcal{A}}^*$ *defined on the power sigma algebra* $\sigma(P(\mathbb{R}))$. *By Remark I.5.17, for* $A \in \sigma(P(\mathbb{R}))$:

$$\mu_{\mathcal{A}}^*(A) \equiv \inf \left\{ \sum_n \mu_{\mathcal{A}}((a_n, b_n]) \mid A \subset \bigcup_n (a_n, b_n] \right\},$$

where $\{(a_n, b_n]\} \subset \mathcal{A}'$ *can be taken to be disjoint. Now* $\mathcal{M}_{\mu_F}(\mathbb{R})$ *is the collection of Carathéodory measurable sets with respect to* $\mu_{\mathcal{A}}^*$, $\mathcal{M}_{\mu_F}(\mathbb{R}) \subset \sigma(P(\mathbb{R}))$ *by definition. Also,* $\mu_{\mathcal{A}}^*$ *is a measure on* $\mathcal{M}_{\mu_F}(\mathbb{R})$, *and we defined* $\mu_F \equiv \mu_{\mathcal{A}}^*$.

Thus to prove (4) is to prove that every set $A \in \sigma(P(\mathbb{R}))$ *is Carathéodory measurable with respect to* $\mu_{\mathcal{A}}^*$. *That is, for any set* $E \subset \mathbb{R}$:

$$\mu_{\mathcal{A}}^*(E) = \mu_{\mathcal{A}}^* \left(A \bigcap E \right) + \mu_{\mathcal{A}}^* \left(\tilde{A} \bigcap E \right).$$

For this, it is enough to prove that

$$\mu_{\mathcal{A}}^* = \mu, \ \ on \ \sigma(P(\mathbb{R})), \tag{5}$$

with μ *given as in (1.2), because* μ *is by definition additive on disjoint sets.*

First, by (2) and countable additivity and monotonicity of μ:

$$
\begin{aligned}
\mu_{\mathcal{A}}^*(A) &\equiv \inf \left\{ \sum_n \mu((a_n, b_n]) \mid A \subset \bigcup_n (a_n, b_n] \right\} \\
&= \inf \left\{ \mu \left(\bigcup_n (a_n, b_n] \right) \mid A \subset \bigcup_n (a_n, b_n] \right\} \\
&\geq \mu(A).
\end{aligned}
$$

To prove equality, we construct a sequence of collections which provide the correct infimum. Let $\{(a_n, b_n]\}$ *be given with* $A \subset \bigcup_n (a_n, b_n]$ *and:*

$$\mu(A) < \mu \left(\bigcup_n (a_n, b_n] \right).$$

If no such collection exists then we are done. Otherwise, by definition of μ *there exists* $\{x_j\} \subset \bigcup_n (a_n, b_n] - A$, *so choose* x_m *with maximum* $\mu(\{x_j\})$ *from this collection. Since* $\sum_j \mu(\{x_k\}) \leq 1$, $\max\{\mu(\{x_j\})\}$ *is well defined and there can be at most finitely many* x_j *at this maximum, so choose any one.*

By the disjointedness of $\{(a_n, b_n]\}$ there is a unique interval with $x_m \in (a_{n_m}, b_{n_m}]$. Construct the countable collection $\{(y_i, y_{i+1}]\}$ where $y_1 = a_{n_m}$ and $\{y_i\}$ is an increasing sequence with $y_i \to b_{n_m}$. Now replace $(a_{n_m}, b_{n_m}]$ in the collection by $\{\{(y_i, y_{i+1}]\}, (x_m, b_{n_m}]\}$ if $x_m < b_{n_m}$, or by $\{(y_i, y_{i+1}]\}$ if $x_m = b_{n_m}$. Then with $\{(a'_n, b'_n]\}$ denoting the new collection, $A \subset \bigcup_n (a'_n, b'_n]$ and:

$$\mu(A) \leq \mu\left(\bigcup_n (a'_n, b'_n]\right) = \mu\left(\bigcup_n (a_n, b_n]\right) - p_m.$$

This process can be continued for each such x_m, of which there are at most countably many. The infimum of these constructed collections converges to $\mu(A)$, since μ only has positive measure on such x_m. Thus (5) is proved, which proves (4). Combining with (3) we obtain (1).

Three important and useful **discrete probability measures** are introduced here and will facilitate examples below.

1. **Discrete Rectangular Probability Measure**

   Perhaps the simplest probability measure that can be imagined is one which assumes the same value on every one of a finite number of sample points. The domain of this distribution is arbitrary but conventionally taken as $\{x_i\} = \{j/n\}_{j=1}^n$ or sometimes $\{j/n\}_{j=0}^n$, so in either case $\{x_i\} \subset [0,1]$. For given n, the **discrete rectangular measure on $[0,1]$**, sometimes called the **discrete uniform measure,** is defined on $\{j/n\}_{j=1}^n$ by:

   $$\mu_R(j/n) = 1/n, \quad j = 1, 2, .., n, \tag{1.5}$$

   or on $\{j/n\}_{j=0}^n$ by:

   $$\mu_R(j/n) = 1/(n+1), \quad j = 0, 1, 2, .., n. \tag{1.6}$$

   Using the simple transformation:

   $$x_i \to (b-a)x_i + a,$$

   provides corresponding discrete rectangular measures on the interval $[a, b]$.

2. **Binomial Probability Measure**

   For given p, $0 < p < 1$, the **standard binomial measure** is defined on $\{x_i\} = \{0, 1\}$. This is often economically expressed, defining $p' \equiv 1 - p$:

   $$\mu_B(j) = \begin{cases} p, & j = 1, \\ p', & j = 0. \end{cases} \tag{1.7}$$

   This probability measure can be defined on a space $Y = \{T, H\}$, representing a "tail" or "head" in a coin flip, and these results conventionally identified with 0 and 1, respectively.

   This measure can be extended to accommodate sample spaces defined with respect to the sum of n **independent** standard binomials, where independence is defined below. This produces the **general binomial measure** which now has 2 parameters, p and $n \in \mathbb{N}$. That is, $\{x_i\} = \{0, 1, 2, ..., n\}$, and the associated probabilities are given by:

   $$\mu_B(j) = \binom{n}{j} p^j (1-p)^{n-j}, \ j = 0, 1, .., n. \tag{1.8}$$

The formula in (1.8) will be justified in the next section once independence is defined. Here, the **binomial coefficient** $\binom{n}{j}$ is defined for integers $0 \leq j \leq n$ by:

$$\binom{n}{j} = \frac{n!}{j!(n-j)!}, \tag{1.9}$$

recalling that $n! \equiv n(n-1)...2 \cdot 1$, and $0! \equiv 1$. Binomial coefficients are useful in counting subsets of a set of n distinct objects, and are often notationally represented as $_nC_j$, and read as "n choose j," where C stands for "combination."

Exercise 1.10 (On $\binom{n}{j}$) *Show that $\binom{n}{j}$ equals the number of distinct j-element subsets that can be chosen from a set of n distinct objects, $j \leq n$, where by distinct subset is implied that the ordering of elements is ignored.*

Returning to the probability space $Y = \{T, H\}$ above, μ_B in (1.8) is related to an induced measure on the product space Y^n defined as the collection of all 2^n n-tuples of Ts and Hs. There, for $y \in Y^n$, $\mu_B(y) = p^j(1-p)^{n-j}$ where j denotes the number of Hs in y and $n-j$ the corresponding number of Ts. This probability measure then induces a measure on $\{0, 1, 2, ..., n\}$ by defining a **random variable** on Y^n:

$$X : Y^n \to \mathbb{R},$$

for which $X(y)$ is defined by the number of Hs in y. Under this mapping, the probabilities in (1.8) then equal the probabilities of events, $\{y | X(y) = j\} = X^{-1}(j)$.

That the formula for μ_B produces a probability measure on $\{x_i\} = \{0, 1, ..., n\}$ follows from the **binomial theorem**, which provides an explicit expression for the expansion of the binomial $(a+b)^n$ with n a positive integer:

$$(a+b)^n = \sum_{j=0}^{n} \binom{n}{j} a^j b^{n-j}. \tag{1.10}$$

Letting $a = p$ and $b = p'$, this formula produces

$$1 = \sum_{j=0}^{n} \binom{n}{j} p^j(1-p)^{n-j}.$$

The proof of the binomial theorem follows from the observation that the expansion of $(a+b)^n$ is a sum of terms of the form $\{a^j b^{n-j}\}$, so it is only the coefficients that need to be addressed. This coefficient identifies the number terms in this product equal to $a^j b^{n-j}$. Now for given j, there are:

$$n!/(n-j)! = n(n-1)...(n-j+1),$$

ways to identify j of the n binomial factors from which to select an a, which then by default identifies the $n-j$ factors from which to select a b. However, this count overstates this coefficient. Any such collection of j factors will be counted $j!$ times by this process, representing all the orders in which these factors could be selected. Hence the coefficient of $a^j b^{n-j}$ is $\binom{n}{j}$ as stated in (1.10).

3. **Poisson Probability Measure**

The Poisson distribution is named for **Siméon-Denis Poisson** (1781–1840) who discovered this measure and studied its properties. This measure is characterized by a single parameter $\lambda > 0$, and is defined on the nonnegative integers $\{x_i\} = \{0, 1, 2, ...\}$ by:

$$\mu_P(j) = e^{-\lambda}\frac{\lambda^j}{j!}, \; j = 0, 1, 2, ... \tag{1.11}$$

That $\sum_{j=0}^{\infty} f^P(j) = 1$ is an application of the **Taylor series expansion** for the exponential function:

$$e^x = \sum_{j=0}^{\infty} \frac{x^j}{j!}, \tag{1.12}$$

which is convergent for all $x \in \mathbb{R}$. For more on Taylor series, see Chapter 9 of **Reitano** (2010) or any book on calculus.

One important application of the Poisson distribution is that it provides a good approximation to the binomial distribution when the binomial parameter p is "small," and the binomial parameter n is "large." Specifically, with fixed $\lambda = np$, then for p "small" and n "large":

$$\binom{n}{j}p^j(1-p)^{n-j} \simeq e^{-np}\frac{(np)^j}{j!}. \tag{1.13}$$

The requirement that p be "small" is typically understood as the condition that $p < 0.1$, while n "large" is understood as $n \geq 100$ or so.

The approximation in (1.13) was of critical value in pre-computer days, and comes from the following result. Here it is assumed that as n increases, p decreases so that the product np is fixed and equal to λ.

Proposition 1.11 (Poisson Limit theorem) *For $\lambda = np$ fixed:*

$$\binom{n}{j}p^j(1-p)^{n-j} \longrightarrow e^{-\lambda}\frac{\lambda^j}{j!}, \tag{1.14}$$

for all j as $n \to \infty$.
Proof. *First off:*

$$\binom{n}{j}p^j(1-p)^{n-j} = \frac{n(n-1)...(n-j+1)}{j!}\left(\frac{\lambda}{n}\right)^j\left(1-\frac{\lambda}{n}\right)^n\left(1-\frac{\lambda}{n}\right)^{-j}$$

$$= \frac{n(n-1)...(n-j+1)}{n^j}\frac{\lambda^j}{j!}\left(1-\frac{\lambda}{n}\right)^n\left(1-\frac{\lambda}{n}\right)^{-j}.$$

The second term, $\lambda^j/j!$, is fixed as a function of n and part of the final result, while the last term, $\left(1-\frac{\lambda}{n}\right)^{-j}$, converges to 1 as $n \to \infty$ because the exponent $-j$ is fixed. Similarly, the first term equals the product of j-terms: $\prod_{k=0}^{j-1}(1-\frac{k}{n})$, which also converges to 1 as $n \to \infty$.
  The only subtlety here is to prove that for any real number λ, $(1-\lambda/n)^n \longrightarrow e^{-\lambda}$ as $n \to \infty$. This is equivalent to proving that $n\ln(1-\lambda/n) \longrightarrow -\lambda$, and this follows from the Taylor series expansion for $\ln(1-x)$, which is convergent for $-1 \leq x < 1$:

$$\ln(1-x) = -\sum_{j=1}^{\infty} x^j/j. \tag{1.15}$$

Taylor series theory also allows a finite sum expression with remainder term:

$$\ln(1-x) = -x - \xi_x^2/2,$$

where for positive x, $0 < \xi_x < x$.

Hence, $\ln(1 - \lambda/n) = -\lambda/n - \xi_n/2$, *where* $0 < \xi_n < \lambda/n$, *so:*

$$n\ln(1 - \lambda/n) = -\lambda - n\xi_n^2/2.$$

Here $0 < n\xi_n^2 < \lambda^2/n$, *and the desired limit follows.* ∎

Exercise 1.12 *Show that the limit* $n\ln(1 - \lambda/n) \longrightarrow -\lambda$ *also follows from the definition of the derivative of* $f(x) = \ln(1 - \lambda x)$ *at* $x = 0$.

1.3.2   Measures from Continuous Probability Theory

As for discrete probability theory, there are also infinitely many probability measures on $\mathbb{R}$ under the continuous theory, all of which are then Borel measures. In each example below, the construction is the same, following Chapter I.5:

1. Identify a nonnegative continuous function $f(x)$ for which $\int_{-\infty}^{\infty} f(x)dx = 1$, defined as a Riemann integral, where:

$$\int_{-\infty}^{\infty} f(x)dx = \lim_{N,M\to\infty} \int_{-M}^{N} f(x)dx.$$

    More generally, if $f(x)$ equals 0 outside an interval or disjoint union of such intervals, then it is required that $f(x)$ be continuous on these intervals. The Riemann integral is then well defined over every bounded interval by Proposition I.1.5, since $f(x)$ is continuous almost everywhere.

2. Define the increasing function:

$$F(x) = \int_{-\infty}^{x} f(y)dy.$$

    Note that $F(x)$ is continuous:

$$F(x \pm \epsilon) \to F(x),$$

    as positive $\epsilon \to 0$, since $F(x)$ is bounded on $[x - \epsilon, x + \epsilon]$. For example:

$$|F(x + \epsilon) - F(x)| \leq \int_{x}^{x+\epsilon} |f(y)| \, dy \to 0.$$

3. By Chapter I.5, construct on $\mathbb{R}$ the Borel measure μ_F, and complete sigma algebra $\mathcal{M}_{\mu_F}(\mathbb{R})$, producing the complete probability space denoted $(\mathbb{R}, \mathcal{M}_{\mu_F}(\mathbb{R}), \mu_F)$.

Definition 1.13 (Defining functions of continuous probability theory) *In continuous probability theory the function* $f(x)$ *is called a* **probability function (p.f.)**, *or* **probability density function (p.d.f.)**, *and* $F(x)$ *the associated* **distribution function (d.f.)**, *or sometimes the* **cumulative distribution function (c.d.f.)**. *These are related by:*

$$f(x) = F'(x), \tag{1.16}$$

$$F(x) = \int_{-\infty}^{x} f(y)dy. \tag{1.17}$$

Equation (1.17) is the definition of $F(x)$, *while (1.16) follows from the fundamental theorem of calculus.*

In this model one can explicitly assign probabilities to right semi-closed intervals as in Chapter I.5:

$$\mu_F\left[(a, b]\right] = F(b) - F(a) = \int_a^b f(y)dy.$$

By continuity it follows that the same result is produced for closed, open and left semi-closed intervals. From this it also follows that $\mu\left[\{a\}\right] = 0$ for any point a.

Remark 1.14 (More general density/distribution functions) *As a small introduction to what is to come in later books, Lebesgue integration will be studied in Book III. Then with the assumption that $f(x)$ is nonnegative, Lebesgue integrable, and with $\int_{-\infty}^{\infty} f(x)dx = 1$, one can again define the associated distribution function $F(x)$ as in (1.17) above. It will turn out that $F(x)$ is again continuous, is differentiable almost everywhere, and with (1.16) again true almost everywhere.*

Finally, we will see in Book V that the associated measure μ_F induced by $F(x)$ can be expressed in terms of integrals of $f(x)$, generalizing $\mu_F\left[(a, b]\right]$ above. Specifically, it will be seen that for all Borel sets $A \in \mathcal{B}(\mathbb{R})$:

$$\mu_F[A] = \int_A f(x)dx.$$

Within the Riemann integration theory, while we know from Book I that the associated μ_F is well defined on $\mathcal{M}_{\mu_F}(\mathbb{R})$ and thus also on $\mathcal{B}(\mathbb{R}) \subset \mathcal{M}_{\mu_F}(\mathbb{R})$, we cannot represent $\mu_F[A]$ as above except for the simplest Borel sets, such as intervals. The reason for this is simply that the associated Riemann integrals may not be defined.

*Specifically, with $\chi_A(x)$ denoting the **characteristic function** of A, defined as $\chi_A(x) = 1$ for $x \in A$ and $\chi_A(x) = 0$ otherwise, we will define:*

$$\int_A f(x)dx = \int f(x)\chi_A(x)dx.$$

Thus within the Riemann theory, if $f(x)$ is continuous, this integral will only exist if A is a set for which $\chi_A(x)$ is continuous almost everywhere.

A few popular examples of continuous probability measures follow.

1. **Continuous Uniform Probability Measure**

   Perhaps the simplest continuous probability function that can be imagined is one which is constant. The domain of this distribution is arbitrary but necessarily bounded, and is conventionally denoted as the interval $[a, b]$. The probability density function of the **continuous uniform measure**, sometimes called the **continuous rectangular measure**, is defined by:

   $$f_U(x) = \begin{cases} 1/(b-a), & x \in [a, b], \\ 0, & x \notin [a, b]. \end{cases} \tag{1.18}$$

   Thus $F_U(x)$ is given by:

   $$F_U(x) = \begin{cases} 0, & x \leq a, \\ (x-a)/(b-a), & a \leq x \leq b, \\ 1, & x \geq b. \end{cases} \tag{1.19}$$

2. Exponential Probability Measure

The probability density function of the **exponential probability measure** is non-zero on $[0, \infty)$, and defined with a single scale parameter $\lambda > 0$ by:

$$f_E(x) = \begin{cases} 0, & x < 0, \\ \lambda e^{-\lambda x}, & x \geq 0. \end{cases} \tag{1.20}$$

For any $\lambda > 0$, $f_E(x)$ is strictly decreasing on $[0, \infty)$, and $\int_0^\infty f_E(x)dx = 1$ defined as an improper integral.

The associated distribution function $F_E(x)$ is given:

$$F_E(x) = \begin{cases} 0, & x < 0, \\ 1 - e^{-\lambda x}, & x \geq 0. \end{cases} \tag{1.21}$$

3. Normal Probability Measure

The probability density function of the **normal probability measure** is strictly positive on $(-\infty, \infty)$, depends on a location parameter $\mu \in \mathbb{R}$ and a scale parameter $\sigma > 0$, and is defined by the following with $\exp A \equiv e^A$ to simplify notation:

$$f_N(x) = \frac{1}{\sigma\sqrt{2\pi}} \exp\left(-(x-\mu)^2/(2\sigma^2)\right). \tag{1.22}$$

This is often called the **Gaussian probability measure** after **Carl Friedrich Gauss** (1777–1855), who was one of the codiscoverers of this formula.

In Book IV, μ will be seen to be the **mean** of this distribution, and σ^2 the **variance.**

When $\mu = 0$ and $\sigma = 1$, this is known as the **unit normal** or **standard normal probability density,** and often denoted $\phi(x)$:

$$\phi(x) = \frac{1}{\sqrt{2\pi}} \exp\left(-x^2/2\right). \tag{1.23}$$

The associated distribution functions are then denoted $F_N(x)$ and $\Phi(x)$, respectively, though these have no closed form expressions. For example:

$$\Phi(x) \equiv \int_{-\infty}^x \phi(y)dy. \tag{1.24}$$

A simple substitution into the integral demonstrates that $f_N(x)$ and $\phi(x)$ have the same Riemann integral over $\mathbb{R}$. Also, for any N:

$$\exp\left(-x^2/2\right) < |x|^{-N}, \text{ as } |x| \to \infty.$$

It follows that $\int_{-\infty}^\infty \phi(y)dy < \infty$ as an improper integral. However, there is no elementary derivation that verifies that this function in fact integrates to $\sqrt{2\pi}$.

The simplest justification follows, and even this requires a deep result on multivariate integrals such as **Fubini's theorem,** named for **Guido Fubini** (1879–1943), or **Tonelli's theorem,** named for **Leonida Tonelli** (1885–1946), as well as results on substitution in integrals. These results will be derived in Book V, and we assume these for now since in the current Riemann integral context these assumptions are intuitively reasonable and are likely familiar to the reader.

Consider the square of this integral, which is finite as noted above:

$$\left[\int_{-\infty}^{\infty} \phi(y)dy\right]^2 = \frac{1}{\sqrt{2\pi}}\int_{-\infty}^{\infty} \exp\left(-x^2/2\right)dx \cdot \frac{1}{\sqrt{2\pi}}\int_{-\infty}^{\infty} \exp\left(-y^2/2\right)dy$$

$$= \frac{1}{2\pi}\int_{-\infty}^{\infty}\int_{-\infty}^{\infty} \exp\left[-\left(x^2+y^2\right)/2\right]dxdy.$$

We use **polar coordinates** to evaluate this integral, making the substitution:

$$x = r\cos\theta;\ y = r\sin\theta,$$

where r denotes the standard distance from (x,y) to the origin $(0,0)$, and θ the angle in radian measure that the segment from $(0,0)$ to (x,y) makes with the positive x-axis, so $0 \le \theta < 2\pi$.

This substitution will lead to $dxdy = rdrd\theta$, and an integration over $0 \le r < \infty$ and then $0 \le \theta \le 2\pi$ obtains:

$$\left[\int_{-\infty}^{\infty} \phi(y)dy\right]^2 = \frac{1}{2\pi}\int_0^{2\pi}\int_0^{\infty} \exp\left(-r^2/2\right)rdrd\theta = 1.$$

Admittedly, this derivation requires a lot of faith that the various manipulations are justifiable, but the reader will eventually see in Book V that this is so.

4. **Lognormal Probability Measure**

The probability density function of the **lognormal probability measure** is defined on $[0,\infty)$, depends on a location parameter $\mu \in \mathbb{R}$ and a shape parameter $\sigma > 0$, and unsurprisingly is intimately related to the normal probability measure discussed above. However, to some the name "lognormal" appears to be opposite of the relationship that exists. Stated one way, the variable x has a lognormal probability function with parameters (μ,σ) if $x = e^z$, where z has a normal probability function with the same parameters. So lognormal x can be understood as an **exponentiated normal**. Stated another way, a variable x has a lognormal probability function with parameters (μ,σ) if $\ln x$ has a normal probability function with the same parameters. The name comes from the second statement, in that the log of a lognormal variable is normal.

The probability density function of the lognormal is defined on $[0,\infty)$ as follows, again using $\exp A \equiv e^A$ to simplify notation:

$$f_L(x) = \frac{1}{\sigma x\sqrt{2\pi}}\exp\left(-\left(\ln x-\mu\right)^2/2\sigma^2\right),\ x \ge 0, \tag{1.25}$$

while $f_L(x) = 0$ for $x < 0$. The substitution $y = \left(\ln x - \mu\right)/\sigma$ produces:

$$\int_0^{\infty} f_L(x)dx = \frac{1}{\sqrt{2\pi}}\int_{-\infty}^{\infty} \exp\left(-y^2/2\right)dy$$

$$= \int_{-\infty}^{\infty} \phi(y)dy.$$

Thus, subject to the above "derivation," the integral of the lognormal density over $[0,\infty)$ equals 1.

This density function is well defined and continuous at $x = 0$ with $f_L(0) = 0$. To see this, let $x = e^{-y}$ and consider $y \to \infty$. With this transformation:

$$f_L(e^{-y}) = \frac{e^y}{\sigma\sqrt{2\pi}} \exp\left(-(y+\mu)^2/2\sigma^2\right)$$

$$= \frac{1}{\sigma\sqrt{2\pi}} \exp\left(y - (y+\mu)^2/2\sigma^2\right).$$

As $y \to \infty$, note that $\left[y - (y+\mu)^2/2\sigma^2\right] \to -\infty$, and so $f_L(e^{-y}) \to 0$.

1.3.3 More General Probability Measures on $\mathbb{R}$

There are more general probability measures on $\mathbb{R}$ than those given above by the discrete and continuous models. For example, the constructions outlined above for discrete and continuous measures work equally well if $f(x)$ is non-negative and **continuous almost everywhere (a.e.),** which is to say, continuous except on a set of measure zero. The Riemann integral is then well defined, and if $\int_{-\infty}^{\infty} f(x)dx = 1$, this function produces a continuous increasing $F(x)$ by (1.17), which again provides the basis for a Borel measure space. In Book III we will see that such a distribution function F is differentiable almost everywhere (a.e.), and that $F'(x) = f(x)$ a.e.

In Book III we will also study Lebesgue integration as noted in Remark 1.14, which extends the definition of an integral to many functions which are not Riemann integrable. If such a function is non-negative and satisfies $(\mathcal{L})\int_{-\infty}^{\infty} f(x)dx = 1$, where $(\mathcal{L})$ denotes a Lebesgue integral, then (1.17) produces an increasing continuous $F(x)$, which again provides the basis for a Borel measure space. Such a distribution function is then differentiable a.e., and $F'(x) = f(x)$ a.e.

Even more generally, we can begin with any increasing right continuous function $F(x)$ for which $F(x) \to 0$ as $x \to -\infty$, and $F(x) \to 1$ as $x \to \infty$, and the Borel construction of Book I can be applied. Once again we will see that $F'(x)$ must exist almost everywhere. In this general case, perhaps surprisingly, it need not be true that there is a function $f(x)$ for which (1.17) holds. In other words, not all distribution functions have associated density functions.

This subject will be addressed in Chapter IV.1, where distribution functions on $\mathbb{R}$ will be characterized.

1.4 Independent Events

An important notion in a probability space, which has little or no compelling application in the more general context of a measure space, is that of **independent events**. Also of interest are **independent classes** and **independent sigma algebras**. In this section we introduce these notions and study some of their properties. This study will be continued in Section 3.4 and Section 5.2.4, in the context of independent random variables and associated sigma algebras.

Definition 1.15 (Independent events and classes) *If* $(\mathcal{S}, \mathcal{E}, \mu)$ *is a probability space, then* $\{A_n\}_{n=1}^N \subset \mathcal{E}$ *with* $N < \infty$ *is said to be a collection of* ***independent events,*** *or* ***are mutually independent,*** *if given any subset* $J \subset \{1, 2, ..., N\}$ *with* $J \equiv \{j_1, j_2, ...j_k\}$:

$$\mu\left(\bigcap_{i=1}^k A_{j_i}\right) = \prod_{i=1}^k \mu\left(A_{j_i}\right). \tag{1.26}$$

These events are said to be ***pairwise independent*** *if (1.26) remains valid for any subset* J *with* $k = 2$.

An infinite collection of events $\{A_\alpha\}_{\alpha \in I} \subset \mathcal{E}$, whether countable or uncountable, is said to be a collection of independent events or pairwise independent, respectively, if every **finite subset** of events is independent or pairwise independent, respectively.

Given a finite, countable, or uncountably infinite collection $\{\mathsf{A}_\alpha\}_{\alpha \in I}$, with each $\mathsf{A}_\alpha \subset \mathcal{E}$ a finite or infinite **class** of measurable sets or a **sigma algebra,** we say that $\{\mathsf{A}_\alpha\}_{\alpha \in I}$ is a collection of **independent classes** or **independent sigma algebras,** or simply **are mutually independent,** if given any **finite subset** $J \subset I$ with $J \equiv \{\alpha_1, \alpha_2, ...\alpha_k\}$, and any $\{A_{\alpha_i}\}_{i=1}^k$ with $A_{\alpha_i} \in \mathsf{A}_{\alpha_i}$, (1.26) is satisfied.

For finite and countable collections, the notation $\{\mathsf{A}_j\}_{j=1}^N$ is more convenient, where N is finite or infinite.

Remark 1.16 (On equation (1.26)) *Since $\bigcap_{i=1}^k A_{j_i} \subset A_{j_i}$ for every j_i, it follows by monotonicity that in any measure space:*

$$\mu\left(\bigcap_{i=1}^k A_{j_i}\right) \le \min \mu\left(A_{j_i}\right).$$

In any probability space, since $\mu\left(A_{j_i}\right) \le 1$:

$$\prod_{i=1}^k \mu\left(A_{j_i}\right) \le \min \mu\left(A_{j_i}\right).$$

So the equality in (1.26) is at least qualitatively plausible in a probability space, though apparently not so in a general measure space where the second inequality will typically fail except for "small" measurable sets with $\mu\left(A_{j_i}\right) \le 1$.

*Also, note that the criterion in (1.26) can be restated as follows: $\{A_n\}_{n=1}^N \subset \mathcal{E}$ with $N < \infty$ is a collection of **independent events** if:*

$$\mu\left(\bigcap_{n=1}^N B_n\right) = \prod_{n=1}^N \mu\left(B_n\right),$$

for each of the 2^N expressions where each $B_n = A_n$ or $B_n = \mathcal{S}$. See Proposition 1.19 for another characterization.

Example 1.17 (Independent events)

1. **General Binomial and independent coin flips:** *Recall the discussion above on the general binomial measure μ^B defined on $\{0, 1, 2, ..., n\}$ by (1.8). These probability values were deemed to result from the summing of n "independent" standard binomials, each of which takes value from $\{0, 1\}$.*

   Define the sample space $\mathcal{S}$ of n-tuples $(x_1, x_2, ..., x_n)$ of results, each point of which identifies the ordered outcome of n standard binomials. There are 2^n distinct such n-tuples. Define the 2n events:

   $$A_{i(k)} = \{(x_1, x_2, ..., x_n) | x_i = k\},$$

   where $k = 0, 1$, and note that $A_{i(k)}$ is the collection of n-tuples with the ith component specified to equal k. More visually, it is the collection of n-flips with the ith flip equal to H if $k = 1$, or T if $k = 0$. The probability measure of these events is given by:

   $$\mu\left(A_{i(k)}\right) = \begin{cases} p, & k = 1, \\ p', & k = 0, \end{cases} \tag{1}$$

   where $p' \equiv 1 - p$. This is because $A_{i(k)}$ contains all n-tuples with $x_i = k$ and thus this can be modeled as the full collection of $(n-1)$-flips, which has probability 1, and then an nth flip with probability p or p'.

The events $A_{i(0)}$ and $A_{i(1)}$ cannot be independent for any i since $A_{i(0)} \cap A_{i(1)} = \emptyset$ and thus (1.26) will not be satisfied since $\mu\left(A_{i(k_i)}\right) > 0$. We now investigate the implications of more general independence of these sets since this independence characterizes the independence of the standard binomials noted above. For example, that the ith and jth coin flip are independent standard binomials for $i \neq j$, means that $A_{i(k_i)}$ and $A_{j(k_j)}$ are independent events for any choice of outcomes k_i and k_j.

More generally, given any collection $\{k_i\}_{i=1}^{n}$, the n events $\{A_{i(k_i)}\}_{i=1}^{n}$ are independent if:

$$\mu\left(\bigcap_{i=1}^{n} A_{i(k_i)}\right) = \prod_{i=1}^{n} \mu\left(A_{i(k_i)}\right).$$

Now $\bigcap_{i=1}^{n} A_{i(k_i)} = (k_1, k_2, ..., k_n)$, a single sample point, and so independence requires by (1) that:

$$\mu(k_1, k_2, ..., k_n) = p^j (1-p)^{n-j},$$

where j denotes the number of indexes with $k_i = 1$, and $n - j$ the number with $k_i = 0$. Thus independence of $\{A_{i(k_i)}\}_{i=1}^{n}$ implies that the probability of $(k_1, k_2, ..., k_n)$ depends only on j, the number of indexes with $k_i = 1$, and not the order.

There are $\binom{n}{j}$ of the $A_{i(k)}$-events that have exactly j indexes with $k_i = 1$, and these are mutually disjoint, meaning no sample point can be in any two of these events. Define B_j as the union of all the sample points in these $\binom{n}{j}$ events. That is, B_j is the event containing all sample points with j values equal to 1. It then follows by finite additivity that:

$$\mu(B_j) = \binom{n}{j} p^j (1-p)^{n-j}.$$

This is equivalent to the probability measure in (1.8), and so $\mu_B(j) = \mu(B_j)$.

In other words, the probabilities assigned by (1.8) are consistent with the assumption of n independent flips.

It is left as an exercise to show that given any collection $\{k_i\}_{i=1}^{m}$ for $2 \leq m < n$, the m events $\{A_{i(k_i)}\}_{i=1}^{m}$ are independent. Hint: Consider the argument demonstrating (1).

2. **Independent vs. pairwise independent:** *While independent events are pairwise independent by definition, pairwise independence does not imply independence. For example, let $\mathcal{S}$ denote the probability space of outcomes of 3 independent rolls of a fair die. Thus $\mathcal{S} = \{(r_1, r_2, r_3)\}$, where $r_j \in \{1, 2, ..., 6\}$ is the value on roll j, and $\mu((r_1, r_2, r_3)) = 1/6^3$ for every outcome. Let $\mathcal{E} = \sigma(P(\mathcal{S}))$, the power set of $\mathcal{S}$, and define μ on $\mathcal{E}$ additively. If $A_{12} = \{(r_1, r_2, r_3) | r_1 = r_2\}$, with A_{13} and A_{23} analogously defined, then $A_{12}, A_{13},$ and A_{23} are pairwise disjoint. Each event has probability $1/6$, while $\mu(A_{12} \cap A_{13}) = 1/6^2$, and similarly for the other intersection sets. However, $\mu(A_{12} \cap A_{13} \cap A_{23}) = 1/6^2$, since $A_{12} \cap A_{13} \subset A_{23}$, and so (1.26) fails for $k = 3$.*

3. **Independence and subcollections:** *Requiring (1.26) to be satisfied only for the maximal subset $J = \{1, 2, ..., N\}$ does not ensure independence or pairwise independence. Let $\mathcal{S} = \{r_1\}$ be the sample space of the outcome of one roll of a fair die, with $\mu(r_1) = 1/6$ for all values, and $\mathcal{E} = \sigma(P(\mathcal{S}))$ with μ again extended by additivity. Let $A_1 = \{1, 2, 3, 4\}$, and $A_2 = A_3 = \{1, 5, 6\}$. Then these events have respective probabilities of $2/3$, $1/2$, and $1/2$, and the full intersection event has probability $1/6$. So (1.26) is satisfied with $J = \{1, 2, 3\}$. However, these events are not independent since they are not pairwise independent.*

Next are a few simple yet useful results on independent events. The first addresses "extreme" events A, meaning $\mu(A) \in \{0, 1\}$.

Proposition 1.18 (Extreme events) *If $A \in \mathcal{E}$ satisfies $\mu(A) \in \{0,1\}$, then A is independent of every event $B \in \mathcal{E}$ including itself.*

If $A \in \mathcal{E}$ is independent of itself, then $\mu(A) \in \{0,1\}$.

Proof. *If $\mu(A) = 0$, then A is independent of all events B since $A \cap B \subset A$ and hence $\mu(A \cap B) = 0$ by subadditivity. If $\mu(A) = 1$, then $\mu(\widetilde{A}) = 0$ and since $B = (B \cap A) \cup \left(B \cap \widetilde{A}\right)$, independence follows from finite additivity. Specifically, $\mu(B) = \mu(A \cap B)$, which is (1.26) since $\mu(A) = 1$.*

If A is independent of itself, then (1.26) implies that $\mu(A) = (\mu(A))^2$ and hence $\mu(A) \in \{0,1\}$. ∎

The next result provides an alternative characterization of independence.

Proposition 1.19 (Alternative characterization of independence) *Let $(\mathcal{S}, \mathcal{E}, \mu)$ be a probability space. Then $\{A_n\}_{n=1}^N \subset \mathcal{E}$ are independent events if and only if:*

$$\mu\left(\bigcap_{n=1}^N B_n\right) = \prod_{n=1}^N \mu(B_n), \tag{1.27}$$

for each of the 2^N expressions where each $B_n = A_n$ or $B_n = \widetilde{A}_n$.

Proof. *Assume that (1.27) is satisfied for $\{A_n\}_{n=1}^N$, then so too is (1.26) for $k = N$. Consider the subcollection $\{A_n\}_{n=1}^{N-1}$, noting that the labeling is sequential for notational convenience. Then by finite additivity and (1.27), (1.26) is satisfied for $k = N - 1$:*

$$\begin{aligned}
\mu\left(\bigcap_{n=1}^{N-1} A_n\right) &= \mu\left(\bigcap_{n=1}^{N-1} A_n \bigcap A_N\right) + \mu\left(\bigcap_{n=1}^{N-1} A_n \bigcap \widetilde{A}_N\right) \\
&= \prod_{n=1}^N \mu(A_n) + \prod_{n=1}^{N-1} \mu(A_n)(1 - \mu(A_N)) \\
&= \prod_{n=1}^{N-1} \mu(A_n).
\end{aligned} \tag{1}$$

By iteration this proves that (1.26) is satisfied for all k.

Conversely, assume that (1.26) is satisfied for $\{A_n\}_{n=1}^N$, then so too is (1.27) for $B_n = A_n$ for all n. Consider $\{B_n\}_{n=1}^N$ with $B_n = A_n$ for all $n < N$ and $B_N = \widetilde{A}_N$. As noted above, labeling is sequential for notational convenience. Using the setup in (1), and the fact that (1.26) is satisfied for $\{A_n\}_{n=1}^{N-1}$:

$$\begin{aligned}
\mu\left(\bigcap_{n=1}^{N-1} A_n \bigcap \widetilde{A}_N\right) &= \mu\left(\bigcap_{n=1}^{N-1} A_n\right) - \mu\left(\bigcap_{n=1}^{N-1} A_n \bigcap A_N\right) \\
&= \prod_{n=1}^{N-1} \mu(A_n)(1 - \mu(A_N)) \\
&= \prod_{n=1}^N \mu(B_n).
\end{aligned}$$

This now can be iterated to prove (1.27). For example, since $\{A_n\}_{n=1}^{N-1}$ satisfies (1.26), the prior step can also be applied to this collection:

$$\begin{aligned}
&\mu\left(\bigcap_{n=1}^{N-2} A_n \bigcap \widetilde{A}_{N-1} \bigcap \widetilde{A}_N\right) \\
&= \mu\left(\bigcap_{n=1}^{N-2} A_n \bigcap \widetilde{A}_{N-1}\right) - \mu\left(\bigcap_{n=1}^{N-2} A_n \bigcap \widetilde{A}_{N-1} \bigcap A_N\right) \\
&= \prod_{n=1}^{N-2} \mu(A_n)(1 - \mu(A_{N-1})) - \prod_{n=1}^{N-2} \mu(A_n)(1 - \mu(A_{N-1}))\mu(A_N) \\
&= \prod_{n=1}^N \mu(B_n).
\end{aligned}$$

∎

Corollary 1.20 *Let $(\mathcal{S}, \mathcal{E}, \mu)$ be a probability space, then $\{A_n\}_{n=1}^N \subset \mathcal{E}$ are independent by Definition 1.15 if and only if $\{B_n\}_{n=1}^N$ are independent for each of the 2^N choices where $B_n = A_n$ or $B_n = \widetilde{A_n}$.*

Proof. *This follows from Proposition 1.19.* ∎

1.4.1 Independent Classes and Associated Sigma Algebras

In the section below on Kolmogorov's Zero-One law, we will need results on properties of **independent classes** as defined in Definition 1.15. In particular, we will need to know when independent classes give rise to independent sigma algebras, where these sigma algebras are generated by these classes. It is perhaps natural to expect that if $\{A_j\}_{j=1}^N$ are independent classes by the above definition, then the smallest sigma algebras generated by these classes, $\{\sigma(A_j)\}_{j=1}^N$, would also be independent. But this is not the case and the problem is with the intersection sets.

Example 1.21 (Independent classes $\not\Rightarrow$ independent sigma algebras) *Consider the infinite product probability space of Example I.9.10 associated with fair coin flips, $(Y^{\mathbb{N}}, \sigma(Y^{\mathbb{N}}), \mu_{\mathbb{N}})$. Here $Y = \{H, T\}$ and $\mu_{\mathbb{N}}$ is the measure induced by the probability measure on Y defined by $p(H) = p(T) = 1/2$. Let $A_1 = \{A_{12}, A_{13}\}$ and $A_2 = \{A_{23}\}$ where $A_{jk} = \{y \in Y^{\mathbb{N}} | y_j = y_k\}$. Note that $\mu_{\mathbb{N}}(A_{jk}) = 1/2$ for any $j \neq k$.*

Then A_1 and A_2 are mutually independent classes. For example:

$$\mu_{\mathbb{N}}(A_{12} \cap A_{23}) = 1/4 = \mu_{\mathbb{N}}(A_{12})\mu_{\mathbb{N}}(A_{23}).$$

However, $A_{12} \cap A_{13} \in \sigma(A_1)$, and this set is not independent of $A_{23} \in \sigma(A_2)$ since:

$$\mu_{\mathbb{N}}([A_{12} \cap A_{13}] \cap A_{23}) = 1/4 \neq \mu_{\mathbb{N}}(A_{12} \cap A_{13})\mu_{\mathbb{N}}(A_{23}) = 1/8.$$

The problem here is that while A_1 and A_2 are independent classes, the sets A_{12}, A_{13}, A_{23} are only pairwise independent and not mutually independent:

$$\mu_{\mathbb{N}}(A_{12} \cap A_{13} \cap A_{23}) = 1/4 \neq \mu_{\mathbb{N}}(A_{12})\mu_{\mathbb{N}}(A_{13})\mu_{\mathbb{N}}(A_{23}) = 1/8.$$

The next result shows that if we begin with a collection of mutually independent sets and partition this collection into classes, the above problem cannot occur and finite intersection sets will again be independent. The most complicated part of this next result is the necessary notation. So to simplify this, we represent the initial collection of mutually independent sets as a matrix of sets which may have a finite or infinite number of rows, and each row a finite or infinite number of elements. We will create classes by grouping rows of sets.

The goal of the next few results is to show that these classes generate independent sigma algebras, and we do this in steps.

Proposition 1.22 (Independent classes) *Given a probability space $(\mathcal{S}, \mathcal{E}, \mu)$, let $A = \{A_{jk}\} \subset \mathcal{E}$ be a finite or infinite class of mutually independent sets with $1 \leq j \leq N$ and $1 \leq k \leq M_j$ where N and any M_j can be finite or infinite. Partition this class "by rows" into subclasses $\{A_j\}_{j=1}^N$, with $A_j = \{A_{jk}\}_{k=1}^{M_j}$. Define A_j' as the collection of sets from A_j and all possible finite intersections of sets from A_j.*

Then $\{A_j'\}_{j=1}^N$ are mutually independent classes.

Proof. *First note that $\{A_j\}_{j=1}^N$ are independent classes since A is a collection of independent sets and $\{A_j\}_{j=1}^N$ simply partitions this collection. In detail, given a finite subset $J \subset (1, ..., N)$ with $J \equiv \{j_1, j_2, ...j_n\}$ and any $\{A_{j_i k_i}\}_{i=1}^n$ with $A_{j_i k_i} \in A_{j_i}$, then (1.26) is satisfied because $\{A_{j_i k_i}\}_{i=1}^n \subset A$. Thus only the independence of the intersection sets from $\{A_j'\}_{j=1}^N$ need be checked.*

Choose $\{B_j\}$ with $B_j \in \mathsf{A}'_j$ for a finite index collection $J = \{j_1, ..., j_n\}$, which by re-ordering rows and relabeling for simplicity we can assume that $J = \{1, 2, ..., n\}$. Each B_j is by assumption a finite intersection set so $B_j = \bigcap_{K_j} A_{jk}$ with $A_{jk} \in \mathsf{A}_j$, and where each K_j is a finite index set which by re-ordering we can again assume that $K_j = \{1, 2, ..., k_j\}$. Of course, if $k_j = 1$ for any j, then $B_j = A_{j1}$.

Recalling that $\{A_{jk}\}_{j,k}$ are mutually independent:

$$
\begin{aligned}
\mu\left(\bigcap_{j=1}^{n} B_j\right) &= \mu\left(\bigcap_{j=1}^{n}\bigcap_{k=1}^{k_j} A_{jk}\right) \\
&= \prod_{j=1}^{n}\prod_{k=1}^{k_j} \mu\left(A_{jk}\right) \\
&= \prod_{j=1}^{n} \mu\left(\bigcap_{k=1}^{k_j} A_{jk}\right) \\
&= \prod_{j=1}^{n} \mu\left(B_j\right).
\end{aligned}
$$

■

Corollary 1.23 (Independent semi-algebras) *Given the assumptions and notation of the prior proposition, if $\mathcal{A}'(\mathsf{A}_j)$ denotes the smallest semi-algebra generated by A_j, then $\{\mathcal{A}'(\mathsf{A}_j)\}_{j=1}^{N}$ are mutually independent classes.*

Proof. *The smallest semi-algebra that contains A_j is contained in the smallest semi-algebra that contains A_j but expanded to include $\mathcal{S}$, $\emptyset$, and $\widetilde{A}_{jk}$ for every $A_{jk} \in \mathsf{A}_j$. These expanded classes, which we label $\{\overline{\mathsf{A}}_j\}_{j=1}^{N}$, continue to be mutually independent since $\mathcal{S}$ and $\emptyset$ are independent of every set by Proposition 1.18, and the result for complements follows from Proposition 1.19.*

Hence, the prior proposition applies to these expanded classes and assures that $\{\overline{\mathsf{A}}'_j\}_{j=1}^{N}$, defined as collections of sets from $\overline{\mathsf{A}}_j$ and all possible finite intersections of sets from $\overline{\mathsf{A}}_j$, are again mutually independent classes. We claim that each $\overline{\mathsf{A}}'_j$ is a semi-algebra.

First, $\overline{\mathsf{A}}'_j$ is closed under finite intersections since each set in $\overline{\mathsf{A}}'_j$ is a finite intersection of sets from the expanded $\overline{\mathsf{A}}_j$, and hence intersecting any finite number of such sets produces another such finite intersection set. Consider next the complement of a set $B_j \in \overline{\mathsf{A}}'_j$, say $B_j = \bigcap_{k=1}^{n} B_{jk}$, where each B_{jk} is an element of the expanded $\overline{\mathsf{A}}_j$ class. That is, each B_{jk} equals one of A_{jk}, $\widetilde{A}_{jk}$, $\mathcal{S}$ or $\emptyset$. Then $\widetilde{B}_j = \bigcup_{k=1}^{n} \widetilde{B}_{jk}$, a finite union of sets from the expanded $\overline{\mathsf{A}}_j$ class.

Hence $\overline{\mathsf{A}}'_j$ is a semi-algebra, which by construction contains A_j, and so also contains the smallest semi-algebra generated by A_j. That is, $\mathcal{A}'(\mathsf{A}_j) \subset \overline{\mathsf{A}}'_j$, and thus $\{\mathcal{A}'(\mathsf{A}_j)\}_{j=1}^{N}$ are mutually independent classes. ■

Remark 1.24 ($\mathcal{A}'(\mathsf{A}_j) \neq \overline{\mathsf{A}}'_j$) *We cannot conclude that $\mathcal{A}'(\mathsf{A}_j) = \overline{\mathsf{A}}'_j$ since the latter class contains $\widetilde{A}_{jk}$ for every original $A_{jk} \in \mathsf{A}_j$, while $\mathcal{A}'(\mathsf{A}_j)$ needs only have such sets expressible as a finite union of other A_j-sets.*

Our final goal is to expand the conclusion of this corollary to assert that the smallest sigma algebras generated by these classes, $\{\sigma(\mathsf{A}_j)\}_{j=1}^{N}$, are mutually independent classes. We then extend this result.

Proposition 1.25 (Independent sigma algebras 1) *Given a probability space $(\mathcal{S}, \mathcal{E}, \mu)$, let $\mathsf{A} = \{A_{jk}\} \subset \mathcal{E}$ be a finite or infinite class of mutually independent sets with $1 \leq j \leq N$ and $1 \leq k \leq M_j$, partitioned into subclasses $\{\mathsf{A}_j\}_{j=1}^{N}$ with $\mathsf{A}_j = \{A_{jk}\}_{k=1}^{M_j}$.*

If $\sigma(A_j)$ denotes the smallest sigma algebra generated by A_j, then $\{\sigma(A_j)\}_{j=1}^N$ are mutually independent.

Proof. *Consider a finite collection of semi-algebras, say $\{\mathcal{A}'(A_j)\}_{j=1}^n$ by relabeling indexes, and which are mutually independent classes by corollary 1.23. Thus given $A_j \in \mathcal{A}'(A_j)$:*

$$\mu\left(\bigcap_{j=1}^n A_j\right) = \prod_{j=1}^n \mu(A_j).$$

Fix $\{A_j\}_{j=2}^n$, and define the class:

$$C(A_1) = \left\{A_1 \in \mathcal{E} \,\middle|\, \mu\left(\bigcap_{j=1}^n A_j\right) = \prod_{j=1}^n \mu(A_j)\right\}.$$

Then $\mathcal{A}'(A_1) \subset C(A_1)$, and we now show that $C(A_1)$ is a sigma algebra.

First, $C(A_1)$ is closed under complementation by Proposition 1.19, so $A \in C(A_1)$ if and only if $\widetilde{A} \in C(A_1)$. Next, consider a countable collection, $\{A_{1k}\}_{k=1}^\infty \subset C(A_1)$ which are disjoint, $A_{1k} \cap A_{1i} = \emptyset$ for $k \neq i$, and we show that $\bigcup_{k=1}^\infty A_{1k} \in C(A_1)$. To this end, note that with $A_1 \equiv \bigcup_{k=1}^\infty A_{1k}$, it follows from De Morgan's laws that:

$$\bigcap_{j=1}^n A_j = \bigcup_{k=1}^\infty \left[A_{1k} \cap \left(\bigcap_{j=2}^n A_j\right)\right],$$

and since $\{A_{1k}\}_{k=1}^\infty$ are disjoint, so too are $\left\{A_{1k} \cap \left(\bigcap_{j=2}^n A_j\right)\right\}_{k=1}^\infty$.

Thus by countable additivity:

$$\begin{aligned}
\mu\left(\bigcap_{j=1}^n A_j\right) &= \sum_{k=1}^\infty \mu\left[A_{1k} \cap \left(\bigcap_{j=2}^n A_j\right)\right] \\
&= \sum_{k=1}^\infty \left[\mu(A_{1k}) \prod_{j=2}^n \mu(A_j)\right] \\
&= \prod_{j=1}^n \mu(A_j).
\end{aligned}$$

Now consider $A_1 \equiv \bigcup_{k=1}^\infty A_{1k}$ where $\{A_{1k}\}_{k=1}^\infty \subset C(A_1)$ are not necessarily disjoint. In this case there exists disjoint $\{A'_{1k}\}_{k=1}^\infty \subset \mathcal{E}$ for which $\bigcup_{k=1}^\infty A_{1k} = \bigcup_{k=1}^\infty A'_{1k}$. For this construction, define $A'_{11} = A_{11}$, and then for $k > 1$:

$$A'_{1k} = A_{1k} \cap \left[\bigcap_{j=1}^{k-1} \widetilde{A}_{1j}\right].$$

As each $A_{1k} \in C(A_1)$, so too is each $\widetilde{A}_{1k} \in C(A_1)$ by (1.27), and thus as a finite intersection, each $A'_{1k} \in C(A_1)$ by Proposition 1.22. Applying the conclusion with disjoint sets and noting that:

$$\sum_{k=1}^\infty \mu(A'_{1k}) = \mu(A_1),$$

by countable additivity, we conclude that $\bigcup_{k=1}^\infty A_{1k} \in C(A_1)$.

Hence $C(A_1)$ is a sigma algebra that contains A_1, and thus by definition contains the smallest such sigma algebra $\sigma(A_1)$. Now the definition of $C(A_1)$ depended on the initial selection of $\{A_j\}_{j=2}^n$, but for any selection the conclusion is the same, that $\sigma(A_1) \subset C(A_1)$, and $\{\sigma(A_1), \mathcal{A}'(A_2), ..., \mathcal{A}'(A_n)\}$ are mutually independent classes.

We can repeat this construction iteratively, next investigating $C(A_2)$ by fixing $A_1 \in \sigma(A_1)$ and $A_j \in \mathcal{A}'(A_j)$ for $3 \leq j \leq n$, and proving $\sigma(A_2)$ to be independent of the other classes. Repeating, we finally conclude that $\{\sigma(A_j)\}_{j=1}^n$ are mutually independent classes. Since this is then true for any finite collection, $\{\sigma(A_j)\}_{j=1}^N$ are mutually independent classes by Definition 1.15. ∎

Remark 1.26 *The above conclusion was framed from the point of view that we began with a class of mutually independent sets* A, *then partitioned this class into subclasses,* $\{A_j\}_{j=1}^N$. *This approach will be seen to fit our needs for Kolmogorov's zero-one law of Section 2.3. But what if we simply began with this latter collection of classes,* $\{A_j\}_{j=1}^N$, *and were given only that these were mutually independent as classes? Could we then conclude that* $\{\sigma(A_j)\}_{j=1}^N$ *are mutually independent classes? In general the answer is "no" as Example 1.21 demonstrates.*

*One might wonder why the "rows" of the mutually independent collection above, which were independent classes, gave rise to independent sigma algebras. The essential ingredient that was added by the assumed mutual independence of all sets, was that this assured **independence of finite intersection sets** from the various* A_j. *In general, intersection sets of independent classes need not be independent as shown in Example 1.21.*

The next result generalizes Proposition 1.25 to reflect this additional ingredient.

Proposition 1.27 (Independent sigma algebras 2) *Assume as given a probability space* $(\mathcal{S}, \mathcal{E}, \mu)$, *and mutually independent classes* $\{A_j\}_{j=1}^N$ *with* $A_j = \{A_{jk}\}_{k=1}^{M_j} \subset \mathcal{E}$, *where* N *and the* M_j *can be finite or infinite.*

If these classes are each closed under finite intersections, then $\{\sigma(A_j)\}_{j=1}^N$ *are mutually independent, where* $\sigma(A_j)$ *denotes the smallest sigma algebra generated by* A_j.

Proof. *Because the* A_j *classes are assumed to be closed under finite intersections, the mutual independence proof of Proposition 1.22 for* $\{A_j'\}_{j=1}^N$ *is unnecessary since now* $A_j = A_j'$. *The remaining steps for Corollary 1.23 and Proposition 1.25 now proceed as above, since these proofs made no further use of the independence of all sets in* A. ■

Remark 1.28 (Beyond countable collections) *Because independence of collections of sets is defined in terms of independence of finite subcollections, all of the above results are true if we begin with an arbitrarily double-indexed collection of independent sets,* A $=$ $\{A_{\alpha\beta}\} \subset \mathcal{E}$ *with* $\alpha \in I$ *and* $\beta \in J_\alpha$. *By arbitrary is here meant not necessarily countable. The conclusions are then:*

1. *If* A $= \{A_{\alpha\beta}\}$ *is a collection of independent sets partitioned into subclasses* $\{A_\alpha\}_{\alpha \in I}$ *with* $A_\alpha \equiv \{A_{\alpha\beta}\}_{\beta \in J_\alpha}$, *and if* $\sigma(A_\alpha)$ *denotes the smallest sigma algebra generated by* A_α, *then* $\{\sigma(A_\alpha)\}$, $\alpha \in I$, *are mutually independent.*

2. *If* $\{A_\alpha\}_{\alpha \in I}$ *is a collection of independent classes,* $A_\alpha \equiv \{A_{\alpha\beta}\}_{\beta \in J_\alpha}$, *and if these classes are each closed under finite intersections, then* $\{\sigma(A_\alpha)\}$, $\alpha \in I$, *are mutually independent.*

1.5 Conditional Probability Measures

We begin with an exercise, that given a measure space $(\mathcal{S}, \mathcal{E}, \mu)$, every set $B \in \mathcal{E}$ induces a measure denoted $\mu_{|B}$.

Exercise 1.29 (Induced measures) *Given a **measure space** $(\mathcal{S}, \mathcal{E}, \mu)$ and $B \in \mathcal{E}$, prove that:*

1. $(B, \mathcal{E}_B, \mu)$ *is a measure space where* $\mathcal{E}_B \equiv \{A \bigcap B | A \in \mathcal{E}\}$. *In other words, prove that* $\mathcal{E}_B$ *is a sigma algebra and μ is a measure on $\mathcal{E}_B$.*

2. $(\mathcal{S}, \mathcal{E}, \mu_{|B})$ *is a measure space where for* $A \in \mathcal{E}$, $\mu_{|B}$ *is defined by:*

$$\mu_{|B}(A) = \mu(A \cap B).$$

Definition 1.30 (Induced measure spaces) *The measure* $\mu_{|B}$ *is sometimes referred to as the measure **induced** by* μ *and* B, *and the associated measure spaces of Exercise 1.29 are called **induced measure spaces.***

When $(\mathcal{S}, \mathcal{E}, \mu)$ is a **probability space**, to ensure that $\mu_{|B}$ is in fact a probability measure we must modify the above construction. In order to do so we must now require that $\mu(B) > 0$. This modified induced measure is called a **conditional probability measure**. This notion will be substantially generalized in Chapter 5 of Book VI.

Definition 1.31 (Conditional probability measure) *If* $(\mathcal{S}, \mathcal{E}, \mu)$ *is a **probability space** and* $B \in \mathcal{E}$ *with* $\mu(B) > 0$, *the **conditional probability measure** $\mu(\cdot|B)$ is defined on* $A \in \mathcal{E}$ *by:*

$$\mu(A|B) \equiv \mu(A \cap B)/\mu(B). \tag{1.28}$$

The expression $\mu(A|B)$ *is called the **conditional probability of** A, and if additional emphasis on B is needed, the **probability of** A **conditional on** B.*

Exercise 1.32 *Prove that* $\mu(\cdot|B)$ *is a measure on* $\mathcal{E}$ *and thus* $(\mathcal{S}, \mathcal{E}, \mu(\cdot|B))$ *is a probability space. With* $\mathcal{E}_B \equiv \{A \cap B | A \in \mathcal{E}\}$, *a sigma algebra by Exercise 1.29, prove that* $\mu(\cdot|B)$ *is a probability measure on* $\mathcal{E}_B$, *where for* $C \in \mathcal{E}_B$:

$$\mu(C|B) \equiv \mu(C)/\mu(B),$$

and thus $(B, \mathcal{E}_B, \mu(\cdot|B))$ *is a probability space.*

Notation 1.33 *It is common practice to denote the conditional probability measure by* $\mu(\cdot|B)$, *and the conditional probability of a set* $A \in \mathcal{E}$ *by* $\mu(A|B)$. *When probability measures are denoted* P, *the standard notations are* $P(\cdot|B)$ *for the conditional probability measure, and* $P(A|B)$ *for the conditional probability of* A. *We will see below that this latter notation is convenient in formulas which involve multiple conditional probabilities.*

A transparent result that links the notion of independent events and conditional probabilities is the following.

Proposition 1.34 (Conditional probability and pairwise independence) *If* $(\mathcal{S}, \mathcal{E}, \mu)$ *is a probability space and* $B \in \mathcal{E}$ *with* $\mu(B) > 0$, *then a set* $A \in \mathcal{E}$ *is pairwise independent of* B *if and only if:*

$$\mu(A|B) = \mu(A). \tag{1.29}$$

Proof. *Immediate from the definitions.* ∎

1.5.1   Law of Total Probability

A relatively simple-to-prove result, but one of great application, is the following. In essence it states that the probability of an event can be evaluated from a "complete" collection of conditional probabilities. This result will also be seen as useful in the evaluation of "moments" of random variables as will be addressed in Book IV.

Proposition 1.35 *(Law of Total Probability) Given a probability space $(\mathcal{S}, \mathcal{E}, \mu)$ and a finite or countable collection of disjoint sets $\{B_i\} \subset \mathcal{E}$ with $\mu(B_i) > 0$ and $\bigcup_i B_i = \mathcal{S}$, then for any $A \in \mathcal{E}$:*

$$\mu(A) = \sum_i \mu(A|B_i)\mu(B_i). \tag{1.30}$$

Proof. *Because $\{A \cap B_i\}$ is a finite or countable disjoint collection of measurable sets with $\bigcup_i(A \cap B_i) = A$, it follows by finite or countable additivity that:*

$$\mu(A) = \sum_i \mu(A \cap B_i).$$

Then (1.30) follows from (1.28). ■

Remark 1.36 (Small generalizations) *The result in (1.30) and proof are equally valid for disjoint $\{B_i\}$ if $\bigcup_i B_i \subset \mathcal{S}$ and $\mu(\mathcal{S} - \bigcup B_i) = 0$, and/or, if the collection $\{B_i\}$ are not disjoint but $\mu(B_i \cap B_j) = 0$ for $i \neq j$. But this observation is more of theoretical interest than practical consequence.*

The law of total probability is very useful for evaluating the probability of events which are defined by the sequential outcome of simpler random events. A simple traditional probability example follows, as well as a finance example.

Example 1.37

1. **Coins and dice:** *A pair of fair dice is rolled with outcomes assumed independent, and then n coins are flipped where n denotes the total score on the dice. Letting A denote the event that 6 heads result, we seek $\mu(A)$. We can thus model the probability space in terms of the collection of events:*

$$\mathcal{S} = \{(n, m) | 2 \leq n \leq 12, \ 0 \leq m \leq n\},$$

and $A = \{(n, 6) | 2 \leq n \leq 12\}$.

Let B_n denote the event that n is the score on the dice, so $B_n = \{(n, m) | 0 \leq m \leq n\}$. Then $\mu(A|B_n) = 0$ for $2 \leq n \leq 5$ since $A \cap B_n = \emptyset$. For $6 \leq n \leq 12$:

$$\mu(A|B_n) = \binom{n}{6}(1/2)^n.$$

Here we are using the binomial probability measure from (1.8) since conditional on B_n, the number of heads has a binomial probability measure with the given n and $p = 1/2$, and so $\mu(A|B_n) = \mu_B(6)$. Also, $\mu(B_n)$ is given by:

$$\mu(B_n) = (13 - n)/36, \text{ for } 7 \leq n \leq 12,$$
$$\mu(B_n) = \mu(B_{14-n}), \text{ for } 2 \leq n \leq 6.$$

This follows because each die has a discrete rectangular probability measure, and the dice outcomes are assumed independent. For example, $\mu(B_7)$ is the probability of the event:

$$\mu(B_7) = \{(1, 6) \cup (2, 5) \cup (3, 4) \cup (4, 3) \cup (5, 2) \cup (6, 1)\}.$$

By finite additivity, the probability of this union is the sum of the probabilities, while by independence the probability of each pair is $1/36$ and hence $\mu(B_7) = 6/36$.

Evaluating $\mu(A)$ by (1.30):

$$\mu(A) = 5(1/2)^6/36 + \sum_{n=7}^{12} \binom{n}{6}(13-n)/[36(2)^n]$$

$$= \frac{2971}{36\,864}$$

$$\approx 0.080594.$$

One could similarly evaluate $\mu(A)$ for other events such as the event that 2 or 5 heads result, or that more than 9 or less than 4 heads result, or that the number of heads is even.

2. ***Group life insurance:*** *A group life insurer is contemplating offering a fixed amount life policy to a senior citizen group using a mail solicitation. Offering a fixed $5000 death benefit, it believes it could enroll 2500 individuals, split 60% female and 40% male. Due to the minimal underwriting involved it will use population-based mortality rates and age distribution, with which it has derived claim rates for the first year of 0.025 for males and 0.018 for females. The CFO requested the probability that claims will exceed $0.25 million in the first year.*

    As above, the probability space is given:

    $$\mathcal{S} = \{(m, f)|0 \le m \le 1000, \ 0 \le f \le 1500\}.$$

    The probability requested is the same as the probability that the total number of claims exceeds 50. Hence if A is the event that claims exceed $0.25 million:

    $$A = \{(m, f)|m + f > 50\}.$$

    Letting B_j denote the event of exactly j total claims:

    $$B_j = \{(m, f)|m + f = j\},$$

    obtains that:

    $$\mu(A) = 1 - \mu(\widetilde{A}) = 1 - \sum_{j=0}^{50} \mu(B_j).$$

    This formula is stated in terms of $\mu(\widetilde{A})$ to avoid the computational effort of evaluating B_j for $51 \le j \le 2500$, and more importantly the complexity associated with the observation that we cannot have more than 1500 female claims nor more than 1000 male claims.

    Now for any j, $\mu(B_j)$ will depend on the male/female mix of claimants. So conditioning on the split between male and female claims, let C_k be the event of k male claims, and thus $j - k$ female claims:

    $$\mu(B_j) = \sum_{k=0}^{j} \mu(B_j|C_k)\mu(C_k).$$

    Now:

    $$\mu(B_j|C_k)\mu(C_k) = \mu(B_j \cap C_k)$$

    is the probability of k male claims and $j - k$ female claims. Using the binomial probability measure:

    $$\mu(B_j|C_k)\mu(C_k) = \binom{1500}{j-k}(0.018)^{j-k}(0.982)^{1500-j+k}$$

    $$\times \binom{1000}{k}(0.025)^k(0.982)^{1000-k}.$$

Thus:

$$\mu(A) = 1 - \sum_{j=0}^{50} \sum_{k=0}^{j} \mu(B_j|C_k)\mu(C_k).$$

Perhaps surprisingly, for large populations, these binomial probabilities can be remarkably difficult to evaluate explicitly without a fair amount of expression manipulation. This is because the evaluation involves products of large n!-type terms, and very small $p^j(1-p)^{n-j}$-type terms. But recall the Poisson approximation with:

$$\begin{aligned} \lambda_1 &= 1500\,(0.018) = 27, \\ \lambda_2 &= 1000\,(0.025) = 25. \end{aligned}$$

Then by (1.13), the above expression can be approximated:

$$\mu(B_j|C_k)\mu(C_k) \approx e^{-52}(27)^{j-k}(25)^k/[(j-k)!k!].$$

Using the binomial theorem:

$$\mu(B_j) = \sum_{k=0}^{j} \mu(B_j|C_k)\mu(C_k) \approx e^{-52}52^j/j!$$

In other words, the measure of j total claims $\mu(B_j)$, is approximated by a single Poisson probability measure with $\lambda = \lambda_1 + \lambda_2$.

Finally,

$$\begin{aligned} \mu(A) &\approx 1 - \sum_{j=0}^{50} e^{-52}52^j/j! \\ &= 1.0621 \times 10^{-3}, \end{aligned}$$

or about 0.1%.

Exercise 1.38

1. **Independent Poissons:** *Generalize the above observation. Assume we are given Poisson probability measures with respective parameters λ_1 and λ_2, so if $B_k^{(i)}$ denotes the event that the outcome is k, then $\mu_P^{(i)}(B_k^{(i)}) = e^{-\lambda_i}\lambda_i^k/k!$, $k = 0, 1, 2, ...$ Letting B_k denote the event that the sum of the outcomes from these distributions is k, derive that the associated probability measure is again Poisson with parameter $\lambda = \lambda_1 + \lambda_2$.*

2. **Auto insurance claims:** *An auto insurance company insures 10,000 vehicles against collision and knows from experience that when a collision occurs that there is a 50% likelihood that the loss will be $5000, and a 50% chance it will be $15,000. It also knows that there is a 10% probability of an accident in a year for any given driver, and that accidents are assumed to be independent of one another since the insurer has a small market share. Use the law of total probability to derive an expression for the probability that the insurer will suffer accident losses in excess of $10,000,000. (Hint: The number of accidents is binomial, but given n accidents, how can one evaluate the probability of different dollar loss levels? Note that if there are then m small losses and n − m large losses, you know both the cost and probability of that event.)*

3. **Bond defaults:** *A bond manager is managing 250 securities in a $1.0 billion par portfolio: 200 medium-risk bonds with $4.5 million par each, and 50 high-risk bonds with $2 million par each. The respective default probabilities in one year are 1.0% and 4.0%, and upon default the manager expects to lose 40% of par on the medium-risk*

bonds, and 75% of par on the high-risk bonds. Use the law of total probability to derive
an expression for the probability that the manager will lose more than $25 million in a
year due to default. (Hint: This is a bit more complicated than exercise 2 since for any
number of defaults, say n, there can be m medium-risk and n − m high-risk defaults
for any logical m. But the probability of such a pair is straightforward, so think about
the loss on each pair.)

1.5.2   Bayes' Theorem

A final result related to conditional probabilities, but again, one with a great many applications, is named for **Thomas Bayes** (c. 1701–1761) who derived the special case of this
result stated in Proposition 1.39 below. This was further developed and popularized over 50
years later by **Pierre-Simon Laplace** (1749–1827). In essence, Bayes' theorem states that
a conditional probability $\mu(A|B)$ can be calculated indirectly from the value of $\mu(B|A)$. It
plays a key role in the theory of **Bayesian statistics** in which probabilities are interpreted
subjectively as reflective of the degree of one's belief in a given outcome, and which can
then be updated with the emergence of data.

Proposition 1.39 (Bayes' Theorem 1) *Given a probability space $(\mathcal{S}, \mathcal{E}, \mu)$ and $\{A, B\} \subset$*
$\mathcal{E}$ sets with $\mu(A) > 0$ and $\mu(B) > 0$, then:

$$\mu(A|B) = \frac{\mu(B|A)\mu(A)}{\mu(B)}. \tag{1.31}$$

Proof. *By (1.28), $\mu(A \cap B) = \mu(A|B)\mu(B) = \mu(B|A)\mu(A)$.* ■

Remark 1.40 (Bayesian statistics) *In the context of Bayesian statistics, $\mu(A|B)$ is in-*
*terpreted as the **posterior probability** of event A conditional on the observation of event*
*B (or "evidence" B), $\mu(A)$ is interpreted as the **prior probability** of event A before the*
*observation of B, while the factor $\mu(B|A)/\mu(B)$ is the **relative likelihood** of observing*
event B given A, where by relative is meant as compared to simply observing B.
   So if $\mu(B|A)/\mu(B) > 1$, meaning the truth of A increases the likelihood of observ-
ing B, then the posterior estimate of the probability of A will exceed the prior estimate:
$\mu(A|B) > \mu(A)$. Similarly, if $\mu(B|A)/\mu(B) < 1$, meaning the truth of A decreases the
likelihood of observing B, then the posterior estimate of the probability of A will be below
the prior estimate: $\mu(A|B) < \mu(A)$. Finally, when $\mu(B|A)/\mu(B) = 1$, meaning that A and
B are independent, then the posterior probability calculation will leave the prior estimate
unchanged.

   In some applications, one has $\{A_i\} \subset \mathcal{E}$, a finite or countable collection of disjoint sets
with $\mu(A_i) > 0$ and $\bigcup_i A_i = \mathcal{S}$, and for which the calculation of $\mu(B|A_i)$ is relatively easy,
where B is an observed outcome. In other words, each A_i represents one of a possible number
of states of the world within which we can calculate the probability of observing the given
outcome B. One is then interested in the probability $\mu(A_j|B)$, that given the observation
B, what is the probability that it was produced by the jth state of the world?

Corollary 1.41 (Bayes' Theorem 2) *If $\{A_i\} \subset \mathcal{E}$ is a finite or countable collection of*
disjoint sets with $\mu(A_i) > 0$ and $\bigcup_i A_i = \mathcal{S}$, then:

$$\mu(A_j|B) = \frac{\mu(B|A_j)\mu(A_j)}{\sum_i \mu(B|A_i)\mu(A_i)}. \tag{1.32}$$

Proof. *Combine (1.31) with (1.30).* ■

Again, we provide a standard probability example below, as well as a finance example.

Example 1.42

1. **Coins and a die:** *Assume that one fair die is rolled, then j fair coins are flipped where j is the score on the die, and n is defined as the number of heads that results. If we observe 2 heads, what is the probability that the score on the die was 3?*

   First, where j denotes the score on the die, and n the number of subsequent heads on the toss of j coins:
   $$S = \{(j,n)|1 \leq j \leq 6, 0 \leq n \leq j\}.$$

   For each pair, $\mu((j,n)) = \binom{j}{n}/(6 \cdot 2^j)$, and we verify by the binomial theorem with $a = b = 1$ that $\mu(S) = 1$.

   Denote the event of 2 heads by $B = \{(j,2)\}$, and the event of a score of i on the die by $A_i = \{(i,n)\}$. Then as a fair die $\mu(A_i) = 1/6$, and this also follows from the event $A_i = \{(i,n)\}$ since the sum of the n-probabilities for $0 \leq n \leq i$ is 1 for all i. Then $\mu(B|A_i)$ is given by:

   $$\mu(B|A_i) = \begin{cases} 0, & i = 1, \\ \binom{i}{2}/2^i, & 2 \leq i \leq 6. \end{cases}$$

   Hence, the probability that the score on the die was 3 given the observed 2 heads is obtained in (1.32) by:

   $$\mu(A_3|B) = \frac{\binom{3}{2}/2^3}{\sum_{i=2}^{6} \binom{i}{2}/2^i} = 8/33.$$

2. **Group life insurance:** *Consider the group life insurer in Example 1.37 above. The year has passed and 40 claims are observed. What is the probability that there were 20 or more male claims?*

   Using the above notation, B_{40} is the event of 40 claims, while C_k is the event of k male claims. The objective is to evaluate:

   $$\sum_{k=20}^{40} \mu(C_k|B_{40}).$$

   Now from (1.31) and the above Poisson calculations,

   $$\begin{aligned} \mu(C_k|B_{40}) &= \mu(B_{40}|C_k)\mu(C_k)/\mu(B_{40}) \\ &\approx \left[e^{-52}(27)^{40-k}(25)^k/[(40-k)!k!]\right] / \left[e^{-52}52^{40}/40!\right] \\ &= \binom{40}{k}(25/52)^{40-k}(25/52)^k. \end{aligned}$$

   Finally,

   $$\sum_{k=20}^{40} \left[\binom{40}{k}(27/52)^{40-k}(25/52)^k\right] = 0.46531,$$

   or about 46.5%.

2

Limit Theorems on Measurable Sets

2.1 Introduction to Limit Sets

Given a measure space $(\mathcal{S}, \mathcal{E}, \mu)$, and a countable collection of sets $\{A_n\}_{n=1}^{\infty} \subset \mathcal{E}$, one is sometimes interested in the properties of sets related to $\bigcap_n A_n$. Of course $\bigcap A_n \in \mathcal{E}$, but it can turn out that $\bigcap A_n = \emptyset$ even when this collection has other significant non-empty intersection properties. For example, there may exist points x which are in infinitely many of these sets, and we could distinguish between the following:

1. Points x which are in infinitely many of these sets, and outside only finitely many.

2. Points x which are in infinitely many of these sets, and outside infinitely many.

The **limit superior** of the collection $\{A_n\}_{n=1}^{\infty}$ is defined as the set of points contained in infinitely many of these sets, and so includes groups 1 and 2, while the **limit inferior** is defined as the subset defined in 1, and that is, as the set of points contained in all but finitely many of these sets.

The formal definitions follow, but will require a little thought to reveal these interpretations. We state these notions within a general measure space, but we will largely be interested in probability spaces.

Definition 2.1 ($\liminf A_n$, $\limsup A_n$, $\lim A_n$) *Given a measure space $(\mathcal{S}, \mathcal{E}, \mu)$ and a countable collection of sets $\{A_n\}_{n=1}^{\infty} \subset \mathcal{E}$, define:*

1. *Limit superior:*

$$\limsup A_n = \bigcap_{n=1}^{\infty} \bigcup_{k=n}^{\infty} A_k. \tag{2.1}$$

2. *Limit inferior:*

$$\liminf A_n = \bigcup_{n=1}^{\infty} \bigcap_{k=n}^{\infty} A_k. \tag{2.2}$$

3. *Limit: If $\limsup A_n = \liminf A_n = A$, define:*

$$\lim A_n \equiv A. \tag{2.3}$$

Remark 2.2 (Discussion of $\liminf A_n$, $\limsup A_n$, $\lim A_n$) *To connect these formal definitions with the introductory comments, note that:*

- *If $x \in \limsup A_n$, then $x \in \bigcup_{k=n}^{\infty} A_k$ for every n, and hence, x is an element of infinitely many sets.*

- *If $x \in \liminf A_n$, then $x \in \bigcap_{k=n}^{\infty} A_k$ for some n, and hence x is an element of infinitely many sets, and not an element of at most the finite collection $\{A_j\}_{j=1}^{n-1}$.*

By this informal analysis:

$$\liminf A_n \subset \limsup A_n, \tag{2.4}$$

but this can also be formalized.

Observe that for any n, m, and any $j \geq \max\{n, m\}$:

$$\bigcap\nolimits_{k=m}^{\infty} A_k \subset A_j \subset \bigcup\nolimits_{k=n}^{\infty} A_k. \tag{1}$$

Consequently, for any n:

$$\bigcup\nolimits_{m=1}^{\infty} \bigcap\nolimits_{k=m}^{\infty} A_k \subset \bigcup\nolimits_{k=n}^{\infty} A_k$$

and thus:

$$\bigcup\nolimits_{n=1}^{\infty} \bigcap\nolimits_{k=n}^{\infty} A_k \subset \bigcap\nolimits_{n=1}^{\infty} \bigcup\nolimits_{k=n}^{\infty} A_k,$$

which is (2.4).

Finally, if $\lim A_n$ exists, then (2.4) implies that $\limsup A_n \subset \liminf A_n$. Thus, if x is an element of infinitely many sets, then it is an element of all but finitely many sets. Put another way, when $\lim A_n$ exists, there are no points which are both inside infinitely many sets, and outside infinitely many sets.

Notation 2.3 (lim inf **and** lim sup **of function sequences**) *It may well not be apparent initially how the notions of $\limsup$ and $\liminf$ here relate to these same notions in the context of a sequence of functions $\{f_n(x)\}$ as defined in Definition I.3.42. This terminology is also used for numerical sequences $\{y_n\}$. But the connection between numerical and function sequences is transparent because $\limsup$ and $\liminf$ of a function sequence are defined pointwise for each x, and for each such fixed x the sequence $\{f_n(x)\}$ is a numerical sequence.*

Recall that $\limsup f_n(x)$ is defined by:

$$\limsup f_n(x) \equiv \inf_n \sup_{k \geq n} f_k(x),$$

and $\limsup A_n$ is defined in (2.1) by:

$$\limsup A_n = \bigcap\nolimits_{n=1}^{\infty} \bigcup\nolimits_{k=n}^{\infty} A_k.$$

Thus for any m:

$$\limsup f_n(x) \leq \sup_{k \geq m} f_k(x), \qquad \limsup A_n \subset \bigcup\nolimits_{k=m}^{\infty} A_k.$$

In other words, both notions of $\limsup$ create a decreasing sequence of function values $\left\{\sup_{k \geq m} f_k(x)\right\}_{m=1}^{\infty}$, respectively sets $\left\{\bigcup_{k=m}^{\infty} A_k\right\}_{m=1}^{\infty}$, that approach the final result from above because:

$$\sup_{k \geq m+1} f_k(x) \leq \sup_{k \geq m} f_k(x), \qquad \bigcup\nolimits_{k=m+1}^{\infty} A_k \subset \bigcup\nolimits_{k=m}^{\infty} A_k.$$

*Thus in both cases, $\limsup$ is a **limit** defined by a sequence of values or sets which approach the final result from above. In other words, both final results are a limit of "superior" objects of the same type.*

Using the same analysis, it can be seen that in both cases the $\liminf$ *is a limit defined by a sequence of "inferior" function values* $\{\inf_{k \geq m} f_k(x)\}_{m=1}^{\infty}$ *or sets* $\left\{\bigcap_{k=m}^{\infty} A_k\right\}_{m=1}^{\infty}$, *which approach the final result from below. This follows from the definitions:*

$$\liminf f_n(x) \equiv \sup_n \inf_{k \geq n} f_k(x),$$

$$\liminf A_n \equiv \bigcup_{n=1}^{\infty} \bigcap_{k=n}^{\infty} A_k,$$

and the observation that for any m:

$$\inf_{k \geq m} f_k(x) \leq \inf_{k \geq m+1} f_k(x) \leq \liminf f_n(x),$$

$$\bigcap_{k=m}^{\infty} A_k \subset \bigcap_{k=m+1}^{\infty} A_k \subset \liminf A_n.$$

Finally, the existence of $\lim f_n(x)$ *is equivalent to* $\liminf f_n(x) = \limsup f_n(x)$, *and then all 3 values agree pointwise. For sets, if* $\liminf A_n = \limsup A_n$ *we define* $\lim A_n$ *to equal this common value.*

The next result has a simple proof, owing to the robustness of sigma algebras to set operations. This simple proposition justifies an inquiry into the values of $\mu(\liminf A_n)$ and $\mu(\limsup A_n)$.

Proposition 2.4 (Measurability of $\liminf A_n$ **and** $\limsup A_n$**)** *Given a measure space* $(S, \mathcal{E}, \mu)$ *and a countable collection of sets* $\{A_n\}_{n=1}^{\infty} \subset \mathcal{E}$:

$$\liminf A_n \in \mathcal{E} \ , \qquad \limsup A_n \in \mathcal{E}.$$

Thus $\lim A_n \in \mathcal{E}$ *when this limit exists.*
Proof. *Both results follow from the fact that* $\mathcal{E}$ *is a sigma algebra which is closed under countable unions and intersections, while the last statement is true by definition.* ∎

The next proposition provides a general set of inequalities which follow from the above remark and the continuity results for measures identified in Proposition I.2.45. The second result actually generalizes these continuity results beyond nested collections of sets.

This result is stated for probability spaces, but is **true for measure spaces** under the added assumption that:

$$\mu\left(\bigcup_{k=1}^{\infty} A_k\right) < \infty.$$

This assumption is needed for the last inequality of (2.5), to justify the application of continuity from above of measures, and thus also needed for (2.6).

Proposition 2.5 (On $\mu(\liminf A_n)$ **and** $\mu(\limsup A_n)$**)** *Given a probability space* $(S, \mathcal{E}, \mu)$ *and a countable collection of sets* $\{A_n\}_{n=1}^{\infty} \subset \mathcal{E}$:

1. *Measure inequalities:*

$$
\begin{aligned}
\mu(\liminf A_n) &\leq \liminf \mu(A_n) \qquad\qquad (2.5) \\
&\leq \limsup \mu(A_n) \leq \mu(\limsup A_n).
\end{aligned}
$$

2. *If* $\lim A_n$ *exists, then:*

$$\mu(\lim A_n) = \lim \mu(A_n). \qquad\qquad (2.6)$$

Proof. *For result 1, the middle inequality follows from the definitions, that for any numerical sequence x_n:*

$$\liminf x_n \leq \limsup x_n.$$

Indeed, $\inf_{k \geq n} x_k \leq \sup_{k \geq n} x_k$ *for any* n, *while the sequence on the left is increasing and that on the right is decreasing, so the respective limits satisfy the same inequality.*

As noted in (1) of Remark 2.2, with $j = m = n$:

$$\bigcap_{k=m}^{\infty} A_k \subset A_m \subset \bigcup_{k=m}^{\infty} A_k.$$

Now $B_m \equiv \bigcap_{k=m}^{\infty} A_k$ *is a nested collection of measurable sets,* $B_m \subset B_{m+1}$, *with* $\bigcup_{m=1}^{\infty} B_m = \liminf A_n$. *Thus by continuity from below of Proposition I.2.45,* $\{\mu(B_m)\}$ *has a limiting value and:*

$$\mu(\liminf A_n) = \lim \mu(B_m).$$

Now $B_m \subset A_m$ *implies* $\mu(B_m) \leq \mu(A_m)$ *by monotonicity of measures, thus* $\liminf \mu(B_m) \leq \liminf \mu(A_m)$. *This proves the left-most inequality since* $\liminf \mu(B_m) = \lim \mu(B_m)$.

The argument is analogous for the right-most inequality of (2.5), defining $B_m \equiv \bigcup_{k=m}^{\infty} A_k$ *and using continuity from above.*

The result in 2 is a corollary of (2.5) and the definition of limit set. ∎

2.2   The Borel-Cantelli Lemma

This section introduces a critically important tool for demonstrating that a set, defined as the lim sup of a set sequence, has measure 0 or 1. It will be referenced in many investigations in future books where such an outcome is of interest. This important theorem is universally called the Borel-Cantelli lemma, named for **Émile Borel** (1871–1956) and **Francesco Cantelli** (1875–1966).

The Cantelli part of the result is applicable in general measure spaces, while the Borel part, due to the requirement of independence of sets, is a probability space result.

Proposition 2.6 (Borel-Cantelli Lemma) *Given a probability space* $(\mathcal{S}, \mathcal{E}, \mu)$ *and a countable collection of sets* $\{A_n\}_{n=1}^{\infty} \subset \mathcal{E}$:

1. *(Cantelli) If* $\sum_{n=1}^{\infty} \mu(A_n) < \infty$, *then:*

$$\mu(\limsup A_n) = 0.$$

2. *(Borel) If* $\sum_{n=1}^{\infty} \mu(A_n) = \infty$ *and* $\{A_n\}_{n=1}^{\infty}$ *are mutually independent, then:*

$$\mu(\limsup A_n) = 1.$$

Proof.

1. *By (2.1),* $\limsup A_n \subset \bigcup_{k \geq m}^{\infty} A_k$ *for all* m, *and hence by monotonicity and subadditivity of* μ:

$$\mu(\limsup A_n) \leq \mu\left(\bigcup_{k=m}^{\infty} A_k\right) \leq \sum_{k=m}^{\infty} \mu(A_k).$$

The convergence of $\sum_{n=1}^{\infty} \mu(A_n)$ *then implies that this upper bound can be made as small as desired.*

2. *For the Borel result we prove that:*

$$\mu\left(\widetilde{\limsup A_n}\right) = 0. \tag{1}$$

Using De Morgan's laws:

$$\mu\left(\widetilde{\limsup A_n}\right) \equiv \mu\left(\bigcup_{n=1}^{\infty}\bigcap_{k=n}^{\infty}\widetilde{A}_k\right),$$

and by countable subadditivity:

$$\mu\left(\bigcup_{n=1}^{\infty}\bigcap_{k=n}^{\infty}\widetilde{A}_k\right) \leq \sum_{n=1}^{\infty}\mu\left(\bigcap_{k=n}^{\infty}\widetilde{A}_k\right).$$

The result in (1) is then proved by demonstrating that for every n:

$$\mu\left(\bigcap_{k=n}^{\infty}\widetilde{A}_k\right) = 0. \tag{2}$$

To this end, Corollary 1.20 implies that $\{\widetilde{A}_k\}$ are also mutually independent and hence:

$$\mu\left(\bigcap_{k=n}^{N}\widetilde{A}_k\right) = \prod_{k=n}^{N}\mu\left(\widetilde{A}_k\right) = \prod_{k=n}^{N}(1 - \mu(A_k)).$$

Applying the Taylor series in (1.12) with a remainder term as in the proof of the Poisson limit theorem, obtains for $x > 0$:

$$e^{-x} = 1 - x + \xi^2/2 > 1 - x,$$

where $0 < \xi < x$. Hence, for every N:

$$\prod_{k=n}^{N}(1 - \mu(A_k)) < \exp\left[-\sum_{k=n}^{N}\mu(A_k)\right].$$

Monotonicity of μ assures that $\mu\left(\bigcap_{k=n}^{\infty}\widetilde{A}_k\right) \leq \mu\left(\bigcap_{k=n}^{N}\widetilde{A}_k\right)$ for all n, N, and thus:

$$\mu\left(\bigcap_{k=n}^{\infty}\widetilde{A}_k\right) < \exp\left[-\sum_{k=n}^{N}\mu(A_k)\right].$$

The divergence of $\sum_{n=1}^{\infty}\mu(A_n)$ now implies that for $\mu\left(\bigcap_{k=n}^{\infty}\widetilde{A}_k\right) = 0$ for all n, which is (2).

∎

Remark 2.7 *Two comments on the assumptions underlying the above results:*

1. *(**Cantelli**) The weaker version of the Cantelli assumption, that simply $\mu(A_n) \to 0$, is enough to imply that $\mu(\liminf A_n) = 0$ by (2.5). The result here is the stronger conclusion that $\mu(\limsup A_n) = 0$ because of the stronger assumption that $\mu(A_n) \to 0$ fast enough to ensure the convergence of $\sum_{n=1}^{\infty}\mu(A_n)$.*

2. *(**Borel**) In any case with $\sum_{n=1}^{\infty}\mu(A_n) = \infty$ and $\limsup \mu(A_n) > 0$, (2.5) provides the conclusion that $\mu(\limsup A_n) > 0$. But in a case of divergence with $\limsup \mu(A_n) = 0$, (2.5) provides no such assurance. The Borel result provides a strong conclusion from the divergence of $\sum_{n=1}^{\infty}\mu(A_n)$, that $\mu(\limsup A_n) = 1$, because of the strong assumption on mutual independence.*

The following corollary of the Borel-Cantelli lemma is remarkable, and yet is a special case of **Kolmogorov's zero-one law** addressed in the next section.

Corollary 2.8 (Borel Zero-One Law) *Let* $\{A_n\}_{n=1}^{\infty}$ *be mutually independent. Then* $\mu(\limsup A_n)$ *can assume only two values, 0 or 1, and it does so based on the convergence or divergence of the series* $\sum_{n=1}^{\infty} \mu(A_n)$, *respectively.*
Proof. *Immediate from Borel-Cantelli.* ∎

Example 2.9 *Here we explore examples of cases addressed by the Borel-Cantelli theorem, as well as cases not addressed.*

1. *(**Cantelli**) The Cantelli result is silent on the value of* $\mu(\limsup A_n)$ *for the general case where* $\sum_{n=1}^{\infty} \mu(A_n) = \infty$. *If* $\{A_n\}_{n=1}^{\infty}$ *are mutually independent, then the Borel result obtains that* $\mu(\limsup A_n) = 1$, *but what about the general case? Here are examples of both extremes.*

    (a) *On the Borel probability space* $([0,1], \mathcal{B}[0,1], m)$ *with Lebesgue measure, let* $A_n = [0, 1/n]$. *Then since* $\mu(A_n) = 1/n$, *it is clear that* $\sum_{n=1}^{\infty} \mu(A_n)$ *is divergent. But:*

    $$\limsup A_n = \bigcap_{n=1}^{\infty} \bigcup_{k=n}^{\infty} A_k = \bigcap_{n=1}^{\infty} [0, 1/n] = [0],$$

    and so $\mu(\limsup A_n) = 0$.
    Of course $\{A_n\}_{n=1}^{\infty}$ *cannot be mutually independent by Borel's result, but also by definition since with* $m < n$:

    $$m(A_n \cap A_m) = 1/n \neq m(A_n)m(A_m).$$

    (b) *On the same probability space, let* $B_n = [1/n, 1]$. *Then* $\mu(B_n) = (n-1)/n$ *and it follows that* $\sum_{n=1}^{\infty} \mu(B_n)$ *is divergent. But*

    $$\limsup B_n = \bigcap_{n=1}^{\infty} \bigcup_{k=n}^{\infty} B_k = \bigcap_{n=1}^{\infty} (0, 1] = (0, 1],$$

    and so $\mu(\limsup B_n) = 1$.
    In this case $\{B_n\}_{n=1}^{\infty}$ *are not mutually independent since with* $m < n$:

    $$m(B_n \cap B_m) = (m-1)/m \neq m(B_n)m(B_m).$$

2. *(**Cantelli**) From the Borel zero-one law, for independent* $\{A_n\}_{n=1}^{\infty}$ *the value of* $\mu(\limsup A_n)$ *can only be 0 or 1, and this is determined by the convergence or divergence of the series* $\sum_{n=1}^{\infty} \mu(A_n)$, *respectively. But what about the general case?*

    For general $\{A_n\}_{n=1}^{\infty}$, *Cantelli applies and assures that* $\mu(\limsup A_n) = 0$ *in the convergence case. By example 1,* $\mu(\limsup A_n)$ *can attain either a value of 0 and 1 in the divergence case.*

Exercise 2.10 *Identify examples* $\{A_n\}_{n=1}^{\infty}$ *in the space* $([0,1], \mathcal{B}[0,1], m)$ *above, which will necessarily not be mutually independent, for which* $\mu(\limsup A_n) = r$ *for any* r *with* $0 < r < 1$.

3. *(**Borel**) Recalling Chapter 9 of Book I, consider the infinite product probability space associated with fair coin flips* $(Y^{\mathbb{N}}, \sigma(Y^{\mathbb{N}}), \mu_{\mathbb{N}})$ *where* $Y = \{H, T\}$, *and* $\mu_{\mathbb{N}}$ *is the measure induced by the probability measure* μ^B *on* Y *as defined in (1.7) with*

$\mu^B(H) = \mu^B(T) = 1/2$. *Each point* $y \in Y^{\mathbb{N}}$ *can be envisioned as an infinite sequence of* H*s and* T*s. Letting* $\{n_j\}_{j=1}^{\infty}$ *be an arbitrary sequence of integers,* $n_j \geq 0$*, define* $N_0 = 0$ *and:*

$$N_j = \sum_{i=1}^{j} n_i.$$

Define A_j *by*

$$A_j = \{y \in Y^{\mathbb{N}} | y_i = H \text{ for } N_{j-1} + 1 \leq i \leq N_j\},$$

so A_j *is the set of points which has a run of* H*s in the* y_i*-positions for* $N_{j-1} + 1 \leq i \leq N_j$*. If* $n_j = 0$*, then* $A_j = \emptyset$*.*

For $n_j \geq 1$*, it is an exercise in the Book I construction that:*

$$\mu_{\mathbb{N}}(A_j) = 1/2^{n_j}.$$

Further, $\{A_n\}_{n=1}^{\infty}$ *are pairwise independent since for non-empty sets:*

$$\mu_{\mathbb{N}}(A_j \cap A_k) = 1/2^{n_j + n_k} = \mu_{\mathbb{N}}(A_j)\mu_{\mathbb{N}}(A_k),$$

and the same is true if one or both of the sets is empty. This independence result extends to any finite collection of distinct sets, and thus $\{A_n\}_{n=1}^{\infty}$ *are independent sets.*

Now for any k*:*

$$\bigcup_{j \geq k} A_j = \{y \in Y^{\mathbb{N}} | y_i = H \text{ for } N_{j-1} + 1 \leq i \leq N_j, \text{ some } j \geq k\}.$$

Thus $\bigcup_{j \geq k} A_j$ *is the set of points which have a run of* H*s in the* y_i*-positions for* $N_{j-1} + 1 \leq i \leq N_j$ *for some* $j \geq k$*.*

Now consider

$$\limsup A_n \equiv \bigcap_{k=1}^{\infty} \bigcup_{j \geq k} A_j.$$

If $\{N_j\}$ *is bounded, then* $\limsup A_n = \emptyset$ *since* $\bigcup_{j \geq k} A_j = \emptyset$ *for* $k > \max N_j$*.*

If $\{N_j\}$ *is unbounded, meaning infinitely many* $n_j \geq 1$*, then* $y \in \limsup A_n$ *implies that* y *has an infinite run sequence. That is, there exists increasing* $\{j_n\}_{n=1}^{\infty}$ *so that* $y_i = H$ *for* $N_{j_n - 1} + 1 \leq i \leq N_{j_n}$ *and all* n*. Otherwise there exists* J*, so that* y *has no run for some index interval* $N_{j-1} + 1 \leq i \leq N_j$ *for* $j \geq J$*. But then such* $y \notin \bigcup_{j \geq J} A_j$ *and hence* $y \notin \limsup_n A_n$*, a contradiction. Further, for any* y *defined by such an infinite sequence of runs,* $y \in \limsup A_n$ *since then* $y \in \bigcup_{j \geq k} A_j$ *for all* k*. In other words,* $\limsup A_n$ *is characterized by the collection of points with an infinite sequence of runs.*

By the Borel zero-one law, this set has probability 0 or 1, and we can determine which by looking to the series $\sum_{j=1}^{\infty} \mu_{\mathbb{N}}(A_j)$*. When* $\{N_j\}$ *is bounded,* $\sum_{j=1}^{\infty} \mu_{\mathbb{N}}(A_j)$ *is a sum of finitely many terms and hence converges, so* $\mu_{\mathbb{N}}(\limsup A_n) = 0$ *consistent with the observation that* $\limsup A_n = \emptyset$*. Eliminating the empty sets in the case of unbounded* $\{N_j\}$*, which does not change this summation, we have by the above calculation:*

$$\sum_{j=1}^{\infty} \mu_{\mathbb{N}}(A_j) = \sum_{j=1}^{\infty} 1/2^{n_j},$$

where now all $n_j \geq 1$*.*

Some conclusions:

(a) *If* $\{n_j\}_{j=1}^{\infty}$ *is bounded, say* $n_j \leq n$ *for all* j*, then this series diverges and hence* $\mu(\limsup A_n) = 1$*.*

(b) *If $\{n_j\}_{j=1}^{\infty}$ is unbounded, then either conclusion can follow. For example:*

    i If $n_j = j$, the series is convergent and hence $\mu_{\mathbb{N}}(\limsup_n A_n) = 0$.

    *ii If $n_j = \lfloor \log_2 j \rfloor$, where this symbol denotes the **greatest integer** less than or equal to $\log_2 j$, then the series diverges. This follows since:*

$$\log_2 j - 1 < \lfloor \log_2 j \rfloor \le \log_2 j,$$

and this implies that:

$$1/j \le 1/2^{n_j} < 2/j.$$

Hence the series is divergent and $\mu_{\mathbb{N}}(\limsup A_n) = 1$.

Remark 2.11 (On 3. of Example 2.9) *It may appear that in none of the cases of Example 3 would the value of $\mu(\limsup A_n)$ be readily derivable from the $\limsup$ definition. But for this example, the conclusion is derivable directly based on an application of the Borel proof.*

First, since $\{A_j\}_{n=1}^{\infty}$ are independent, so too are $\{\widetilde{A}_j\}_{n=1}^{\infty}$ by Corollary 1.20, and thus:

$$
\begin{aligned}
\mu_{\mathbb{N}}\left(\bigcup_{j \ge k} A_j\right) &= 1 - \mu_{\mathbb{N}}\left(\widetilde{\bigcup_{j \ge k} A_j}\right) \\
&= 1 - \mu_{\mathbb{N}}\left(\bigcap_{j \ge k} \widetilde{A}_j\right) \\
&= 1 - \prod_{j \ge k}^{\infty} (1 - \mu(A_j)) \\
&= 1 - \prod_{j \ge k}^{\infty} (1 - 1/2^{n_j}).
\end{aligned}
$$

By the nested property:

$$\bigcup_{j \ge k+1} A_j \subset \bigcup_{j \ge k} A_j,$$

and continuity of $\mu_{\mathbb{N}}$ from above obtains:

$$\mu_{\mathbb{N}}(\limsup A_n) = \lim_{k \to \infty} \mu_{\mathbb{N}}\left(\bigcup_{j \ge k} A_j\right) = \lim_{k \to \infty}\left[1 - \prod_{j \ge k}(1 - 1/2^{n_j})\right].$$

For a divergent series $\{1/2^{n_j}\}_{j=1}^{\infty}$:

$$\prod_{j \ge k}(1 - 1/2^{n_j}) \to 0,$$

as $k \to \infty$ and hence $\mu_{\mathbb{N}}(\limsup A_n) = 1$. This follows by taking a logarithm of the infinite product and approximating each term $\ln\left[1 - 1/2^{n_j}\right]$ using a Taylor series expansion with remainder as in the proof of the Poisson limit theorem. As the log of the product diverges to $-\infty$, the result follows by continuity of the exponential function, noting that:

$$\prod_{j=k}^{\infty}(1 - 1/2^{n_j}) = \exp\left[\sum_{j=k}^{\infty} \ln(1 - 1/2^{n_j})\right].$$

For a convergent series $\{1/2^{n_j}\}_{j=1}^{\infty}$, a similar analysis obtains that:

$$\prod_{j \ge k}(1 - 1/2^{n_j}) \to 1,$$

as $k \to \infty$ and thus $\mu_{\mathbb{N}}(\limsup A_n) = 0$. The derivation is similar but adding the observation that if $\{x_j\}_{j=1}^{\infty}$ is convergent, so too is $\{x_j^2\}_{j=1}^{\infty}$.

2.3 Kolmogorov's Zero-One Law

Kolmogorov's zero-one law is named for **Andrey Kolmogorov** (1903–1987). In effect it says that the Borel-Cantelli theorem, as summarized in the Borel zero-one law, is in fact a special case of a very general result. And this is so because $\limsup A_n$ is a special case of a so-called **tail event**. Kolmogorov proved that the probability of every tail event defined in terms of mutually independent $\{A_n\}_{n=1}^\infty$, is either 0 or 1.

For the following definition, the sigma algebra $\sigma(A_n, A_{n+1}, A_{n+2}, ...)$ generated by $\{A_j\}_{j=n}^\infty$ is defined as the smallest sigma algebra that contains these sets. Recalling Proposition I.2.8, this in turn is defined as the intersection of all sigma algebras that contain these sets, a definition that is not vacuous since $\mathcal{E}$ is one such sigma algebra.

Definition 2.12 ($\mathcal{T} \equiv \mathcal{T}(\{A_n\}_{n=1}^\infty)$ **and tail events**) *Given a probability space* $(\mathcal{S}, \mathcal{E}, \mu)$ *and a countable collection of sets* $\{A_n\}_{n=1}^\infty \subset \mathcal{E}$, *the* **tail sigma algebra associated with** $\{A_n\}_{n=1}^\infty$, *denoted:*

$$\mathcal{T} \equiv \mathcal{T}(\{A_n\}_{n=1}^\infty),$$

is defined by:

$$\mathcal{T} = \bigcap_{n=1}^\infty \sigma(A_n, A_{n+1}, A_{n+2}, ...), \tag{2.7}$$

where $\sigma(A_n, A_{n+1}, A_{n+2}, ...)$ *is defined as the* **sigma algebra generated by** $\{A_j\}_{j=n}^\infty$. *A* **tail event** *is any set* $A \in \mathcal{T}$.

Intuitively, a tail event is any event $A \in \mathcal{E}$ which depends only on $\{A_n\}$ for arbitrarily large n, and this notion is formalized in the following way. For any finite N:

$$\mathcal{T}(\{A_n\}_{n=N}^\infty) = \mathcal{T}(\{A_n\}_{n=1}^\infty),$$

because $\{\sigma(A_n, A_{n+1}, A_{n+2}, ...)\}_{n=1}^\infty$ is a nested collection of sigma algebras:

$$\sigma(A_{n+1}, A_{n+2}, A_{n+3}, ...) \subset \sigma(A_n, A_{n+1}, A_{n+2}, ...).$$

Thus the definition in (2.7) is equivalent to:

$$\mathcal{T} = \bigcap_{n=N}^\infty \sigma(A_n, A_{n+1}, A_{n+2}, ...),$$

for any N.

Remark 2.13 ($\mathcal{T}$ **is a sigma algebra**) *Note that* $\mathcal{T}$ *is indeed a sigma algebra by Proposition I.2.8 since it is the intersection of a countable collection of sigma algebras. Further,* $\mathcal{T} \subset \mathcal{E}$ *since* $\sigma(A_n, A_{n+1}, A_{n+2}, ...) \subset \mathcal{E}$ *for all* n. *Hence,* μ *is well defined on* $\mathcal{T}$ *and it therefore makes sense to seek the value of* $\mu(A)$ *for tail events. Kolmogorov's zero-one law addresses the probability of tail events in the special case where the collection of sets,* $\{A_n\}_{n=1}^\infty$, *are mutually independent.*

Example 2.14 ($\liminf A_n$, $\limsup A_n \in \mathcal{T}$) *Given a probability space* $(\mathcal{S}, \mathcal{E}, \mu)$ *and a countable collection of sets* $\{A_n\}_{n=1}^\infty \subset \mathcal{E}$, *both* $\limsup A_n$ *and* $\liminf A_n$ *are tail events:*

$$\limsup A_n \in \mathcal{T}, \qquad \liminf A_n \in \mathcal{T}.$$

For $\limsup A_n$, *note that for any* m:

$$\limsup A_n \equiv \bigcap_{j=1}^\infty \bigcup_{k \geq j}^\infty A_k$$

$$= \bigcap_{j=m}^\infty \bigcup_{k \geq j}^\infty A_k,$$

by nesting of the intersected sets:

$$\bigcup\nolimits_{k \geq j+1}^{\infty} A_k \subset \bigcup\nolimits_{k \geq j}^{\infty} A_k.$$

Thus for any m

$$\limsup\nolimits_n A_n = \bigcap\nolimits_{j=m}^{\infty} \bigcup\nolimits_{k \geq j}^{\infty} A_k \in \sigma(A_m, A_{m+1}, A_{m+2}, ...),$$

and the result follows.

The result for $\liminf A_n$ *follows analogously and is left as an exercise.*

We now state and prove **Kolmogorov's zero-one law.** The proof here is relatively easy because all the technical results are provided by Section 1.4.1 on Independent Classes and Associated Sigma Algebras. The key to the result is to prove that a tail event is independent of itself.

Proposition 2.15 (Kolmogorov's zero-one law) *Given a probability space* $(\mathcal{S}, \mathcal{E}, \mu)$ *and mutually independent events* $\{A_j\}_{j=1}^{\infty}$, *if* $A \in \mathcal{T}(\{A_n\}_{n=1}^{\infty})$, *then:*

$$\mu(A) \in \{0, 1\}. \tag{2.8}$$

Proof. *The result will follow from Proposition 1.18 once it is proved that every such A is independent of itself. To this end, Proposition 1.25 obtains that for any n:*

$$\sigma(A_1), \sigma(A_2), ..., \sigma(A_{n-1}), \sigma(A_n, A_{n+1}, A_{n+2}, ...),$$

are independent classes. Since $A \in \sigma(A_n, A_{n+1}, A_{n+2}, ...)$, *it follows that* $A, A_1, A_2, ..., A_{n-1}$ *is an independent collection for any n, and thus by Definition 1.15,* $\{A\} \cup \{A_j\}_{j=1}^{\infty}$ *is a mutually independent collection. Another application of Proposition 1.25 yields that* $\sigma(A)$ *and* $\sigma(A_1, A_2, ...)$ *are mutually independent classes.*

Now since $A \in \mathcal{T} \subset \sigma(A_1, A_2, ...)$, *and obviously* $A \in \sigma(A)$, *we conclude that A is independent of itself and the result follows.* ∎

Remark 2.16 (0 or 1?) *Kolmogorov's conclusion is very powerful in limiting the potential values of* $\mu(A)$ *for* $A \in \mathcal{T}$, *but it provides no insight in a given application as to which outcome is correct, nor how one can even begin to determine which outcome is correct.*

While the Borel-Cantelli theorem is on the one hand a special case of Kolmogorov's more general result, it is also very much more informative for the special case of the tail event $\limsup A_n$ *for independent* $\{A_n\}$. *The Borel-Cantelli result not only specifies that* $\mu(A) = 0$ *or* $\mu(A) = 1$, *but also provides the criterion which identifies when either happens based on the convergence or divergence of* $\sum \mu(A_n)$.

For other tail events it is often difficult to determine criteria on which to base the final conclusion. The following is a tractable example where a conclusion can be reached.

Example 2.17 (Example 2.9, cont'd) *Recall the analysis of* $\limsup A_n$ *in 3 of Example 2.9, but now adapted to investigate* $\liminf A_n$, *for which we have no Borel-Cantelli-type result. Given a sequence of integers* $\{n_j\}_{j=1}^{\infty}$ *with* $n_j \geq 0$, *we defined* $N_0 = 0$ *and:*

$$N_j = \sum\nolimits_{i=1}^{j} n_i,$$

and then:

$$A_j \equiv \{y \in Y^{\mathbb{N}} | y_i = H \text{ for } N_{j-1} + 1 \leq i \leq N_j\}.$$

For given $n_j \geq 1$, A_j is the set of points which have a run of Hs in the y_i-positions for $N_{j-1} + 1 \leq i \leq N_j$ and $\mu_{\mathbb{N}}(A_j) = 1/2^{n_j}$, while $A_j = \emptyset$ for $n_j = 0$. Also proved above, $\{A_n\}_{n=1}^{\infty}$ is a mutually independent collection.

Now:

$$\bigcap_{k \geq j} A_k = \{y \in Y^{\mathbb{N}} | y_i = H \text{ for } i \geq N_{j-1} + 1\}.$$

That is, $\bigcap_{k \geq j} A_k$ is the set of points which have all Hs in the y_i-positions for $i \geq N_{j-1} + 1$.

By definition:

$$\liminf A_n = \bigcup_{j=1}^{\infty} \bigcap_{k \geq j} A_k,$$

and there is a relatively simple way to characterize the points that are in this set. That is, $y \in \liminf A_n$ if and only if there exists N_{j-1} so that $y_i = H$ for $i \geq N_{j-1} + 1$. As $\liminf A_n \in \mathcal{T}$, the Kolmogorov zero-one law restricts the possibilities for $\mu_{\mathbb{N}}(\liminf A_n)$ to 0 or 1, *but gives no resolution as to which is the correct value, so we investigate a direct evaluation.*

1. First, assume all $n_i \geq 1$. Fixing j, define for $n \geq j$:

$$B_n = \bigcap_{k=j}^{n} A_k.$$

Then $\{B_n\}$ is a nested sequence of sets, $B_{n+1} \subset B_n$, with $\mu_{\mathbb{N}}(B_n) = 1/2^{M_j(n)}$ where $M_j(n) \equiv \sum_{i=j}^{n} n_i$, and $\bigcap_{k \geq j} A_k = \bigcap_{n \geq j} B_n$. Hence by continuity from above of $\mu_{\mathbb{N}}$:

$$\mu_{\mathbb{N}}\left(\bigcap_{k \geq j} A_k\right) = \lim_{n \to \infty} 1/2^{M_j(n)}.$$

Since $n_i \geq 1$ for all i, it follows that $M_j(n)$ is unbounded in n for every j, and thus $\mu_{\mathbb{N}}\left(\bigcap_{k \geq j} A_k\right) = 0$.

2. If there exists I so that $n_i \geq 1$ for $i \geq I + 1$, then again $\mu_{\mathbb{N}}\left(\bigcap_{k \geq j} A_k\right) = 0$ for every j. For $j \leq I$, $\bigcap_{k \geq j} A_k = \emptyset$ if there exists i with $j \leq i \leq I$ and $n_i = 0$, while otherwise $n_i \geq 1$ for $i \geq j$. This latter case and that for $j > I$ follow the derivation in part 1 to prove that $\mu_{\mathbb{N}}\left(\bigcap_{k \geq j} A_k\right) = 0$.

3. Finally, if there does not exist I so that $n_i \geq 1$ for $i \geq I + 1$, then $\bigcap_{k \geq j} A_k = \emptyset$ for all j.

Thus for this example we can conclude that in all cases:

$$\mu_{\mathbb{N}}(\liminf A_n) = 0,$$

as the countable union of sets of measure 0.

3

Random Variables and Distribution Functions

3.1   Introduction and Definitions

As noted above, a probability space is a finite measure space by another name, but one that typically reflects a model and a story. Similarly, a random variable is a measurable function by another name, and one that again often reflects a model or a story. However, a random variable is a measurable function with a very different notational convention.

The following can be compared with that for Borel measurability in Definition I.3.5.

Definition 3.1 (Random variable) *Given a probability space* $(\mathcal{S}, \mathcal{E}, \mu)$, *a **random variable (r.v.)** is a real-valued function:*

$$X : \mathcal{S} \longrightarrow \mathbb{R},$$

such that for any bounded or unbounded interval, $(a, b) \subset \mathbb{R}$:

$$X^{-1}(a, b) \in \mathcal{E}.$$

*The **distribution function (d.f.)**, or **cumulative distribution function (c.d.f.)** associated with* X, *denoted* F *or* F_X, *is defined on* $\mathbb{R}$ *by:*

$$F(x) = \mu[X^{-1}(-\infty, x]]. \tag{3.1}$$

Remark 3.2 *Note that* $F(x)$ *is well defined since:*

$$X^{-1}(-\infty, x] = \bigcap_{n=1}^{\infty} X^{-1}(-\infty, x + 1/n),$$

and thus $X^{-1}(-\infty, x] \in \mathcal{E}$.

It is common to state that $F(x)$ ***equals the probability that*** X ***is less than or equal to*** x. *As* X *is a function on* $\mathcal{S}$ *which takes values in* $\mathbb{R}$, *this terminology identifies the probability of* $X(s) \leq x$ *with the probability of the event in* $\mathcal{S}$ *which produces this outcome:*

$$X^{-1}(-\infty, x] = \{s \in \mathcal{S} | X(s) \leq x\}.$$

This terminology also reflects the intuitive model of "random" sampling from the space $\mathcal{S}$, *discussed in Chapter 4, whereby the probability of obtaining a sample from any given event is defined as the probability of that event. So this terminology for* $F(x)$ *can also be understood from this perspective, that* $F(x)$ *equals the probability that on random sampling from* $\mathcal{S}$, *one will obtain a variate s for which* $X(s) \leq x$.

More specifically, a random variable is a **Borel measurable function** on $(\mathcal{S}, \mathcal{E}, \mu)$.

Exercise 3.3 (X is Borel measurable) *Show that if* $X : \mathcal{S} \longrightarrow \mathbb{R}$ *is a random variable, then* $X^{-1}(A) \in \mathcal{E}$ *for every Borel set* $A \in \mathcal{B}(\mathbb{R})$. *Hint: Recall that* $\mathcal{B}(\mathbb{R})$ *is generated by* $\{(a, b)\}$ *(Definition I.2.13) and* $X^{-1}((a, b)) \in \mathcal{E}$ *for all open intervals* (a, b). *Show that* $\{X^{-1}(A) | A \in \mathcal{B}(\mathbb{R})\}$ *is a sigma algebra because* X^{-1} *preserves unions and intersections (or see Corollary I.3.27).*

Proposition 3.4 (On $\mu[X^{-1}(a,b]]$**)** *For any right semi-closed interval* $(a,b]$ *with finite* $a < b$:

$$\mu[X^{-1}(a,b]] = F(b) - F(a). \tag{3.2}$$

If $b = \infty$ *and* $(a,\infty] \equiv (a,\infty)$:

$$\mu[X^{-1}(a,\infty)] = 1 - F(a).$$

Finally, for all $b \in \mathbb{R}$:

$$\mu[X^{-1}[b]] = F(b) - F(b^-),$$

where $F(b^-)$ *is the left limit of* $F(x)$ *at* b.
Proof. *Because* $(-\infty, a] \cup (a,b] = (-\infty, b]$:

$$X^{-1}((-\infty,b]) = X^{-1}((-\infty,a]) \bigcup X^{-1}((a,b]).$$

As $X^{-1}((-\infty,a])$ *and* $X^{-1}((a,b])$ *are disjoint, (3.2) follows by finite additivity of* μ. *The case* $b = \infty$ *follows from this since* $X^{-1}((-\infty,\infty)) = \mathcal{S}$.

For $b \in \mathbb{R}$, $F(b^-)$ *exists by Proposition I.3.60, and this final result follows from (3.2) by letting* $a \to b^-$. ■

Notation 3.5 (Book I distribution functions) *The definition of a distribution function in (3.1) is identical to that given in Definition I.3.56, but with a dramatic change in notation. In real analysis and measure theory, measure spaces are often denoted* $(X, \sigma(X), \mu)$, *and measurable functions almost always denoted as* f, g, *etc., with* $f : X \to \overline{\mathbb{R}}$, *where the extended reals* $\overline{\mathbb{R}}$ *are defined to include* $\{-\infty, \infty\}$ *(Definition I.3.1). Correspondingly, the* **distribution function associated with** $f(x)$, *denoted* $F(y)$ *for* $y \in \overline{\mathbb{R}}$, *was defined in terms of the* **cumulative level sets** *of* $f(x)$ *by:*

$$F(y) \equiv \mu[L_f(y)] \equiv \mu[f^{-1}(-\infty, x]]. \tag{3.3}$$

In probability theory, a probability space might be denoted $(\mathcal{S}, \mathcal{E}, \mu)$, *and random variables almost surely denoted as* X, Y, *etc., with* $X : \mathcal{S} \to \mathbb{R}$. *Correspondingly, the* **distribution function associated with** X, *denoted* $F(x)$ *for* $x \in \mathbb{R}$, *and sometimes* $F_X(x)$ *when* X *needs to be identified, is defined as in (3.1).*

Thus every random variable X *defined on a probability space* $(\mathcal{S}, \mathcal{E}, \mu)$ *gives rise to a Book I distribution function* $F(x)$. *By Proposition I.3.60 and Remark I.3.61,* $F(x)$ *is increasing, right continuous and Borel measurable, has left limits everywhere, and satisfies* $F(-\infty) = 0$ *and* $F(\infty) = 1$ *defined as limits. Such* $F(x)$ *is then continuous except for at most countably many points, and thus continuity points are dense in* $\mathbb{R}$. *This is summarized in Proposition 6.1 below for completeness.*

The next proposition states that given any such function $F(x)$, there is a probability space $(\mathcal{S}, \mathcal{E}, \mu)$ and random variable X so that $F(x) = \mu[X^{-1}(-\infty, x]]$, and thus $F(x)$ is the distribution function of X. This result will be generalized in Chapter 6, and a related general result will be developed with the aid of **Skorokhod's Representation theorem** in Section 8.4.

Proposition 3.6 (Identifying a distribution function) *Let* $F(x)$ *be an increasing function which is right continuous and satisfies* $F(-\infty) = 0$ *and* $F(\infty) = 1$, *defined as limits.*

Then there exists a probability space $(\mathcal{S}, \mathcal{E}, \mu)$ *and random variable* X *so that* $F(x) = \mu[X^{-1}(-\infty, x]]$. *In other words, every such function is the distribution function of a random variable.*

Proof. *Recalling Exercise 1.29, let* $\mathcal{S} = (0,1)$, $\mathcal{E} = \mathcal{B}((0,1))$, *the sigma algebra of Borel subsets of* $\mathcal{S}$, *and* $\mu = m$, *Lebesgue measure on* $\mathcal{E}$. *Define:*

$$X : ((0,1), \mathcal{B}(0,1), m) \to (\mathbb{R}, \mathcal{B}(\mathbb{R}), m)$$

by:

$$X(s) \equiv \inf\{x | F(x) \geq s\}.$$

Now for any $s \in (0,1)$, $\{x | F(x) \geq s\}$ *is an unbounded interval because* $F(x)$ *is increasing. Also,* $F(x)$ *is continuous from the right and* $F(x) \geq s$ *for* $x > X(s)$. *This implies that* $F(X(s)) \geq s$ *and* $X(s) \in \{x | F(x) \geq s\}$, *and thus this interval must be of the form* $[X(s), \infty)$. *Hence for any* $s \in (0,1)$, $\{x | F(x) \geq s\} = [X(s), \infty)$ *and so* $x \geq X(s)$ *if and only if* $F(x) \geq s$.
Equivalently:

$$X^{-1}[(-\infty, F(x)]] = (-\infty, x],$$

and so X *is Borel measurable and hence a random variable on* $((0,1), \mathcal{B}(0,1), m)$. *Further, since Lebesgue measure preserves interval length:*

$$m[\{X(s) \leq x\}] = m[\{s \leq F(x)\}] = m[(0, F(x)]] = F(x),$$

and thus F *is the distribution function of* X. ∎

Corollary 3.7 (On $\mu[X^{-1}(x)] = 0$**)** *For the measure space* $(\mathcal{S}, \mathcal{E}, \mu)$ *and random variable* X *of Proposition 3.6:*

$$\mu(\{X(s) = x\}) = F(x) - \lim_{y \to x-} F(y).$$

Thus $\mu(X^{-1}(x)) = 0$ *if and only if* $F(x)$ *is left continuous at* x.
Proof. *If* $y < x$, *(3.2) yields:*

$$F(x) - F(y) = m[\{y < X(s) \leq x\}].$$

Restricting y *to rationals and applying continuity from above of* m:

$$\lim_{y \to x-} m[\{y < X(s) \leq x\}] = m\left[\bigcap_{y < x} \{y < X(s) \leq x\}\right] = m(\{X(s) = x\}).$$

∎

Remark 3.8 (On the definition of X**)** *It should be noted that a key result of the above proof generalizes to an **arbitrary increasing and right continuous function** $F(x)$ that satisfies* $F(-\infty) = a$ *and* $F(\infty) = b$, *defined as limits, with either or both infinite. Specifically, if* $y \in (a,b)$, *then with* X *defined as above as:*

$$X(s) = \inf\{x | F(x) \geq s\},$$

one again obtains:

$$\{x | F(x) \geq y\} = [X(y), \infty),$$

and so $x \geq X(y)$ *if and only if* $F(x) \geq y$.
In Section 3.2 below, this definition for X *will be taken as the definition of the **left continuous inverse of** F, and denoted* F^*, *where* F *is a distribution function or more generally an increasing function. There we define:*

$$F^*(s) \equiv \inf\{x | F(x) \geq s\}.$$

The above proof shows that for $y \in (a,b)$:

$$\{x | F(x) \geq y\} = [F^*(y), \infty),$$

and so:

$$x \geq F^*(y) \text{ if and only if } F(x) \geq y.$$

This is noted below in (3.8).
Of course, that $F^*(s)$ *is indeed left continuous remains to be proved.*

3.1.1   Bond Loss Example (Continued)

Example 3.9 (Bond default r.v.) *We continue with* $(\mathcal{S}, \mathcal{E}, \mu)$ *as developed in part 3 of Example 1.5, where* f_j *denoted the loan amount of the jth bond or loan of N such bonds or loans,* $\{B_j\}_{j=1}^N$. *Recall:*

$$\mathcal{S} = \{(B_J, B_{J'}) | J \subset \{1, 2, ..., N\}\},$$

where $B_J \equiv \{B_{j_1}, ..., B_{j_M}\}$ *denotes the defaulting subset, and* $B_{J'}$ *the complementary non-defaulting subset. Then* $\mathcal{E} = \sigma(P(\mathcal{S}))$, *the power sigma algebra of subsets of* $\mathcal{S}$, *and* μ *is defined on* $A \equiv (B_J, B_{J'}) \in \mathcal{E}$ *by:*

$$\mu(A) \equiv \prod_{j \in J} q(B_j) \prod_{j \in J'} (1 - q(B_j)),$$

and extended additively to other sets in $\mathcal{E}$.

For each j, $1 \le j \le N$, *define a binomial "default" function* $D_j : \mathcal{S} \to \mathbb{R}$ *on* $A \equiv (B_J, B_{J'}) \in \mathcal{S}$ *by:*

$$D_j(A) = \begin{cases} 1, & B_j \in B_J, \\ 0, & B_j \notin B_J. \end{cases}$$

This default function identifies the elements of $\mathcal{S}$ *that contain the bond* B_j *in the default collection, and is in fact a measurable function and thus a random variable on this probability space.*

To see this, define $A_j \in \mathcal{E}$ *by:*

$$A_j = \{A \in \mathcal{S} | B_j \in B_J\}.$$

Then for any interval (a, b):

$$D_j^{-1}((a,b)) = \begin{cases} \emptyset, & 0, 1 \notin (a, b), \\ \mathcal{S} - A_j, & 0 \in (a, b), 1 \notin (a, b), \\ A_j, & 0 \notin (a, b), 1 \in (a, b), \\ \mathcal{S} & 0, 1 \in (a, b). \end{cases}$$

Hence, D_j *is measurable for all* j.

Further, $\mu(A_j) = q(B_j)$ *and consequently* $\mu(\mathcal{S} - A_j) = 1 - q(B_j)$. *That* $\mu(A_j) = q(B_j)$ *follows from the definition of* μ *above. If* $A \in A_j$:

$$\begin{aligned} \mu(A) &\equiv \prod_{j \in J} q(B_j) \prod_{j \in J'} (1 - q(B_j)) \\ &= q(B_j) \prod_{k \in J, k \ne j} q(B_k) \prod_{k \in J'} (1 - q(B_k)). \end{aligned}$$

Then $\mu(A_j)$ *is defined additively over all such sets* $A \in A_j$, *and so:*

$$\mu(A_j) = q(B_j) \sum_J \prod_{k \in J, k \ne j} q(B_k) \prod_{k \in J'} (1 - q(B_k)).$$

This summation is over all 2^{N-1} *J-subsets which contain* j, *and so:*

$$\sum_J \prod_{k \in J, k \ne j} q(B_k) \prod_{k \in J'} (1 - q(B_k)) = \prod_{k=1, \; k \ne j}^N (q(B_k) + [1 - q(B_k)]) = 1.$$

Thus $\mu(A_j) = q(B_j)$.

The distribution function for D_j, *denoted* $F_j^D(x)$ *say, is then defined by (3.1) and given as:*

$$F_j^D(x) = \begin{cases} 0, & x < 0, \\ 1 - q(B_j), & 0 \le x < 1, \\ 1, & 1 \le x. \end{cases}$$

Remark 3.10 *In the language of probability theory, it is often said that $D_j = 1$ with probability $q_j \equiv q(B_j)$, and $D_j = 0$ with probability $1 - q_j$. This language is unique to the probability branch of measure theory and is suggestive of the interpretation of random variables as representing outcomes of experiments or simulations. So in the current application one can envision the default of B_j as the outcome of an H on the flip of a biased coin, where $\Pr(H) = q_j$. In that case, the language is explicitly justified.*

But it is important to remember that while colorful and often useful in understanding a given application, the probabilities attributed to random variables are simply the measures of the pre-image domain sets, in this case, $q_j = D_j^{-1}(1)$, so by definition:

$$\Pr[D_j = 1] = \mu\left[D_j^{-1}(1)\right].$$

*For a general measurable function $f(x)$, say $f : X \to \mathbb{R}$ defined on the measure space $(X, \sigma(X), \mu)$, the μ-measure of the pre-image sets of f also play an important role in Book III, in defining the **Lebesgue integral** of $f(x)$, and named for by **Henri Lebesgue** (1875–1941). In that book the Lebesgue theory is developed for $X = \mathbb{R}^n$ and applied to probability theory in Book IV. This development will be generalized to other measures and spaces in Book V, and as will be seen in Book VI, such generalized integrals play a key role in probability theory.*

Example 3.11 (Bond loss r.v.) *Returning to Example 1.7, the **loss given default** (L.G.D.) or **loss ratio** for each loan can be denoted L_j, say. That is, L_j is a function with range in the interval $[0, 1]$, so $L_j : \mathcal{S} \to \mathbb{R}$, and specifies the proportion of the loan amount f_j lost to the lender on the default of loan B_j. The **loan recovery** function R_j represents the relative amount recovered by the lender given a default of loan B_j, and as noted above, $L_j + R_j \geq 1$.*

If $L_j = l_j$ is modeled as a constant, then for each j, $1 \leq j \leq N$, the loss function $L_j : \mathcal{S} \to \mathbb{R}$ is defined for $A \equiv (B_J, B_{J'}) \in \mathcal{S}$ by:

$$L_j(A) = \left\{ \begin{array}{ll} l_j f_j, & B_j \in B_J, \\ 0, & B_j \notin B_J. \end{array} \right.$$

Then with $A_j = \{A \in \mathcal{S} | B_j \in B_J\}$ as above and (a, b) an arbitrary open interval:

$$L_j^{-1}((a, b)) = \left\{ \begin{array}{ll} \emptyset, & 0, \; l_j f_j \notin (a, b), \\ \mathcal{S} - A_j, & 0 \in (a, b), \; l_j f_j \notin (a, b), \\ A_j, & 0 \notin (a, b), \; l_j f_j \in (a, b), \\ \mathcal{S} & 0, \; l_j f_j \in (a, b). \end{array} \right.$$

Consequently such L_j is a random variable on $(\mathcal{S}, \mathcal{E}, \mu)$.

*More generally if $L_j : \mathcal{S} \to [0, 1]$ is not constant, then the above sigma algebra $\mathcal{E}$ is too coarse to make this function measurable. For example, imagine as in Example 1.7 that on default, every L_j can assume 3 values: $L_j = 0.25, 0.50, 1.00$, with given probabilities. Then $L_j^{-1}((0.3, 0.6))$ should logically be a subset of A_j, since this interval implies that a default did occur on bond B_j. But with the current definition of $\mathcal{E}$ there is no way to identify as an event the subset of A_j for which bond B_j defaulted **and** incurred a loss: $0.3 < L_j < 0.6$. What is needed is a refinement of $\mathcal{S}$ and $\mathcal{E}$. The example above in effect split event $A \in \mathcal{S}$ into finer events which identified the loss levels incurred by the defaulted bonds which A identifies.*

Specifically, a probability space $(\mathcal{S}', \mathcal{E}', \mu')$ was introduced as follows, reflecting the requirement that upon default of any B_j, the associated L_j could assume 3 values: $L_j = 0.25$, 0.50, 1.00. If $A \equiv (B_J, B_{J'}) \in \mathcal{S}$ is a point where J contains M of the N bonds with index *set $J = (j_1, j_2, ..., j_M)$, then this point becomes 3^M points in $\mathcal{S}'$. Each point in $\mathcal{S}'$ represents one of the 3^M possible loss outcomes for these M bonds given that each can incur 3 possible loss levels. These 3^M points in $\mathcal{S}'$ will be of the form $A^K \equiv (B_J^K, B_{J'})$ with:*

$$B_J^K \equiv (B_{j_1}^{k_1}, B_{j_2}^{k_2}, ..., B_{j_M}^{k_M}).$$

Here $K = (k_1, k_2, ..., k_M)$ denotes one of the 3^M M-tuples of points where each k_i equals $1, 2, 3$, and with respective $L_j^{k_j}$ equal to 0.25, 0.50, or 1.00. Now with $\mathcal{E}' \equiv \sigma(P(\mathcal{S}'))$, L_j is a random variable on this space.

*With such L_j, one could similarly define a random variable L to equal **total losses**. Using the so-called **individual loss model**:*

$$L = \sum_j f_j D_j L_j, \tag{3.4}$$

and is defined on A^K above by:

$$L\left(A^K\right) = \sum_J f_j L_j^{k_j}.$$

As an algebraic combination of random variables, L is again a random variable on $(\mathcal{S}', \mathcal{E}', \mu')$. Now $L^{-1}((a,b))$ identifies the points of $\mathcal{S}'$ with total losses in this interval, and the collection of these points is an event in $\mathcal{E}'$ by construction.

For more general loss random variables, the probability space models become increasingly difficult to explicitly specify. Fortunately, such a specification is typically unnecessary in practice.

3.2 "Inverse" of a Distribution Function

The title of this section has inverse in quotes because distribution functions, while increasing, need not be strictly increasing and therefore need not have inverses that are definable as functions. Of course if $F : X \to Y$, then F^{-1} is always definable as a **set function** on any set $B \subset Y$ by:

$$F^{-1}(B) = \{x \in X | F(x) \in B\}.$$

But for F^{-1} to be definable as an inverse function requires that for all $x \in X$ and all $y \in Y$:

$$F^{-1}(F(x)) = x, \qquad F(F^{-1}(y)) = y.$$

Thus in order for F^{-1} to be well defined as a function, it is necessary and sufficient for F to be:

1. **One-to-one,** meaning $F(x) = F(x')$ implies $x = x'$.

2. **Onto,** meaning for all $y \in Y$ there is an $x \in X$ so that $F(x) = y$.

If F is one-to-one but not onto, then F^{-1} is well defined as a function for $y \in Rng\,[F] \subset Y$, the range of f.

The "problem" with distribution functions is that they are in general neither one-to-one nor onto when considered as a function $F : \mathbb{R} \to [0,1]$. In general there can be intervals (recall right continuity) $[a, b)$ on which $F(x) = F(a)$, a **constant**. Visually, the graph of F has **flat spots** like a step function. This occurs with discrete distribution functions such as the Poisson distribution defined by (1.11), but can also happen with a continuous distribution for which such an interval is outside of the range of the underlying random variable. In such cases, F is clearly not one-to-one and hence F^{-1} cannot be defined in the traditional sense even for $y \in Rng\,[F]$.

Another problem with distribution functions is the existence of **discontinuities** or **jumps** in the graph of F. In this case, F is not onto, so $Rng\,[F] \subsetneq [0,1]$, though this problem is easily overcome at least definitionally as noted above. Any increasing function can have at most countably many such discontinuities, and also at most countably many flat spots, since each such identifies an interval which of necessity contains a different rational number.

However, even in the most general cases there is a notion related to the inverse of a function which has important applications in Chapters 4 and 9, as well as in later books. This notion is that of a **"left-continuous"** inverse, denoted F^* and defined below. That F^* is indeed left continuous is something that needs to be proved and this will be done in Proposition 3.16, after which time the quotes will be dropped.

We define the "left continuous" inverse function for an arbitrary increasing function $F(x)$, not necessarily right continuous. By Proposition I.5.8, any such function has left and right limits everywhere, and these limits agree except at most on a countable collection of points. In other words, any such function has at most countably many discontinuities.

Note that $F^*(y)$ is defined identically to the random variable $X(s)$ constructed on the measure space $((0,1), \mathcal{B}(0,1), m)$ in the proof of Proposition 3.6.

Definition 3.12 ("Left continuous" inverse) *Let $F(x)$ be an increasing function, $F : \mathbb{R} \to \mathbb{R}$. The "left-continuous" inverse of F, denoted F^*, is defined by:*

$$F^*(y) = \inf\{x|F(x) \geq y\}. \tag{3.5}$$

Thus if $\{x|F(x) \geq y\} = \mathbb{R}$, then $F^*(y) = -\infty$. By convention, we define $F^*(y) = \infty$ if $\{x|F(x) \geq y\} = \emptyset$.

Notation 3.13 *F^* is sometimes called the **generalized inverse of** F, but the more common name above will be justified in Proposition 3.16.*

Remark 3.14 (Well-definedness of F^*) *Note that F^* is well defined when F is increasing since if $F(x) \geq y$, then $F(x') \geq y$ for all $x' > x$. Thus with $\langle$ denoting either $[$ or $($:*

$$\{x|F(x) \geq y\} = \langle F^*(y), \infty).$$

This interval is open if $F\,[F^*(y)] < y$ and closed if $F\,[F^*(y)] \geq y$. These respective cases occur if F is left continuous but not right continuous at $F^*(y)$, and if F is right continuous at $F^*(y)$. If F is neither left nor right continuous, then either case can result.

Example 3.15 (On properties of F^*) *Define a distribution function F by:*

$$F(x) = \begin{cases} 0, & x < 0, \\ 0.5x, & 0 \leq x < 1, \\ 0.5, & 1 \leq x < 2, \\ 1, & x \geq 2. \end{cases}$$

Then F has both of the problems noted above, which preclude the existence of a traditional inverse function. Specifically, F is neither one-to-one, with $F(x) = 0.50$ for $1 \leq x < 2$, nor onto, with $F(\mathbb{R}) = [0, 0.5] \cup \{1\}$.

From the definition $F^*(y) = \inf\{x|F(x) \geq y\}$:

$$F^*(y) = \begin{cases} -\infty, & y \leq 0, \\ 2y, & 0 < y \leq 0.5, \\ 2.0, & 0.5 < y \leq 1.0. \\ \infty, & y > 1. \end{cases}$$

Note that $F^*(y)$ is left continuous in this case, and that $F^*(y) = F^{-1}(y)$ on $[0, 0.5]$, where F is one-to-one and onto as a function $F : [0, 1] \rightarrow [0, 0.5]$. Also, F^* has both flat spots and jumps. In fact, we see that the jump in F at $x = 2$ produces a flat spot in F^* over $0.5 < y \leq 1.0$, while the flat spot in F over $1 \leq x < 2$ produces a jump in F^* at $y = 0.5$.

This example illustrates the definition of F^* for all y. However, it is common to restrict the analysis of distribution function to $y \in (0, 1)$, and similarly restrict y for increasing right continuous functions. See Remark 3.17.

3.2.1 Properties of F^*

We begin with a proposition which identifies general properties of F^* in the general case of increasing F. For some of the more detailed properties below, it will also be assumed that F is continuous, or, right continuous as is always true for distribution functions.

For the next result, recall the notion of one-sided limits:

1. **Left limit:**

$$F\left(z^-\right) \equiv \lim_{x \to z^-} F(x),$$

where $x \to z^-$ implies that all $x < z$.

2. **Right limit:**

$$F\left(z^+\right) \equiv \lim_{x \to z^+} F(x),$$

where $x \to z^+$ implies that all $x > z$.

Proposition 3.16 (Basic properties of F^*) *Given an **increasing function** F defined on (a, b) for $-\infty \leq a < b \leq \infty$:*

1. *The function F^* is increasing, and hence Borel measurable.*

2. *F^* is left continuous, and has right limits on $\{y|F^*(y) \in (-\infty, \infty)\}$.*

3. *Both $F(x)$ and $F^*(y)$ are continuous on (a, b) and $\{y|F^*(y) \in (-\infty, \infty)\}$, respectively, except for at most countably many points.*

4. *For all y such that $F^*(y) \in (-\infty, \infty)$:*

$$F\left(F^*(y)^-\right) \leq y \leq F\left(F^*(y)^+\right). \tag{3.6}$$

5. *For all x such that $F(x) \in \{y|F^*(y) \in (-\infty, \infty)\}$:*

$$F^*\left(F(x)\right) \leq x \leq F^*\left(F(x)^+\right). \tag{3.7}$$

Proof.

1. *That $F^*(y)$ is an increasing function follows since F is increasing. If $y < y'$, then:*

$$\{x|F(x) \geq y'\} \subset \{x|F(x) \geq y\},$$

and so $F^*(y) \leq F^*(y')$.

Then given (c, d):

$$(F^*)^{-1}(c, d) \equiv \{y|F^*(y) \in (c, d)\},$$

and so $(F^*)^{-1}(c, d)$ is an interval: open, closed or semi-closed, because F^* is an increasing function. Thus F^* is Borel measurable.

2. *We first show that any increasing function G has left and right limits everywhere. This then applies to increasing F on (a, b) and to F^* on $\{y|F^*(y) \in (-\infty, \infty)\}$, the domain on which F^* is real valued. To this end, if G is increasing, for any z let:*

$$G_1(z) \equiv \sup\{G(z')|z' < z\}.$$

This set is bounded from above by $G(z)$ and so the supremum is finite. This supremum then equals the left limit $G(z^-)$ since given increasing $z_n \to z$, it follows that $G(z_n)$ is also increasing.

Similarly, letting:

$$G_2(z) \equiv \inf\{G(z')|z' > z\},$$

this infimum is finite and equals the right limit $G(z^+)$.

Returning to F^*, if y_0 is given with $F^*(y_0) \in (-\infty, \infty)$, let $\{y_n\}_{n=1}^{\infty}$ be an increasing sequence with $y_n \to y_0$. To prove left continuity we show that $F^*(y_n) \to F^*(y_0)$. With $x_n \equiv F^*(y_n)$, it follows from monotonicity that $\{x_n\}_{n=1}^{\infty}$ is an increasing sequence and bounded above by $x_0 \equiv F^*(y_0)$. Hence as above, $\{x_n\}_{n=1}^{\infty}$ is convergent, say $x_n \to x' \leq x_0$. To prove that $x' = x_0$, we argue by contradiction.

By definition of $x_n = \inf\{x|F(x) \geq y_n\}$, it follows that for all $\epsilon > 0$ and all $n \geq 0$:

$$F(x_n - \epsilon) < y_n \leq F(x_n + \epsilon).$$

If $x' < x_0$, let $\epsilon = (x_0 - x')/2$. Then $x_n + \epsilon \leq x' + \epsilon = x_0 - \epsilon$, and thus:

$$y_n \leq F(x_n + \epsilon) \leq F(x_0 - \epsilon) < y_0.$$

Letting $n \to \infty$:

$$y_0 = \lim_{n \to \infty} y_n \leq F(x_0 - \epsilon) < y_0,$$

a contradiction, and so $x' = x_0$. Hence, $F^*(y_n) \to F^*(y_0)$ and F^* is left continuous on $\{y|F^*(y) \in (-\infty, \infty)\}$.

3. *From the proof in part 2 it follows that for increasing G, that $G(z^-)$ and $G(z^+)$ are well defined for all z, and $G(z^-) \leq G(z) \leq G(z^+)$. For any z for which $G(z^-) \neq G(z^+)$, the interval $[G(z^-), G(z^+)]$ contains a rational number, and different intervals must contain different rationals. Thus there can be at most countably many such intervals and discontinuities. As both F and F^* are increasing, the result follows.*

4. *For any $\epsilon > 0$:*
$$F\left(F^*(y) - \epsilon\right) < y < F\left(F^*(y) + \epsilon\right). \tag{1}$$

If $F\left(F^*(y) - \epsilon\right) \geq y$, then by (3.5) it follows that:

$$F^*(y) \equiv \inf\{x|F(x) \geq y\} \leq F^*(y) - \epsilon,$$

a contradiction. Similarly, if $y \geq F\left(F^*(y) + \epsilon\right)$, then $F^*(y) \geq F^*(y) + \epsilon$ since F is increasing. The bounds in (3.6) now follow from (1) by letting $\epsilon \to 0$. These limits are well defined since F has left and right limits by part 2.

5. *For any $\epsilon > 0$:*
$$F^*\left(F(x) - \epsilon\right) < x < F^*\left(F(x) + \epsilon\right). \tag{2}$$

If $F^*\left(F(x) - \epsilon\right) \geq x$, then since F is increasing and:

$$F^*\left(F(x) - \epsilon\right) = \inf\{x'|F(x') \geq F(x) - \epsilon\},$$

we obtain that $F(x) < F(x) - \epsilon$. Similarly, if $x \geq F^*\left(F(x) + \epsilon\right)$, then $F(x) \geq F(x) + \epsilon$. The result in (3.7) again follows by letting $\epsilon \to 0$ and recalling that F^* is left continuous.

∎

Remark 3.17 (On distribution functions) *The proof of Proposition 3.16 and comments in Remark 3.14 provide important additional insights on $F^*(y)$ when $F(x)$ is a distribution function, or more generally, any increasing right continuous function:*

1. *If $F(x)$ is a **distribution function**, then $F^*(y)$ is a random variable defined from $((0,1), \mathcal{B}(0,1), m)$ to $(\mathbb{R}, \mathcal{B}(\mathbb{R}), m)$ as noted in the proof of Proposition 3.6. This random variable will be important in Chapter 4.*

2. *If $F(x)$ is a **distribution function**, then for any $y \in (0,1)$:*

$$\{x|F(x) \geq y\} = [F^*(y), \infty).$$

First, $\{x|F(x) \geq y\}$ is a left closed, right unbounded interval. Right unboundedness follows from the fact that $F(x)$ is increasing. Since right continuous, if $F(x_n) \geq y$, $x_n > x$, and $x_n \to x$, then $F(x) \geq y$. The result now follows from the definition of $F^*(y)$.

*In general, if $F(x)$ is an **increasing and right continuous** function, this result holds for $y \in IntConv\left[Rng(F)\right]$, the **interior of the convex hull** of $Rng(F)$, the range of F. In general, the convex hull is defined as the smallest convex set which contains $Rng(F)$, but on $\mathbb{R}$ this is equivalent to the smallest interval. The interior of this set is then the largest open interval contained in this interval.*

3. *If $F(x)$ is a **distribution function**, it follows from part 2 that for any $y \in (0,1)$:*

$$\{x|F(x) \geq y\} = \{x|x \geq F^*(y)\},$$

and thus:

$$y \leq F(x) \text{ if and only if } F^*(y) \leq x. \tag{3.8}$$

This is equivalent to the observation that:

$$y > F(x) \text{ if and only if } F^*(y) > x. \tag{3.9}$$

*In general, if $F(x)$ is an **increasing right continuous** function, these results hold for $y \in IntConv\left[Rng(F)\right]$.*

Exercise 3.18 (On F/F^* inequalities) *The inequalities in (3.8) and (3.9) are important, and especially because they do not generalize. Show by examples that:*

$$y < F(x) \not\Rightarrow F^*(y) < x,$$

$$F^*(y) < x \not\Rightarrow y < F(x).$$

These examples then also show that $y \geq F(x)$ does not imply $F^*(y) \geq x$, and $F^*(y) \geq x$ does not imply $y \geq F(x)$.

The next proposition identifies the extent to which F^* is truly the inverse of F.

Proposition 3.19 (Additional properties for distribution F) *Let F be a distribution function and F^* its left continuous inverse.*

1. *For any $y \in (0,1)$:*
$$F\left(F^*(y)^-\right) \leq y \leq F\left(F^*(y)\right), \tag{3.10}$$

and thus except for at most countably many points:

$$y = F\left(F^*(y)\right).$$

2. *For any x with $0 < F(x) < 1$, then by (3.7):*

$$F^*\left(F(x)\right) \leq x \leq F^*\left(F(x)^+\right),$$

and thus except for at most countably many points:

$$x = F^*\left(F(x)\right).$$

Proof.

1. *The bounds in (3.10) follow from (3.6) and right continuity of F. Alternatively, the inequality on the right follows from (3.8) since $y \leq F\left(F^*(y)\right)$ if and only if $F^*(y) \leq F^*(y)$. Similarly, if $x < F^*(y)$, then $F(x) < y$ from (3.9), and taking a supremum over all such $x < F^*(y)$ obtains $F\left(F^*(y)^-\right) \leq y$. Thus $y = F\left(F^*(y)\right)$ at all points of continuity of F, of which there can be at most countably many exceptions.*

2. *The bounds from (3.7) can also be proved as follows. The inequality on the left follows from (3.8) since $F^*\left(F(x)\right) \leq x$ if and only if $F(x) \leq F(x)$. For the inequality on the right, if $y > F(x)$, then $F^*(y) > x$ from (3.9). Taking an infimum over all such $y > F(x)$ produces $F^*\left(F(x)^+\right) \geq x$. The statement on $x = F^*\left(F(x)\right)$ follows as above.*

■

Corollary 3.20 (Increasing, right continuous F) *If F is an increasing, right continuous function on $\mathbb{R}$ with $F(-\infty) = m$ and $F(\infty) = M$ defined as limits, then the above results apply with $y \in (m, M)$ in part 1, and $m < F(x) < M$ in part 2.*
Proof. The above proof only relied on (3.8) and (3.9), which remain valid in this context since $(m, M) = IntConv\left[Rng(F)\right]$ in the notation of Remark 3.17. ■

Example 3.21 *Looking back to Example 3.15, note that (3.10) is satisfied with strict inequality for $0.5 < y < 1.0$. On this interval, $F^*(y) = 2.0$ so $F\left(F^*(y)^-\right) = 0.5$, while $F\left(F^*(y)\right) = 1.0$. Similarly, (3.7) is satisfied with strict inequality for $1 < x < 2$. Here, $F(x) = 0.5$ so $F^*\left(F(x)\right) = 1.0$, while $F^*\left(F(x)^+\right) = 2.0$.*

The next proposition formalizes the intuition from this example. Namely, strict inequality in (3.10) can only occur when F has jumps, meaning (left) discontinuities, whereas strict inequality in (3.7) can only occur where F has flat spots, meaning where F is not strictly monotonic. For this result, we formally relate F^* to the left and right inverses of F.

Recall that F has a **left inverse** F_L^{-1} if for all $x \in Dmn(F)$, the domain of F:

$$F_L^{-1}(F(x)) = x. \tag{3.11}$$

Similarly, we say that F has a **right inverse** F_R^{-1} if for all $y \in Rng(F)$, the range of F:

$$F(F_R^{-1}(y)) = y. \tag{3.12}$$

When $F_L^{-1} = F_R^{-1}$ we say that F has an **inverse** and this is denoted F^{-1}.

Part 1 of the next result states that a distribution function has a right inverse on $(0,1)$ if and only if F is continuous on:

$$D' \equiv \{F^*(y)|y \in (0,1)\} \subset Dmn(F).$$

Part 2 states that such F has a left inverse on:

$$D \equiv \{x|0 < F(x) < 1\} \subset Dmn(F),$$

if and only if F is strictly increasing on D.

Note that D and D' need not be equal in the general case. For Example 3.15, $D = (0,2)$ while $D' = (0,1] \bigcup \{2\}$. However, if F is continuous and strictly increasing on D, then we prove in part 3 below that $D = D'$.

Corollary 3.20 then generalizes these results to right continuous, increasing functions.

Proposition 3.22 (F^* vs. F^{-1} for distribution functions) *Let F be a distribution function and thus increasing and right continuous, and F^* its increasing and left continuous inverse. Then:*

1. *$F(F^*(y)) = y$ for all $y \in (0,1)$ if and only if F is continuous on $D' \equiv \{F^*(y)|y \in (0,1)\}$.*

2. *$F^*(F(x)) = x$ for all $x \in D \equiv \{x|\ 0 < F(x) < 1\}$ if and only if F is strictly increasing on D.*

3. *If F is both continuous and strictly increasing on D, then $D = D'$ and on $(0,1)$:*

$$F^* = F^{-1}. \tag{3.13}$$

Thus F^* is also continuous and strictly increasing on $(0,1)$.

Proof.

1. *If F is continuous, then $F(F^*(y)) = y$ for all $y \in (0,1)$ by part 1 of Proposition 3.19. Conversely, if F is not continuous at $x \equiv F^*(y)$, then since right continuous with $F(x^+) = F(x)$, it is then not left continuous. Thus:*

$$(F(x^-), F(x)) \equiv (F(F^*(y)^-), F(F^*(y))),$$

is a non-empty open interval. If $y' \in (F(x^-), F(x))$, then since $x \equiv F^*(y)$:

$$F^*(y') = \inf\{x'|F(x') \geq y'\} = F^*(y).$$

Thus for all $y' \in (F(x^-), F(x))$:

$$F\left(F^*(y')\right) = F\left(F^*(y)\right),$$

and so $F\left(F^*(y')\right) = y'$ *for at most one such* y'. *Hence if* $F\left(F^*(y)\right) = y$ *for all* $y \in (0,1)$, *then* F *must be continuous on* $(0,1)$.

2. *While part 2 of Proposition 3.19 obtains that* $F^*\left(F(x)\right) = x$ *if* F^* *is continuous at* $F(x)$, *this is not useful until translated to a property of* F. *To prove that* $F^*\left(F(x)\right) = x$ *for all* $x \in D$ *if and only if* F *is strictly monotonic on* D, *note that by definition:*

$$F^*\left(F(x)\right) = \inf\{x'|F(x') \geq F(x)\}.$$

Now $x \in \{x'|F(x') \geq F(x)\}$, *and if* F *is strictly increasing, there can be no* $x' < x$ *in this set, so* $F^*\left(F(x)\right) = x$. *Conversely, if* F *is not strictly increasing on* D, *then there exists* x, x' *with* $x < x'$ *and* $0 < F(x) = F(x') < 1$. *But then* $F^*\left(F(x)\right) = F^*\left(F(x')\right)$, *and thus at most one can equal the respective* x *or* x'. *Hence if* $F^*\left(F(x)\right) = x$ *for all* $x \in D$, *then* F *must be strictly increasing on* D.

3. *If* F *is both continuous and strictly increasing on* D, *then* $F : D \to (0,1)$ *is one-to-one and onto, and thus* F^{-1} *is well defined. Further,* F^{-1} *is strictly increasing, and thus continuous by Proposition I.3.12, since* $\left(F^{-1}\right)^{-1} = F$ *takes open intervals to open intervals.*

To see that $D = D'$, *let* $x \in D$. *Then* $y \equiv F(x) \in (0,1)$, *and* $F^*(y) \leq x \leq F^*(y^+)$ *by (3.7). We claim that* $F^*(y) = F^*(y^+)$ *and thus* $x \in D'$. *To see this, if* $1 - y > \epsilon > 0$ *then:*

$$F^*(y + \epsilon) = \inf\{x'|F(x') \geq F(x) + \epsilon\}.$$

Since $y + \epsilon \in (0,1)$, *there exists unique* $x'_\epsilon > x$ *so that* $F(x'_\epsilon) = F(x) + \epsilon$ *and then* $F^*(y + \epsilon) = x'_\epsilon$. *Thus* $F^*(y^+) = \lim_{\epsilon \to 0} x'_\epsilon$. *But as* $\epsilon \to 0$, $F(x'_\epsilon) \to F(x)$ *by construction, and then* $x'_\epsilon \to x$ *since* F *is one-to-one.*

Conversely, if $x \in D'$, *then* $x = F^*(y)$ *for* $y \in (0,1)$. *But then* $F(x^-) \leq y \leq F(x)$ *by (3.10). As* F *is continuous,* $F(x) = y$ *and thus* $x \in D$.

With $D = D'$, *(3.13) now follows from parts 1 and 2 by definition of* F^{-1}. ∎

Corollary 3.23 (Increasing, right continuous F**)** *Let* F *be an increasing, right continuous function with* $F(-\infty) = m$ *and* $F(\infty) = M$, *defined as limits. Then:*

1. $F\left(F^*(y)\right) = y$ *for all* $y \in (m, M)$ *if and only if* F *is continuous on* $D' \equiv \{F^*(y)|y \in (m, M)\}$.

2. $F^*\left(F(x)\right) = x$ *for all* $x \in D \equiv \{x| \ m < F(x) < M\}$ *if and only if* F *is strictly increasing on* D.

3. *If* F *is both continuous and strictly increasing on* D, *then (3.13) is satisfied on* (m, M).

Proof. *The proof of 1 and 2 is the same, only using Corollary 3.20. The proof of 3 is just a change of notation.* ∎

Remark 3.24 (Pointwise statements) *The above proof provides some pointwise statements for these results for increasing, right continuous functions, which includes distribution functions. We state these in the general case.*

1. *If F is continuous at $F^*(y)$ for given $y \in (m, M)$, then $F(F^*(y)) = y$ by (3.10). However, $F(F^*(y)) = y$ for given $y \in (m, M)$ does not assure continuity of F at $F^*(y)$. This can be observed in Example 3.15. There for $y = 1$:*

$$F(F^*(1)) = F(2) = 1,$$

    yet F is discontinuous at $x = 2$.

2. *If F is strictly increasing at $x \in D \equiv \{x \mid m < F(x) < M\}$, meaning strictly increasing on an interval $(x - \delta, x + \delta)$ about x, then $F^*(F(x)) = x$ by the proof of part 2 above. However, $F^*(F(x)) = x$ for given $x \in D$ does not assure that F is strictly increasing at x. Again by Example 3.15 with $x = 1$:*

$$F^*(F(1)) = F^*(1/2) = 1,$$

    yet F is not strictly increasing on any interval about $x = 1$.

3.2.2   The Function F^{**}

This section investigates applying the left continuous inverse transformation twice:

$$F \;\to\; F^* \;\to\; F^{**}.$$

This sequence is well defined because Definition 3.12 applies to all increasing functions, and F^* is increasing by Proposition 3.16.

By Proposition 3.22, if F is continuous and strictly increasing, then $F^* = F^{-1}$ and is also continuous and strictly increasing. Another application of this result obtains:

$$F^{**} = \left(F^{-1}\right)^* = \left(F^{-1}\right)^{-1} = F.$$

But this is too much to expect with a general increasing function with potentially countably many discontinuities and countably many flat sections.

Example 3.25 (On $F^{**}(x)$) *Recall Example 3.15, for which:*

$$F(x) = \begin{cases} 0, & x < 0, \\ 0.5x, & 0 \le x < 1, \\ 0.5, & 1 \le x < 2, \\ 1, & x \ge 2. \end{cases}$$

Defining $F^{**}(x) = \inf\{y \mid F^*(y) \ge x\}$ with $F^*(y)$ given above produces:

$$F^{**}(x) = \begin{cases} 0, & x < 0, \\ 0.5x, & 0 \le x < 1, \\ 0.5, & 1 \le x \le 2, \\ 1, & x > 2. \end{cases}$$

Note that $F^{**}(x) = F(x)$ at all continuity points of F, while at the discontinuity of F at $x = 2$, we have $F^{**}(x) = F(x^-)$, the left limit of F at x.

The following proposition generalizes this example and states that if F is a distribution function, then $F^{**}(x) = F(x)$ at all continuity points of F, and otherwise $F^{**}(x) = F(x^-)$. Since F is right continuous, while F^{**} is left continuous, this is indeed the best result possible. At any discontinuity x of F, F^{**} is defined to equal the left limit at this point, and so redefined, F^{**} will be left continuous. The interested reader may compare this observation with the statement and proof of Proposition I.5.8, and reformulate the earlier result to obtain a left continuous increasing function.

Proposition 3.26 (F^{**} vs. F for distributions) *Let F be a distribution function, F^* its left continuous inverse, and F^{**} the left continuous inverse of F^*. Then for all x:*

$$F^{**}(x) = F(x^-). \tag{3.14}$$

Further, the continuity points of F and F^{**} agree, and thus:

$$F^{**}(x) = F(x),$$

at all such continuity points.

Proof. *By definition:*

$$F^{**}(x) \equiv \inf\{y | F^*(y) \geq x\},$$

and F^{**} is left continuous and increasing by Proposition 3.16. Given x, choose $y < F(x^-)$, so then $y \leq F(x-\epsilon)$ for some $\epsilon > 0$. By (3.8) it follows that $F^*(y) \leq x-\epsilon$, and by definition, $y < F^{**}(x)$. As this is true for any $y < F(x^-)$, it follows that $F(x^-) \leq F^{**}(x)$.

If $y > F(x)$, then $F^*(y) > x$ by (3.9), and so by definition $y \geq F^{**}(x)$. As this is true for all $y > F(x)$, it follows that $F(x) \geq F^{**}(x)$. Hence:

$$F(x^-) \leq F^{**}(x) \leq F(x),$$

and in particular,

$$F^{**}(x) = F(x) \tag{1}$$

at all continuity points of F.

If x is a discontinuity point of F, then it is a left discontinuity point, so consider $(F(x^-), F(x)]$. Then F^* is constant on this interval since for any $y \in (F(x^-), F(x)]$:

$$F^*(y) = \inf\{z | F(z) \geq y\} = x.$$

Thus $F^*(y) = x$ for $F(x^-) < y \leq F(x)$, and so for any such discontinuity point x:

$$F^{**}(x) \equiv \inf\{y | F^*(y) \geq x\} = F(x^-). \tag{2}$$

Combining (1) and (2) obtains (3.14).

That F and F^{**} have the same continuity points is proved as follows. If x_0 is a continuity point of F^{**}, then for any $\epsilon > 0$ there exists δ so that $|x - x_0| < \delta$ assures:

$$|F^{**}(x) - F^{**}(x_0)| < \epsilon.$$

This implies by (3.14) that:

$$\left| F(x^-) - F(x_0^-) \right| < \epsilon.$$

If $x < x_0$ and $|x - x_0| < \delta$, choose $\bar{x} > x_0$ and $|\bar{x} - x_0| < \delta$. As F is increasing:

$$
\begin{aligned}
|F(x) - F(x_0)| &\leq \left| F(x^-) - F(x_0) \right| \\
&\leq \left| F(x^-) - F(x_0^-) \right| + \left| F(\bar{x}^-) - F(x_0^-) \right| \\
&< 2\epsilon.
\end{aligned}
$$

If $x > x_0$ and $|x - x_0| < \delta$, choose $\bar{x} > x$ with $|\bar{x} - x_0| < \delta$. As above:

$$
\begin{aligned}
|F(x) - F(x_0)| &\leq \left| F(x) - F(x_0^-) \right| \\
&\leq \left| F(x^-) - F(x_0^-) \right| \\
&< \epsilon.
\end{aligned}
$$

Thus x_0 is a continuity point of F.

Conversely, if x_0 is a continuity point of F, then for any $\epsilon > 0$ there exists δ so that $|x - x_0| < \delta$ assures:

$$\left| F(x) - F(x_0^-) \right| < \epsilon,$$

since $F(x_0) = F(x_0^-)$. If $x > x_0$ and $|x - x_0| < \delta$, then:

$$\left| F(x^-) - F(x_0^-) \right| \leq \left| F(x) - F(x_0^-) \right| < \epsilon,$$

since F is increasing. If $x < x_0$ and $|x - x_0| < \delta$, choose $\bar{x} < x$ with $|\bar{x} - x_0| < \delta$, then:

$$\left| F(x^-) - F(x_0^-) \right| \leq \left| F(\bar{x}) - F(x_0^-) \right| < \epsilon.$$

Thus x_0 is a continuity point of F^{**} by (3.14). ∎

Corollary 3.27 (Increasing, right continuous F) *The above result remains true if F is an increasing, right continuous function.*
Proof. *The above proof applies since (3.8) and (3.9) are valid for such functions by Remark 3.17.* ∎

3.3   Random Vectors and Joint Distribution Functions

The general framework for a random vector X is that it is a vector-valued, Borel measurable function defined on a probability space $(\mathcal{S}, \mathcal{E}, \mu)$. That is, $X : \mathcal{S} \longrightarrow \mathbb{R}^n$ is measurable relative to the Borel sigma algebra $\mathcal{B}(\mathbb{R}^n)$, so $X^{-1}(A) \in \mathcal{E}$ for all $A \in \mathcal{B}(\mathbb{R}^n)$. But we are also interested in random vectors defined in terms of a collection of n random variables $\{X_j\}_{j=1}^n$ defined on $(\mathcal{S}, \mathcal{E}, \mu)$, so $X \equiv (X_1, X_2, ..., X_n)$.

In this section we introduce both notions and prove their equivalence.

Definition 3.28 (Random vector 1) *If $X_j : \mathcal{S} \longrightarrow \mathbb{R}$ are random variables on a probability space $(\mathcal{S}, \mathcal{E}, \mu)$, $j = 1, 2, ..., n$, define the **random vector** $X = (X_1, X_2, ..., X_n)$ as the vector-valued function:*

$$X : \mathcal{S} \longrightarrow \mathbb{R}^n,$$

defined on $s \in \mathcal{S}$ by:

$$X(s) = (X_1(s), X_2(s), ..., X_n(s)).$$

*The **joint distribution function (d.f.)**, or **joint cumulative distribution function (c.d.f.)** associated with X, denoted F or F_X, and often without the qualifier "joint," is defined on $(x_1, x_2, ..., x_n) \in \mathbb{R}^n$ by:*

$$F(x_1, x_2, ..., x_n) = \mu \left[\bigcap_{j=1}^n X_j^{-1}(-\infty, x_j] \right]. \tag{3.15}$$

Remark 3.29 (On $F(x_1, x_2, ..., x_n)$) *Because random variables are Borel measurable by Exercise 3.3, $F(x_1, x_2, ..., x_n)$ in (3.15) is well defined since $\bigcap_{j=1}^n X_j^{-1}(-\infty, x_j] \in \mathcal{E}$.*

*As in the case for random variables, one often uses the terminology that $F(x_1, x_2, ..., x_n)$ **equals the probability that** $X(s) = (X_1(s), X_2(s), ..., X_n(s))$ **is less than or equal to** $x = (x_1, x_2, ..., x_n)$, denoted $X \leq x$. This ordering of vectors is understood to mean:*

$$X \leq x \quad \Longleftrightarrow \quad X_j(s) \leq x_j \text{ all } j.$$

The event $X \leq x$ is well defined in $\mathcal{S}$ by:

$$\{s \in \mathcal{S} | X(s) \leq x\} = \bigcap_{j=1}^{n} X_j^{-1}(-\infty, x_j].$$

Thus the distribution function in (3.15) can be equivalently written:

$$F(x) = \mu\left[X \leq x\right].$$

*This terminology that $F(x)$ **equals the probability that** X **is less than or equal to** x also reflects the intuitive model of a "random" sampling from the space $\mathcal{S}$, the topic of Chapter 4. That is, $F(x)$ equals the probability that on a random sampling from $\mathcal{S}$, one will obtain a variate $s \in \bigcap_{j=1}^{n} X_j^{-1}(-\infty, x_j]$ for which $X(s) \leq x$, and so $X_j(s) \leq x_j$ all j.*

As noted in the introduction, a random vector could have been defined in the apparently more general way as follows.

Definition 3.30 (Random vector 2) *Given a probability space $(\mathcal{S}, \mathcal{E}, \mu)$, a **random vector** is a mapping $X : \mathcal{S} \longrightarrow \mathbb{R}^n$, so that for all $A \in \mathcal{B}(\mathbb{R}^n)$, the sigma algebra of Borel measurable sets on $\mathbb{R}^n$:*

$$X^{-1}(A) \in \mathcal{E}. \tag{3.16}$$

*The **joint distribution function (d.f.)**, or **joint cumulative distribution function (c.d.f.)** associated with X, denoted F or F_X, and often without the qualifier "joint," is defined on $(x_1, x_2, ..., x_n) \in \mathbb{R}^n$ by:*

$$F(x_1, x_2, ..., x_n) = \mu\left[X^{-1}\left(\prod_{j=1}^{n}(-\infty, x_j]\right)\right]. \tag{3.17}$$

Remark 3.31 (Borel sigma algebra) *The Borel sigma algebra $\mathcal{B}(\mathbb{R}^n)$ is introduced in Definition I.2.13 as the smallest sigma algebra that contains all the open sets in $\mathbb{R}^n$. As an exercise it can be checked that this sigma algebra can be equivalently defined as the smallest sigma algebra that contains all the **open balls:***

$$B_r(a) \equiv \{x \in \mathbb{R}^n | \ |x - a| < r\},$$

*or **open rectangles:***

$$R = \prod_{j=1}^{n}(a_i, b_i),$$

or by complementarity, in terms of closed sets. However, the open set characterization is standard.

The following result shows that these definitions are equivalent. Consequently, we can use whichever characterization is more convenient for the applications that follow.

Proposition 3.32 (Equivalence of d.f. definitions) *If X is a random vector on $(\mathcal{S}, \mathcal{E}, \mu)$ by Definition 3.30, then $X = (X_1, X_2, ..., X_n)$ where each $X_j : \mathcal{S} \longrightarrow \mathbb{R}$ is a random variable on $(\mathcal{S}, \mathcal{E}, \mu)$.*

Conversely, if $X = (X_1, X_2, ..., X_n)$ where each $X_j : \mathcal{S} \longrightarrow \mathbb{R}$ is a random variable on $(\mathcal{S}, \mathcal{E}, \mu)$, then $X^{-1}(A) \in \mathcal{E}$ for all $A \in \mathcal{B}(\mathbb{R}^n)$.

Proof. *If $X : \mathcal{S} \longrightarrow \mathbb{R}^n$ is a random vector by Definition 3.30 let $X \equiv (X_1, X_2, ..., X_n)$ where X_j denotes the jth coordinate function. To see that each X_j is a random variable, let $A_j \in \mathcal{B}(\mathbb{R})$ and define with apparent notation $A = \mathbb{R}^{j-1} \times A_j \times \mathbb{R}^{n-j}$. Then $A \in \mathcal{B}(\mathbb{R}^n)$ and hence $X^{-1}(A) \in \mathcal{E}$. However:*

$$X^{-1}(A) = \{s | X_j(s) \in A_j\} = X_j^{-1}(A_j),$$

and so X_j is a random variable on $(\mathcal{S}, \mathcal{E}, \mu)$.

Conversely, let $X = (X_1, X_2, ..., X_n)$ where each $X_j : \mathcal{S} \longrightarrow \mathbb{R}$ is a random variable on $(\mathcal{S}, \mathcal{E}, \mu)$. *With $A_j \in \mathcal{B}(\mathbb{R})$ for all j, define $A = \prod_{j=1}^n A_j$ and note that $A \in \mathcal{B}(\mathbb{R}^n)$. Then:*

$$X^{-1}(A) = \bigcap_{j=1}^n X_j^{-1}(A_j),$$

and so $X^{-1}(A) \in \mathcal{E}$ since $X_j^{-1}(A_j) \in \mathcal{E}$ for all j. By Proposition I.8.1, such sets generate the sigma algebra $\mathcal{B}(\mathbb{R}^n)$. That is, $\mathcal{B}(\mathbb{R}^n)$ is the smallest sigma algebra that contains such sets, and all sets produced by unions, intersections and complements of such sets.

Now X^{-1} preserves unions, intersections and complements. If $\{B, B_1, ..., B_m\} \subset \mathcal{B}(\mathbb{R}^n)$:

$$X^{-1}\left(\bigcup_{j=1}^m B_j\right) = \bigcup_{j=1}^m X^{-1}(B_j),$$

$$X^{-1}\left(\bigcap_{j=1}^m B_j\right) = \bigcap_{j=1}^m X^{-1}(B_j),$$

$$X^{-1}(\widetilde{B}) = \widetilde{X^{-1}(B)},$$

and thus it follows $X^{-1}(A) \in \mathcal{E}$ for all $A \in \mathcal{B}(\mathbb{R}^n)$. ∎

Example 3.33 (Dice triplets) *Define $\mathcal{S}$ as the collection of 3-tuples of scores on rolling a **fair** die **sequentially** three times. So:*

$$\mathcal{S} = \{(f_1, f_2, f_3) \,|\, f_j \in \{1, ..., 6\}.$$

By "fair" is meant that for any roll, the six potential outcomes for f_j are equally likely, meaning each has probability $1/6$.

That these rolls are "sequential" implies that on any given roll, there is no "memory" of the outcomes of prior rolls, and these prior outcomes have no influence on the current outcome. In Example 3.49 we will formalize this notion by relating the outcomes $\{f_j\}_{j=1}^3$ to independent random variables $\{Y_j\}_{j=1}^3$ defined on $\mathcal{S}$. Thus each of the 6^3 potential outcomes is equally likely, and we define μ on $\mathcal{S}$ by the discrete rectangular distribution. That is, for all $(f_1, f_2, f_3) \in \mathcal{S}$:

$$\mu(f_1, f_2, f_3) = 1/6^3.$$

On this space, let $\mathcal{E} = \sigma(P(\mathcal{S}))$, the collection of all subsets of $\mathcal{S}$, to which we extend μ additively.

Define random variables X_1 and X_2 by:

$$\begin{aligned} X_1(f_1, f_2, f_3) &= f_1 + f_2, \\ X_2(f_1, f_2, f_3) &= f_2 + f_3. \end{aligned}$$

Then

$$X \equiv (X_1, X_2) : \mathcal{S} \to \mathbb{R}^2,$$

and the range of X is a proper subset of $\{(j, k) | 2 \le j, k \le 12\}$, since for example, the point $(2, 12)$ is impossible.

The joint distribution function $F(j, k)$ can be evaluated using the law of total probability in (1.30), conditioning on the value of f_2. In other words, define:

$$A = \{(f_1, f_2, f_3) \,|\, f_1 + f_2 \le j \text{ and } f_2 + f_3 \le k\},$$

then $F(j,k) = \mu(A)$. By (1.30):

$$
\begin{aligned}
F(j,k) &= \sum\nolimits_{f_2=1}^{6} \mu(A|f_2)\mu(f_2) \\
&= \sum\nolimits_{f_2=1}^{6} \mu(f_1 \leq j - f_2 \text{ and } f_3 \leq k - f_2)/6. \\
&= \sum\nolimits_{f_2=1}^{6} \mu(f_1 \leq j - f_2)\mu(f_3 \leq k - f_2)/6 \\
&= n(j,k)/6,
\end{aligned}
$$

where

$$
n(j,k) = \sum\nolimits_{f_2=1}^{6} \mu(f_1 \leq j - f_2)\mu(f_3 \leq k - f_2).
$$

*Note that the second-to-last step follows because f_1 and f_3 are **independent random variables** by construction (see the next section).*

Now for $\mu(f_1 \leq j - f_2)$:

$$
\mu(f_1 \leq j - f_2) = \begin{cases} (j - f_2)/6, & f_2 + 1 \leq j \leq f_2 + 6, \\ 1, & j > f_2 + 6, \\ 0, & j < f_2 + 1, \end{cases}
$$

with an identical formula for $\mu(f_3 \leq k - f_2)$.

3.3.1 Marginal Distribution Functions

Given a joint distribution function, one can also define a host of marginal distribution functions, though the notation quickly gets cumbersome. Consequently, we first introduce the idea for $n = 2$, then generalize. While these will be called distribution functions in the following definition, it should be noted that this characterization requires proof.

Definition 3.34 (Marginal distribution function) *Let $X : S \longrightarrow \mathbb{R}^n$ be the **random vector** $X = (X_1, X_2, ..., X_n)$ defined on $(S, \mathcal{E}, \mu)$, where $X_j : S \longrightarrow \mathbb{R}$ are random variables on $(S, \mathcal{E}, \mu)$ for $j = 1, 2, ..., n$.*

1. ***Special Case** $n = 2$: Given the joint distribution function $F(x_1, x_2)$ defined on $\mathbb{R}^2$, define two **marginal distribution functions** on $\mathbb{R}$, $F(x_1)$ and $F(x_2)$, by:*

$$
F_1(x_1) \equiv \lim_{x_2 \to \infty} F(x_1, x_2), \quad F_2(x_2) \equiv \lim_{x_1 \to \infty} F(x_1, x_2). \tag{3.18}
$$

2. ***General Case:** More generally, given $F(x_1, x_2, ..., x_n)$ and $I = \{i_1, ..., i_m\} \subset \{1, 2, ..., n\}$, let $x_J \equiv (x_{j_1}, x_{j_2}, ..., x_{j_{n-m}})$ for $j_k \in J \equiv \tilde{I}$. The **marginal distribution function** $F_I(x_I) \equiv F_I(x_{i_1}, x_{i_2}, ..., x_{i_m})$ is defined on $\mathbb{R}^m$ by:*

$$
F_I(x_I) \equiv \lim_{x_J \to \infty} F(x_1, x_2, ..., x_n). \tag{3.19}
$$

Remark 3.35 (Marginal combinatorics) *Given $F(x_1, x_2, ..., x_n)$, there are $2^n - 2$ **proper** marginal distribution functions defined by the $2^n - 2$ proper subsets of $\{1, 2, ..., n\}$. Of these, the most important in many applications are the n marginal distributions $\{F_j(x_j)\}_{j=1}^{n}$, respectively defined with $I = \{j\}$. See Chapter 7 on Copulas and Sklar's Theorem.*

Because these "distribution" functions are defined as limits, it is necessary to verify that these limits are well defined, and despite what the above terminology implies, that these limiting functions are indeed distribution functions.

Proposition 3.36 ($F_I(x_I)$ is a distribution function) *The function $F_I(x_I) \equiv F_I(x_{i_1}, x_{i_2}, ..., x_{i_m})$ of (3.19) is well defined, and is a distribution function.*

Specifically with $X_I \equiv (X_{i_1}, X_{i_2}, ..., X_{i_m})$:

$$F_I(x_I) = \mu \left[X_I^{-1} \left(\prod_{k=1}^{m} (-\infty, x_{i_k}] \right) \right]. \tag{3.20}$$

Thus the marginal distribution function $F_I(x_I)$ is the joint distribution function of X_I.

Proof. *Given $F(x_1, x_2, ..., x_n)$, note that as a distribution function, $0 \leq F \leq 1$ for all $(x_1, x_2, ..., x_n)$. Also, F is an increasing function in each variable separately since measures are monotonic. To prove that the limit in (3.19) is well defined for any $I = \{i_1, ..., i_m\} \subset \{1, 2, ..., n\}$, we can reorder indexes for notational simplicity and assume that $\{i_1, ..., i_m\} = \{1, 2, ..., m\}$ and so $x_J = (x_{m+1}, x_{m+1}, ..., x_n)$. The **marginal distribution function** $F_I(x_{i_1}, x_{i_2}, ..., x_{i_m})$ in (3.19) is then represented as:*

$$F_I(x_1, x_2, ..., x_m) \equiv \lim_{(x_{m+1}, x_{m+1}, ..., x_n) \to \infty} F(x_1, x_2, ..., x_n).$$

Given $(x_1, x_2, ..., x_m)$, let $\{x^{(k)}\}_{k=1}^{\infty} \subset \mathbb{R}^n$ be defined with $x^{(k)} \equiv (x_1, x_2, ..., x_m, x_{m+1}^{(k)}, ..., x_n^{(k)})$, where for each $j \geq m+1$, $\{x_j^{(k)}\}_{k=1}^{\infty}$ monotonically increases to ∞. There is no loss of generality in this monotonicity assumption since given arbitrary $X_J \to \infty$, we can always choose a monotonic subsequence. Define:

$$A_{x^{(k)}} \equiv \prod_{j=1}^{n} (-\infty, x_j^{(k)}],$$

where for notational simplicity we let $x_j^{(k)} \equiv x_j$ for $j \leq m$.

Thus $F(x^{(k)}) \equiv \mu \left[X^{-1} (A_{x^{(k)}}) \right]$ where $X \equiv (X_1, X_2, ..., X_n)$ is the original random vector defined on $(S, \mathcal{E}, \mu)$. By monotonicity of $(x_{m+1}^{(k)}, ..., x_n^{(k)})$, it follows that $\{A_{x^{(k)}}\}_k \subset S$ is a nested sequence of sets with $A_{x^{(k)}} \subset A_{x^{(k+1)}}$, and thus $X^{-1}(A_{x^{(k)}}) \subset X^{-1}(A_{x^{(k+1)}})$. Further:

$$F_I(x_1, x_2, ..., x_m) \equiv \lim_{k \to \infty} \mu \left[X^{-1} (A_{x^{(k)}}) \right].$$

Now define:

$$A'_x = \bigcup_k A_{x^{(k)}}.$$

Then $A'_x \subset \mathbb{R}^n$ is a Borel set, and is well defined independent of the sequence $\{x^{(k)}\}_{k=1}^{\infty}$. To see this, if $\{x_j^{(k)}\}_{k=1}^{\infty}$ and $\{y_j^{(k)}\}_{k=1}^{\infty}$ are two such sequences, meaning $x_j^{(k)} = y_j^{(k)} = x_j$ for $j \leq m$ and for $j > m$, $x_j^{(k)} \to \infty$ and $y_j^{(k)} \to \infty$ monotonically, then for any k there is a $j(k)$ and $j'(k)$ so that $A_{x^{(k)}} \subset A_{y^{j(k)}}$ and $A_{y^{(k)}} \subset A_{x^{j'(k)}}$. The result is then obtained by nesting.

Thus $\mu \left[X^{-1} (A'_x) \right]$ is well defined because A'_x is a Borel set, and since measures are continuous from below (Proposition I.2.45):

$$\begin{aligned} F_I(x_1, x_2, ..., x_m) &\equiv \lim_{k \to \infty} \mu \left[X^{-1} (A_{x^{(k)}}) \right] \\ &= \mu \left[X^{-1} (A'_x) \right]. \end{aligned}$$

In other words, $F_I(x_1, x_2, ..., x_m)$ is well defined as a function.

For (3.20), it follows by construction that:

$$A'_x = \prod_{j=1}^{m} (-\infty, x_j] \times \mathbb{R}^{n-m} \equiv A_x \times \mathbb{R}^{n-m}.$$

Now $(X_I(s), X_J(s)) \in A_x \times \mathbb{R}^{n-m}$ if and only if $X_I(s) \in A_x$, and so:

$$
\begin{aligned}
F_I(x_1, x_2, ..., x_m) &\equiv \mu\{s \mid (X_I, X_J) \in A_x \times \mathbb{R}^{n-m}\} \\
&= \mu\{s \mid X_I \in A_x\} \\
&\equiv \mu\left[X_I^{-1}(A_x)\right].
\end{aligned}
$$

∎

Remark 3.37 (On memory of X_J) *The significance of this result is that a marginal distribution function $F(X_I)$ obtained from $F(x_1, x_2, ..., x_n) \equiv F(X_I, X_J)$ has no "memory" of the X_J-variates that were originally represented by the joint distribution function. This will be eminently obvious in the case of independent random variables in the next section, but remains true without this assumption.*

Example 3.38 *For Example 3.33 above, the marginal distribution functions of $F(j, k)$ are identical, $F(j) = F(k)$. For example:*

$$
\begin{aligned}
F(j) &= \lim_{k \to \infty} \sum\nolimits_{f_2=1}^{6} \mu(f_1 \le j - f_2)\mu(f_3 \le k - f_2)/6 \\
&= \sum\nolimits_{f_2=1}^{6} \mu(f_1 \le j - f_2)/6.
\end{aligned}
$$

In Chapter 6, we prove that a marginal distribution function induces a well-defined Borel measure on $\mathbb{R}^m$, which of course is a probability measure.

3.3.2 Conditional Distribution Functions

While a more general development of conditional distribution functions requires the general measure and integration theory of Book V and is thus deferred to Book VI, the basic ideas can be introduced here and are often adequate for many applications. In this section we develop this idea in the context of conditional probabilities of Section 1.5. As was the case in the prior section, while these will be called distribution functions in the following definition, it should be noted that this characterization again requires proof.

Definition 3.39 (Conditional distribution function) *Let $X : S \longrightarrow \mathbb{R}^n$ be the random vector $X = (X_1, X_2, ..., X_n)$ defined on $(S, \mathcal{E}, \mu)$, $J \equiv \{j_1, ..., j_m\} \subset \{1, 2, ..., n\}$, and $X_J \equiv (X_{j_1}, X_{j_2}, ..., X_{j_m})$. Given a Borel set $B \in \mathcal{B}(\mathbb{R}^m)$ with $\mu\left[X_J^{-1}(B)\right] \ne 0$, define the **conditional distribution function** of X given $X_J \in B$, denoted $F(x|X_J \in B) \equiv F(x_1, x_2, ..., x_n | X_J \in B)$, in terms of the **conditional probability measure**:*

$$
F(x|X_J \in B) \equiv \mu\left[X^{-1}\left(\prod\nolimits_{i=1}^{n}(-\infty, x_i]\right) \middle| X_J^{-1}(B)\right].
$$

By Definition 1.31:

$$
F(x|X_J \in B) = \mu\left[X^{-1}\left(\prod\nolimits_{i=1}^{n}(-\infty, x_i]\right) \bigcap X_J^{-1}(B)\right] \Big/ \mu\left[X_J^{-1}(B)\right]. \tag{3.21}
$$

This distribution function is sometimes denoted $F_{J|B}(x)$.

Example 3.40

1. *A very important example for extreme value theory of Chapter 9 is the 1-dimensional case and $B = \{X > t\}$. Then $F(x|B) = 0$ for $x \le t$ by (3.21) since then*

$$
X^{-1}((-\infty, x]) \cap X^{-1}((-\infty, t]) = \emptyset.
$$

For $x \ge t$:

$$
F(x|B) = \frac{F(x) - F(t)}{1 - F(t)}.
$$

2. When $B = \mathbb{R}^m$, the conditional distribution equals the joint distribution for any J:

$$F(x|X_J \in B) = F(x).$$

Except in relatively simple cases as in Example 3.40, it may not be apparent that $F(x|X_J \in B)$ is in fact a distribution function by Definition 3.28, despite what it is called in Definition 3.39.

Proposition 3.41 ($F(x|X_J \in B)$ **is a distribution function**) *Given the random vector* $X = (X_1, X_2, ..., X_n)$ *defined on* $(\mathcal{S}, \mathcal{E}, \mu)$, $J \equiv \{j_1, ..., j_m\} \subset \{1, 2, ..., n\}$, $X_J \equiv (X_{j_1}, X_{j_2}, ..., X_{j_m})$, *and a Borel set* $B \in \mathcal{B}(\mathbb{R}^m)$ *with* $\mu\left[X_J^{-1}(B)\right] \neq 0$, *the function* $F(x|X_J \in B)$ *is a distribution function.*
Proof. *Define the set function* $\mu(\cdot|X_J^{-1}(B))$ *on* $A \in \mathcal{E}$ *by:*

$$\mu(A|X_J^{-1}(B)) = \mu(A \cap X_J^{-1}(B))/\mu\left(X_J^{-1}(B)\right).$$

This is a probability measure on $\mathcal{E}$ *by Exercise 1.29 since* $\mu\left[X_J^{-1}(B)\right] \neq 0$, *and thus* $(\mathcal{S}, \mathcal{E}, \mu(\cdot|X_J^{-1}(B)))$ *is a probability space.*
Now define the random vector $X = (X_1, X_2, ..., X_n)$ *on* $(\mathcal{S}, \mathcal{E}, \mu(\cdot|X_J^{-1}(B)))$, *and note that* $F(x|X_J \in B)$ *is the joint distribution function of* $(X_1, X_2, ..., X_n)$ *defined on this probability space. Thus* $F(x|X_J \in B)$ *is a distribution function by Definition 3.28.* ∎

Example 3.42 (Continuous probability theory) *It is likely that the reader has encountered conditional distribution examples for* $n = 2$ *for which* B *is a single point, and for which the above definition makes little sense because* $\mu\left[X_2^{-1}(B)\right] = 0$. *The purpose of this example is to connect the above definition to that case. To do so will require some hopefully familiar mathematics, but nonetheless, mathematics which will be fully justified in later books.*
Assume that the distribution function $F(x, y)$ *has an associated* **continuous density function** $f(x, y) \geq 0$ *so that defined as a Riemann integral:*

$$F(x, y) = \int_{-\infty}^{y} \int_{-\infty}^{x} f(s, t) ds dt.$$

Density functions will be more generally discussed in Books IV and VI, but this discussion is simplified by keeping within the Riemann context. The marginal distribution for y *is then given by:*

$$F(y) = F(\infty, y) = \int_{-\infty}^{y} \int_{-\infty}^{\infty} f(s, t) ds dt.$$

Continuity of f *then assures that both* $F(x, y)$ *and* $F(y)$ *are differentiable with:*

$$\frac{\partial F(x, y)}{\partial y} = \int_{-\infty}^{x} f(s, y) ds, \quad \frac{\partial^2 F(x, y)}{\partial x \partial y} = f(x, y), \quad \frac{\partial F(y)}{\partial y} = f(y),$$

where $f(y) = \int_{-\infty}^{\infty} f(s, y) ds$.
We seek to define the conditional distribution $F(x|y')$, *or in the above notation,* $F(x, y|B)$ *when* $B = \{y'\}$ *for given* y'. *In this case,* $\mu\left[Y^{-1}(B)\right] = 0$. *To see this, by continuity from above of measures (Proposition I.2.45):*

$$\mu\left[Y^{-1}(B)\right] = \lim_{\Delta y \to 0} \mu\left(Y^{-1}(y' - \Delta y, y' + \Delta y]\right).$$

But then:

$$\mu\left(Y^{-1}(y'-\Delta y, y'+\Delta y]\right) = \left(Y^{-1}(-\infty, y'+\Delta y]\right) - \left(Y^{-1}(-\infty, y'-\Delta y]\right)$$
$$= F(y'+\Delta y) - F(y'-\Delta y),$$

and so $\mu\left[Y^{-1}(B)\right] = 0$ *by the continuity of* $F(y)$.

But intuitively, the notion of $F(x|y')$ *would seem to make sense. If it is known that* $Y = y'$, *there must be an associated distribution function of* X *implied by the joint distribution function when restricted to* $\{(X, y')\}$.

Since the above definition cannot be used directly, let $B^+ \equiv [y, y+\Delta y]$, *where we now drop the* y' *notation. Then by (3.21) and continuity of* F:

$$F(x, y+\Delta y|B^+) = \frac{F(x, y+\Delta y) - F(x, y)}{F(y+\Delta y) - F(y)}.$$

Now since $F(x, y)$ *and* $F(y)$ *are differentiable, we can divide numerator and denominator by* Δy *and take a limit of* $\Delta y \to 0$. *We define this limit to be:*

$$F(x|y) \equiv F(x, y|B = \{y\}),$$

and assuming that $\frac{\partial F(y)}{\partial y} = f(y) \neq 0$ *obtain:*

$$F(x|y) = \frac{\partial F(x, y)}{\partial y} \bigg/ f(y).$$

Thus $F(x|y)$ *is differentiable in* x, *and:*

$$\frac{\partial F(x|y)}{\partial x} = \frac{\partial^2 F(x, y)}{\partial x \partial y} \bigg/ f(y)$$
$$= f(x, y)/f(y).$$

It now follows by the fundamental theorem of calculus that when $f(y) \neq 0$ *as assumed, that the function* $F(x|y)$ *has a density function in the sense that:*

$$F(x|y) = \int_{-\infty}^{x} f(s, y)/f(y) ds.$$

This proves that $F(x|y)$ *is a distribution function since it is continuous, increasing, and with limits of 0 and 1 and* $-\infty$ *and* ∞, *respectively.*

Denoting this density function by $f(x|y)$, *the above derivation obtains the perhaps familiar formula:*

$$f(x|y) = f(x, y)/f(y). \tag{3.22}$$

3.4   Independent Random Variables

In this section we discuss the notion of **independent random variables,** and derive an implication for the associated joint and marginal distribution functions. We begin with a discussion on the sigma algebras generated by random variables, and this will provide the connection between independence of random variables as defined here, and independence of sigma algebras as defined in Section 1.4.

3.4.1   Sigma Algebras Generated by R.V.s

Given a probability space $(\mathcal{S}, \mathcal{E}, \mu)$ and random variable (r.v.) X:

$$X : \mathcal{S} \longrightarrow \mathbb{R}.$$

Recall from Exercise 3.3 that $X^{-1}(A) \in \mathcal{E}$ for every Borel set $A \in \mathcal{B}(\mathbb{R})$. Thus:

$$\{X^{-1}(A) | A \in \mathcal{B}(\mathbb{R})\} \subset \mathcal{E},$$

or more succinctly:

$$X^{-1}(\mathcal{B}(\mathbb{R})) \subset \mathcal{E}.$$

In addition, $X^{-1}(\mathcal{B}(\mathbb{R}))\}$ is a sigma algebra by Corollary I.3.27, and hence is a **sigma subalgebra** of $\mathcal{E}$.

This leads to the following definition.

Definition 3.43 (The sigma algebra $\sigma(X)$, etc.) *If $X : \mathcal{S} \longrightarrow \mathbb{R}$ is a random variable on $(\mathcal{S}, \mathcal{E}, \mu)$, **the sigma algebra generated by** X, denoted $\sigma(X)$, is the smallest sigma algebra with respect to which X is Borel measurable.*

*If $\{X_i\}$ is a finite or infinite collection of random variables, **the sigma algebra generated by** $\{X_i\}$, denoted $\sigma(X_1, X_2, ...)$, is the smallest sigma algebra with respect to which each such X_i is Borel measurable.*

*Analogously, if $X : \mathcal{S} \longrightarrow \mathbb{R}^n$ is a random vector on $(\mathcal{S}, \mathcal{E}, \mu)$, **the sigma algebra generated by** X, denoted $\sigma(X)$, is the smallest sigma algebra with respect to which X is Borel measurable.*

Exercise 3.44 (Characterization of $\sigma(X)$) *Show that $\sigma(X)$ is given by:*

$$\sigma(X) = X^{-1}(\mathcal{B}(\mathbb{R}^n)), \tag{3.23}$$

where $n = 1$ for the random variable definition, and general n in the case of a random vector.

Proposition 3.45 (Characterization of $\sigma(X_1, X_2, ...)$) *Given $\{X_i\}$, a finite or infinite collection of random variables defined on $(\mathcal{S}, \mathcal{E}, \mu)$:*

$$\sigma(X_1, X_2, ...) = \sigma(\sigma(X_1), \sigma(X_2), ...), \tag{3.24}$$

The right-hand expression denotes the sigma algebra generated by the collection of sigma algebras, $\{\sigma(X_j)\}$, and defined as the smallest sigma algebra which contains these classes.
Proof. *By defining $\sigma(\sigma(X_1), \sigma(X_2), ...)$ as the smallest sigma algebra which contains these classes, we require a demonstration that there exists at least one such sigma algebra to avoid a vacuous definition. The power sigma algebra $\sigma(P(\mathcal{S}))$, defined as the collection of all subsets of $\mathcal{S}$, is one such example.*

By definition, $\sigma(X_j) \subset \sigma(X_1, X_2, ...)$ for each j since X_j is measurable with respect to $\sigma(X_1, X_2, ...)$. As $\sigma(X_1, X_2, ...)$ is a sigma algebra, we conclude that $\sigma(\sigma(X_1), \sigma(X_2), ...) \subset \sigma(X_1, X_2, ...)$. But $\sigma(X_1, X_2, ...)$ is the smallest sigma algebra with respect to which each X_j is measurable. As each X_j is measurable with respect to $\sigma(\sigma(X_1), \sigma(X_2), ...)$, it follows that $\sigma(X_1, X_2, ...) \subset \sigma(\sigma(X_1), \sigma(X_2), ...)$, and (3.24) is proved. ■

Example 3.46 *Define $\mathcal{S}$ as the collection of n-tuples of heads or tails obtained by flipping a coin sequentially n-times. So $\mathcal{S} = \{(f_1, f_2, ..., f_n)\}$ where each $f_k = H$ or T. Define μ on $(f_1, f_2, ..., f_n)$ consistently with (1.8):*

$$\mu((f_1, f_2, ..., f_n)) = p^j(1 - p)^{n-j},$$

where j denotes the number of heads in this n-tuple, and p denotes the probability of a head. Now let $\mathcal{E} = \sigma(P(\mathcal{S}))$, the collection of all subsets of $\mathcal{S}$ to which we extend μ additively.

Let $X : \mathcal{S} \longrightarrow \mathbb{R}$ be the random variable defined as the number of heads, so $X : \mathcal{S} \longrightarrow \{0, 1, 2, ..., n\}$. With

$$S_j \equiv \{ (f_1, f_2, ..., f_n) \, | X\, (f_1, f_2, ..., f_n) = j \},$$

$\sigma(X)$ equals the sigma algebra generated by $\{S_j\}$, while $\mu\,(S_j)$ is given as in (1.8).

Here $\sigma(X)$ is a proper sigma subalgebra of $\mathcal{E}$:

$$\sigma(X) \subsetneqq \mathcal{E},$$

since the latter sigma algebra contains the measurable set $\{(H, T, T, ..., T)\}$, while the former does not. In $\sigma(X)$, this set is a subset of the measurable set S_1, but is not itself a measurable set in $\sigma(X)$.

3.4.2   Independent Random Variables and Vectors

Given the sigma algebras generated by random variables, the notion of independent classes of sets from Section 1.4 provides the basis for the definition of independent random variables.

Definition 3.47 (Independent random variables/vectors) *If $X_j : \mathcal{S} \longrightarrow \mathbb{R}$ are random variables on $(\mathcal{S}, \mathcal{E}, \mu)$, $j = 1, 2, ..., n$, we say that $\{X_j\}_{j=1}^n$ are **independent random variables** if $\{\sigma(X_j)\}_{j=1}^n$ are independent sigma algebras in the sense of Definition 1.15.*

In other words, given $\{B_j\}_{j=1}^n$ with $B_j \in \sigma(X_j)$:

$$\mu\left(\bigcap\nolimits_{j=1}^n B_j\right) = \prod\nolimits_{j=1}^n \mu\,(B_j). \tag{3.25}$$

Equivalently, given $\{A_j\}_{j=1}^n \subset \mathcal{B}(\mathbb{R})$:

$$\mu\left(\bigcap\nolimits_{j=1}^n X_j^{-1}(A_j)\right) = \prod\nolimits_{j=1}^n \mu\,(X_j^{-1}(A_j)). \tag{3.26}$$

*If $X_j : \mathcal{S} \longrightarrow \mathbb{R}^{n_j}$ are random vectors on $(\mathcal{S}, \mathcal{E}, \mu)$, $j = 1, 2, ..., n$, we say that $\{X_j\}_{j=1}^n$ are **independent random vectors** if $\{\sigma(X_j)\}_{j=1}^n$ are independent sigma algebras in the sense of Definition 1.15. In other words, given $\{B_j\}_{j=1}^n$ with $B_j \in \sigma(X_j)$:*

$$\mu\left(\bigcap\nolimits_{j=1}^n B_j\right) = \prod\nolimits_{j=1}^n \mu\,(B_j). \tag{3.27}$$

Equivalently, given $\{A_j\}_{j=1}^n$ with $A_j \in \mathcal{B}(\mathbb{R}^{n_j})$:

$$\mu\left(\bigcap\nolimits_{j=1}^n X_j^{-1}(A_j)\right) = \prod\nolimits_{j=1}^n \mu\,(X_j^{-1}(A_j)). \tag{3.28}$$

A countable collection of random variables $\{X_j\}_{j=1}^\infty$ defined on $(\mathcal{S}, \mathcal{E}, \mu)$ are said to be **independent random variables** if given any **finite** index subcollection, $J = (j(1), j(2), ..., j(n))$, $\{X_{j(i)}\}_{i=1}^n$ are independent random variables. The analogous definition of independence applies to a countable collection of **random vectors**, as well as to uncountable collections.

Notation 3.48 *In the language and notation of probability theory, (3.26) is often expressed:*

$$\Pr[X_1 \in A_1, ..., X_n \in A_n] = \prod\nolimits_{j=1}^n \Pr[X_j \in A_j], \tag{3.29}$$

and similarly for other such statements.

Example 3.49 (Dice triplets cont'd) *Recall Example 3.33 with $\mathcal{S}$ defined as the collection of 3-tuples of scores on rolling a fair die sequentially three times:*

$$\mathcal{S} \equiv \{(f_1, f_2, f_3)|f_j \in \{1, ..., 6\}\}.$$

The probability measure μ is defined pointwise on all 3-tuples by:

$$\mu(f_1, f_2, f_3) = 1/6^3, \tag{1}$$

and this measure is extended additively to $\mathcal{E} \equiv \sigma(P(\mathcal{S}))$, the power sigma algebra.

If we define random variables $\{Y_j\}_{j=1}^3$ on $\mathcal{S}$ by:

$$Y_j : (f_1, f_2, f_3) = f_j,$$

*we now show that these are **independent random variables** as noted in Example 3.33.*

To see this, first note that for any j the sigma algebra $\sigma(Y_j)$ is the sigma algebra generated by the class of disjoint sets:

$$\sigma(Y_j) = \sigma\left\{\emptyset, B_1^{(j)}, ..., B_6^{(j)}\right\},$$

where:

$$B_k^{(j)} = \{(f_1, f_2, f_3)|f_j = k\}.$$

It is an exercise using (1) to show that for all j, k:

$$\mu\left(B_k^{(j)}\right) = 1/6. \tag{2}$$

*This is the **fair die** condition, that all results on a single roll are equally likely.*

Denoting:

$$B_j \equiv \left\{\emptyset, B_1^{(j)}, ..., B_6^{(j)}\right\},$$

we claim that $\{B_j\}_{j=1}^3$ are mutually independent classes by Definition 1.15. To prove this, we only need to check pairs and triplets of sets from these classes.

For example, given $B_{k_1}^{(j_1)} \in B_{j_1}$ and $B_{k_2}^{(j_2)} \in B_{j_2}$, where $j_1 \neq j_2$:

$$B_{k_1}^{(j_1)} \bigcap B_{k_2}^{(j_2)} = \{(f_1, f_2, f_3)|f_{j_1} = k_1, f_{j_2} = k_2\},$$

and it follows that when both sets are nonempty:

$$\mu\left(B_{k_1}^{(j_1)} \bigcap B_{k_2}^{(j_2)}\right) = 1/6^2 = \mu\left(B_{k_1}^{(j_1)}\right) \mu\left(B_{k_2}^{(j_2)}\right).$$

If one or both sets are empty, these calculations obtain 0. The case of a triplet of sets is left as an exercise. Thus $\{B_j\}_{j=1}^3$ are mutually independent classes, and each is closed under intersections. By Corollary 1.27, $\{\sigma(B_j)\}_{j=1}^3 = \{\sigma(Y_j)\}_{j=1}^3$ are mutually independent sigma algebras.

*That is, $\{Y_j\}_{j=1}^3$ are independent random variables (see **Note** below).*

Introduced in Example 3.33, define the random variable $X_1 \equiv f_1 + f_2$, the sum of the first and second rolls, and $X_2 \equiv f_2 + f_3$, the sum of the second and third rolls. Then X_1 and X_2 are not independent random variables, most apparently because $X_1^{-1}(2) \bigcap X_2^{-1}(12) = \emptyset$ and hence has probability 0, whereas the product of the individual probabilities is $1/36^2$. As another example, $X_1^{-1}(2) \bigcap X_2^{-1}(2) = (1, 1, 1)$, and so has probability $1/6^3$, while the product of the probabilities of these events is again $1/36^2$.

Intuitively, these random variables are not independent because both values depend on the common value of the second roll.

Note: *The random variables $\{Y_j\}_{j=1}^3$ are independent under the probability measure in (1), and for a fair die this is the only probability measure on $\mathcal{S}$ for which this is true. To see this, independence requires that:*

$$\mu\left(B_{k_1}^{(1)} \bigcap B_{k_2}^{(2)} \bigcap B_{k_3}^{(3)}\right) = \mu\left(B_{k_1}^{(1)}\right)\mu\left(B_{k_2}^{(2)}\right)\mu\left(B_{k_3}^{(3)}\right).$$

By definition, $B_{k_1}^{(1)} \bigcap B_{k_2}^{(2)} \bigcap B_{k_3}^{(3)} = (k_1, k_2, k_3)$ and the fair die condition in (2) obtains the result.

It is left as an exercise for the reader to check that this generalizes to any number of rolls of a fair die, and also to biased dice in a natural way.

Exercise 3.50 (Independent vectors from independent variables) *Let $X_j : \mathcal{S} \longrightarrow \mathbb{R}$ be independent random variables on $(\mathcal{S}, \mathcal{E}, \mu)$, $j = 1, 2, ..., n$, and let $\{I_k\}_{k=1}^m$ be disjoint index subsets with $\bigcup_k I_k = \{1, 2, ..., n\}$. If $I_k = \{k(1), ..., k(n_k)\}$, define the random vector $Y_k : \mathcal{S} \longrightarrow \mathbb{R}^{n_k}$, where $Y_k \equiv (X_{k(1)}, ..., X_{k(n_k)})$. In other words, Y_k contains the X_j-variates identified by I_k.*

Prove that $\{Y_k\}_{k=1}^m$ are independent random vectors. Hint: See Proposition 3.51 below, which justifies that it is enough to prove (3.28) with $A_k \in \mathcal{B}(\mathbb{R}^{n_k})$, defined as a right semi-closed rectangle of the form $A_k \equiv \prod_{i=1}^{n_k} (-\infty, b_i^{(k)}]$.

It turns out that independence of random variables can be characterized by a simpler criterion than (3.26) which requires this condition be satisfied for all collections $\{A_j\}_{j=1}^n \subset \mathcal{B}(\mathbb{R})$. The following proposition states that one only needs to verify this condition for a special class of Borel measurable sets and this result generalizes to random vectors.

Proposition 3.51 (Simplifying the test of independence) *If $X_j : \mathcal{S} \longrightarrow \mathbb{R}$ are random variables on $(\mathcal{S}, \mathcal{E}, \mu)$, $j = 1, 2, ..., n$, and (3.26) is satisfied for all sets of the form $A_j = (-\infty, b_j]$, then $\{X_j\}_{j=1}^n$ are independent random variables.*

Similarly, if $X_j : \mathcal{S} \longrightarrow \mathbb{R}^{n_j}$ are random vectors on $(\mathcal{S}, \mathcal{E}, \mu)$, $j = 1, 2, ..., n$, and (3.28) is satisfied for all sets of the form $A_j = \prod_{k=1}^{n_j} (-\infty, b_k^{(j)}]$, then $\{X_j\}_{j=1}^n$ are independent random vectors.

Proof. *Recall Corollary 1.27, where we are given mutually independent classes $\{\mathsf{A}_j\}_{j=1}^N$ with $\mathsf{A}_j = \{A_{jk}\}_{k=1}^{M_j} \subset \mathcal{E}$, and where N, M_j can be finite or infinite (countable or uncountable by Remark 1.28). If these classes are each closed under finite intersections, then $\{\sigma(\mathsf{A}_j)\}_{j=1}^N$ are mutually independent sigma algebras with $\sigma(\mathsf{A}_j)$ denoting the smallest sigma algebra that contains A_j.*

Define $\mathsf{A}_j = \{X_j^{-1}((-\infty, b_k])\}$, the class of pre-images for all real b_k. Note that each A_j is closed under finite intersections:

$$\bigcap_{k=1}^m X_j^{-1}((-\infty, b_k]) = X_j^{-1}\left(\bigcap_{k=1}^m (-\infty, b_k]\right) = X_j^{-1}((-\infty, \min b_k]).$$

Given the assumption in this proposition, that $\{\mathsf{A}_j\}_{j=1}^n$ are mutually independent classes, this then implies the same for $\{\sigma(\mathsf{A}_j)\}_{j=1}^n$. The proof is complete because the collection $\{(-\infty, b_k]\}$ generates the collection $\{(a_k, b_k]\}$, the semi-algebra of right semi-closed intervals, and hence $\sigma(\mathsf{A}_j) = \mathcal{B}(\mathbb{R})$ by Proposition I.8.1.

The proof is similar for random vectors, and left as an exercise. ∎

Remark 3.52 (Other tests) *For random variables or vectors, the above results generalize to any class of sets $\mathsf{A}_j = \{X_j^{-1}(A_k)\}$ with $\{A_k\} \subset \mathcal{B}(\mathbb{R}^{n_j})$, if A_j is closed under finite* *intersections and $\{A_k\}$ generates $\mathcal{B}(\mathbb{R}^{n_j})$. Proposition I.8.1 identifies two possibilities for classes of sets that generate $\mathcal{B}(\mathbb{R}^{n_j})$, both of the form $A_j = \prod_{k=1}^{n_j}(a_k^{(j)}, b_k^{(j)}]$. One class limits these rectangles to be bounded, one does not. Thus, subject to closure of the resultant A_j to finite intersections, $\{A_k\}$ can be taken as any collection of sets that generates these rectangles. For example, these rectangles can be defined with open intervals, closed intervals, or left semi-closed intervals.*

3.4.3   Distribution Functions of Independent R.V.s

The notion of independent random variables can also be characterized in terms of the associated distribution functions. If $X_j : \mathcal{S} \longrightarrow \mathbb{R}$, $j = 1, 2, ..., n$, are random variables on a probability space $(\mathcal{S}, \mathcal{E}, \mu)$, the **random vector** $X = (X_1, X_2, ..., X_n)$ was defined above as the function:

$$X : \mathcal{S} \longrightarrow \mathbb{R}^n,$$

with value on $s \in \mathcal{S}$ given by:

$$X(s) = (X_1(s), X_2(s), ..., X_n(s)).$$

The distribution function of X is defined in (3.15) by:

$$F(x_1, x_2, ..., x_n) = \mu \left[\bigcap_{j=1}^{n} X_j^{-1}(-\infty, x_j] \right].$$

Proposition 3.53 (Characterization of the joint D.F.) *Let $X_j : \mathcal{S} \longrightarrow \mathbb{R}$, $j = 1, 2, ..., n$, be random variables on a probability space $(\mathcal{S}, \mathcal{E}, \mu)$ with distribution functions $F_j(x)$. Let the random vector $X : \mathcal{S} \longrightarrow \mathbb{R}^n$ be defined on this space by $X(s) = (X_1(s), X_2(s), ..., X_n(s))$ and with distribution function $F(x)$.*
Then $\{X_j\}_{j=1}^{n}$ are independent random variables if and only if:

$$F(x_1, x_2, ..., x_n) = \prod_{j=1}^{n} F_j(x_j). \tag{3.30}$$

Countably many random variables $\{X_j\}_{j=1}^{\infty}$ defined on $(\mathcal{S}, \mathcal{E}, \mu)$ are independent random variables if and only if for every finite index subcollection, $J = (j(1), j(2), ..., j(n))$:

$$F(x_{j(1)}, x_{j(2)}, ..., x_{j(n)}) = \prod_{i=1}^{n} F_{j(i)}(x_{j(i)}). \tag{3.31}$$

These results are valid for random vectors $X_j : \mathcal{S} \longrightarrow \mathbb{R}^{n_j}$, noting that $F_j(x_j)$ are joint distribution functions on $\mathbb{R}^{n_j}$ as given in (3.15) or (3.17).
Proof. *We prove the results for random variables and leave the generalization as an exercise.*
If $\{X_j\}_{j=1}^{n}$ are independent random variables, then applying the definition in (3.26) to (3.15) obtains:

$$
\begin{aligned}
F(x_1, x_2, ..., x_n) &\equiv \mu \left[\bigcap_{j=1}^{n} X_j^{-1}(-\infty, x_j] \right] \\
&= \prod_{j=1}^{n} \mu \left(X_j^{-1}(-\infty, x_j] \right) \\
&\equiv \prod_{j=1}^{n} F_j(x_j),
\end{aligned}
$$

where $F_j(x)$ denotes the distribution function of X_j by (3.1).
Conversely, if $\{X_j\}_{j=1}^{n}$ are random variables defined on $(\mathcal{S}, \mathcal{E}, \mu)$ with the property that for all $(x_1, x_2, ..., x_n) \in \mathbb{R}^n$:

$$F(x_1, x_2, ..., x_n) = \prod_{j=1}^{n} F_j(x_j),$$

then (3.26) is seen to be valid for all sets of the form $A_j = (-\infty, b_j]$. Proposition 3.51 then implies that $\{X_j\}_{j=1}^n$ are independent random variables. ∎

Notation 3.54 *In many texts, the identity in (3.30) and similar statements would have been written in terms of $\{F(x_j)\}$ rather than $\{F_j(x_j)\}$:*

$$F(x_1, x_2, ..., x_n) = \prod_{j=1}^{n} F(x_j).$$

It is conventional to denote distribution functions simply by $F(x)$, and it would be understood that $F(x_j)$ was the distribution function for X_j, and this could be the same or different from the distribution function $F(x_k)$ for X_k.

Corollary 3.55 (On the marginal D.F.) *Let $X_j : \mathcal{S} \longrightarrow \mathbb{R}$, $j = 1, 2, ..., n$, be independent random variables on a probability space $(\mathcal{S}, \mathcal{E}, \mu)$. Let $I = \{i_1, ..., i_m\} \subset \{1, 2, ..., n\}$ and $J = \{j_1, ..., j_{n-m}\}$ with $J \equiv \tilde{I}$.*

Then the marginal distribution function $F_I(x_{i_1}, x_{i_2}, ..., x_{i_m})$ defined in (3.19) is given by:

$$F_I(x_{i_1}, x_{i_2}, ..., x_{i_m}) = \prod_{j=1}^{m} F_{i_j}(x_{i_j}). \tag{3.32}$$

Proof. *This result follows from the previous result by the definition of $F_I(x_{i_1}, x_{i_2}, ..., x_{i_m})$ in (3.20):*

$$F_I(x_{i_1}, x_{i_2}, ..., x_{i_m}) = \mu \left[X_I^{-1} \left(\prod_{k=1}^{m} (-\infty, x_{i_k}] \right) \right],$$

since:

$$X_I^{-1} \left(\prod_{k=1}^{m} (-\infty, x_{i_k}] \right) = \bigcap_{k=1}^{m} X_{k_k}^{-1} (-\infty, x_{i_k}].$$

∎

3.4.4   Independence and Transformations

Let $X_j : \mathcal{S} \longrightarrow \mathbb{R}^{n_j}$, $j = 1, 2, ..., n$, be **independent random vectors** on $(\mathcal{S}, \mathcal{E}, \mu)$, which includes the special case of independent random variables when any $n_j = 1$. Let:

$$g_j : \mathbb{R}^{n_j} \to \mathbb{R}^{m_j},$$

be **Borel measurable transformations, meaning**:

$$g_j^{-1} \left(\mathcal{B} \left(\mathbb{R}^{n_j} \right) \right) \subset \mathcal{B} \left(\mathbb{R}^{m_j} \right),$$

which includes the special case of Borel measurable functions when any $m_j = 1$.

The purpose of this section is to investigate the independence of:

$$g_j(X_j) : \mathcal{S} \longrightarrow \mathbb{R}^{m_j}.$$

The next proposition summarizes the general result. While simply stated, it provides for a host of applications.

Proposition 3.56 (Independence survives transformations) *Let $X_j : \mathcal{S} \longrightarrow \mathbb{R}^{n_j}$, $j = 1, 2, ..., n$, be independent random vectors on $(\mathcal{S}, \mathcal{E}, \mu)$, and $g_j : \mathbb{R}^{n_j} \to \mathbb{R}^{m_j}$ Borel measurable transformations.*

Then $g_j(X_j) : \mathcal{S} \longrightarrow \mathbb{R}^{m_j}$ are independent random vectors on $(\mathcal{S}, \mathcal{E}, \mu)$. The same is true for a countable collection $\{X_j\}_{j=1}^{\infty}$.

Proof. *By Definition 3.47 we must prove that $\{\sigma(g_j(X_j))\}_{j=1}^n$ are independent sigma algebras in the sense of Definition 1.15. In other words, given $\{A_j\}_{j=1}^n$ with $A_j \in \mathcal{B}(\mathbb{R}^{m_j})$,*

$$\mu\left(\bigcap_{j=1}^n [g_j(X_j)]^{-1}(A_j)\right) = \prod_{j=1}^n \mu\left([g_j(X_j)]^{-1}(A_j)\right). \tag{1}$$

Now:

$$[g_j(X_j)]^{-1}(A_j) = X_j^{-1}\left[g_j^{-1}(A_j)\right], \tag{2}$$

and by Borel measurability, $g_j^{-1}(A_j) \in \mathcal{B}(\mathbb{R}^{n_j})$ for all $A_j \in \mathcal{B}(\mathbb{R}^{m_j})$. Thus (1) follows from (2) and (3.28).

For a countable collection, independence is defined in terms of finite subsets of random vectors, so this result follows from that just proved. ∎

Exercise 3.57 (Components of independent random vectors) *Apply the above result to show that if $X_j : \mathcal{S} \longrightarrow \mathbb{R}^{n_j}$ are independent random vectors for $j = 1, 2, ..., n$, say $X_j \equiv (X_{j,1}, ..., X_{j,n_j})$, then $\{X_{j,k(j)}\}_{j=1}^n$ are independent random variables for any $\{k(j)\}_{j=1}^n$ with $1 \le k(j) \le n_j$.*

Example 3.58 *In the following, $\{X_j\}$ is a finite or countable collection of random variables/vectors on $(\mathcal{S}, \mathcal{E}, \mu)$.*

1. ***Random variables from random variables:*** *If $X_j : \mathcal{S} \longrightarrow \mathbb{R}$ are independent random variables, then so too are X_j^2, or e^{X_j}, or $g_j(X_j)$ for any Borel measurable functions $g_j : \mathbb{R} \longrightarrow \mathbb{R}$.*

2. ***Random variables from random vectors:*** *If $X_j : \mathcal{S} \longrightarrow \mathbb{R}^{n_j}$ are independent random vectors, say $X_j \equiv (X_{j,1}, ..., X_{j,n_j})$, then $Y_j : \mathcal{S} \longrightarrow \mathbb{R}$ defined by $Y_j = \sum_{k=1}^{n_j} X_{j,k}$ are independent random variables. The same is true for $Y_j = g_j(X_j)$ for any Borel measurable functions $g_j : \mathbb{R}^{n_j} \longrightarrow \mathbb{R}$.*

3. ***Random vectors from random variables:*** *Let $X_j : \mathcal{S} \longrightarrow \mathbb{R}$ be n independent random variables and $\{I_k\}_{k=1}^m$ disjoint index subsets with $I_k = \{k(1), ..., k(n_k)\}$ and $\bigcup_k I_k = \{1, 2, ..., n\}$. Then $Y_k : \mathcal{S} \longrightarrow \mathbb{R}^{n_k}$ defined by $Y_k \equiv (X_{k(1)}, ..., X_{k(n_k)})$ are independent random vectors by Exercise 3.50. It then follows that $g_j(Y_j)$ are independent random vectors (or variables) for Borel measurable $g_j : \mathbb{R}^{n_j} \longrightarrow \mathbb{R}^{m_j}$.*

   Note: *An application of this will be seen in the beginning of Section 9.1. There we will use the fact that if $X_j : \mathcal{S} \longrightarrow \mathbb{R}$ are independent random variables, $j = 1, 2, ..., n + m$, then so too are $\sum_{j=1}^m X_j$ and $\sum_{j=m+1}^{m+n} X_j$.*

4. ***Random vectors from random vectors:*** *Generalizing Exercise 3.57, let $X_j : \mathcal{S} \longrightarrow \mathbb{R}^{n_j}$ be independent random vectors for $j = 1, 2, ..., n$, say $X_j \equiv (X_{j,1}, ..., X_{j,n_j})$. Then $Y_j : \mathcal{S} \longrightarrow \mathbb{R}^{m_j}$ defined by $Y_j \equiv (X_{j,k_j(1)}, ..., X_{j,k_j(m_k)})$ are independent random vectors for any component subsets, $\{X_{j,k_j(i)}\}_{i=1}^{m_j} \subset \{X_{j,i}\}_{i=1}^{n_j}$ for all j. This follows because $Y_j = C_j X_j$ where $C_j \equiv \left(c_{il}^{(j)}\right)$ is an $m_k \times n_j$ matrix with $c_{i,k_j(i)}^{(j)} = 1$ and $c_{il}^{(j)} = 0$ otherwise.*

4

Probability Spaces and i.i.d. RVs

Given a probability space $(\mathcal{S}, \mathcal{E}, \mu)$ and a random variable $X : \mathcal{S} \longrightarrow \mathbb{R}$, one is often interested (see Chapter 5, for example) in investigating properties of a collection of random variables $\{X_j\}_{j=1}^N$, where N can be finite or countably infinite, which are **independent and identically distributed with** X (**i.i.d.-**X or simply **i.i.d.**):

- **Independence** is characterized in Definition 3.47.

- **Identically distributed with** X reflects the notions of Definition 3.1, that all X_j have the same distribution function as X.

While these random variables may be defined on different probability spaces for the distributional requirement, to apply Definition 3.47 on independence requires that they be defined on a common probability space, say $(\mathcal{S}', \mathcal{E}', \mu')$.

It is often tacitly assumed that i.i.d.-X random variables are defined on the initial probability space $(\mathcal{S}, \mathcal{E}, \mu)$, and the investigations pursued then reflect probability calculations relative to μ. This can seem plausible within the context of human experience, where one associates i.i.d.-X random variables with sequential coin flips, or draws from a well-shuffled deck of playing cards with replacement, etc. In these intuitive models, the underlying probability space is at best implicit, and the need to have yet a different space to support the sequence of outcomes does not seem compelling.

While this intuitive framework is often helpful, it disguises the complexity of the above statement. The random variable X defined on $(\mathcal{S}, \mathcal{E}, \mu)$ is simply a measurable function. And the collection $\{X_j\}_{j=1}^N$ we seek is thus a collection of measurable functions with complex technical requirements identified in Definitions 3.1 and 3.47. Thus, this is ultimately an existence question, and is the question investigated in this chapter.

In short, the question is:

Given $(\mathcal{S}, \mathcal{E}, \mu)$ and $X : \mathcal{S} \longrightarrow \mathbb{R}$, can we construct a probability space $(\mathcal{S}', \mathcal{E}', \mu')$ and a collection of random variables $\{X_j\}_{j=1}^N$ defined on this space, which are i.i.d.-X?

We will refer to the collection $\{X_j\}_{j=1}^N$ as an N-**sample** of X, or a **sample** of X when N is implied. The terminology **random sample** from a random variable X or some population also implies the result of some experimental or other empirical process by which **numerical** values $\{X_j(s_j)\}_{j=1}^N$ are generated or otherwise obtained. For this collection to be deemed a "random sample" requires that these variates be independent, and that they each be governed by the distribution underlying X, or the distributional assumption made for some characteristic of the population of interest.

Thus $\{X_j\}_{j=1}^N$ as a collection of i.i.d. random variables, and $\{X_j(s_j)\}_{j=1}^N$ as a random sample, are intimated related. This connection will be discussed somewhat in Example 4.6 below, and more fully investigated in Chapter 4 of Book IV on simulations of random variables. There we will also investigate examples for a variety of distribution functions.

In the next section we provide one construction of $(\mathcal{S}', \mathcal{E}', \mu')$ and $\{X_j\}_{j=1}^N$ in Proposition 4.4, where this probability space will be seen to reflect the distribution function $F(x)$ of X in a fundamental way. We begin this section by formalizing the definitions above.

In the following section we derive an alternative approach to this construction which is useful for empirical sampling underlying "stochastic simulations," also called "Monte Carlo simulations." Utilizing the investigation initiated in Section 3.2, we will, in Proposition 4.13, construct a single probability space and collection of variates, and then transform these variates to possess any distribution desired. This section will also introduce the notion of an essential subspace, which can be useful for reducing a probability space to a more convenient equivalent space.

The results of these two sections are perfectly general and apply to any probability space $(\mathcal{S}, \mathcal{E}, \mu)$ and random variable X. However, these propositions may well appear to be overly general when applied to discrete random variables, producing unfamiliar notational conventions. Indeed there is a simpler and perhaps more compelling model in this case, and that is addressed in the final section, again using a reduction with an essential subspace.

4.1 Probability Space $(\mathcal{S}', \mathcal{E}', \mu')$ and i.i.d. $\{X_j\}_{j=1}^N$

Given a probability space $(\mathcal{S}, \mathcal{E}, \mu)$ and a random variable $X : \mathcal{S} \longrightarrow \mathbb{R}$, we begin by formalizing the definitions underlying the above discussion.

Definition 4.1 (N-Sample 1) *Let a probability space $(\mathcal{S}, \mathcal{E}, \mu)$ and random variable $X : \mathcal{S} \longrightarrow \mathbb{R}$ be given. With N finite or infinite, a collection of random variables $\{X_j\}_{j=1}^N$ defined on a probability space $(\mathcal{S}', \mathcal{E}', \mu')$ is said to be an N-**sample of** X, or a **sample of** X when N is implied, if this collection is **independent, and identically distributed with** X (i.i.d.-X):*

1. *$\{X_j\}_{j=1}^N$ are **independent** if given $(i_1, ..., i_m) \subset (1, 2, ..., N)$ and $\{A_j\}_{j=1}^m \subset \mathcal{B}(\mathbb{R})$:*

$$\mu'\left[\bigcap_{j=1}^m X_{i_j}^{-1}(A_j)\right] = \prod_{j=1}^m \mu'[X_{i_j}^{-1}(A_j)]. \tag{4.1}$$

2. *$\{X_j\}_{j=1}^N$ are **identically distributed with** X if for all j, and all $A \in \mathcal{B}(\mathbb{R})$:*

$$\mu'[X_j^{-1}(A)] = \mu[X^{-1}(A)]. \tag{4.2}$$

Exercise 4.2 *The above definition applies to sets $A = (-\infty, x]$ for any x and thus implies statements about the associated distribution functions. Prove that this definition can be equivalently framed as in Definition 4.3, in terms of distribution functions. Hint: The collection $\{(-\infty, x]\}$ generates $\mathcal{B}(\mathbb{R})$, and thus if the above identities are valid for such A they are valid for all $A \in \mathcal{B}(\mathbb{R})$. See also Proposition 3.53.*

Definition 4.3 (N-Sample 2) *Let a probability space $(\mathcal{S}, \mathcal{E}, \mu)$ and random variable $X : \mathcal{S} \longrightarrow \mathbb{R}$ with distribution function $F(x)$ be given. With N finite or infinite, a collection of random variables $\{X_j\}_{j=1}^N$ defined on a probability space $(\mathcal{S}', \mathcal{E}', \mu')$ with distribution functions $\{F_j(x)\}_{j=1}^N$ is said to be an N-**sample of** X, or a **sample of** X when N is implied, if this collection is **independent, and identically distributed with** X (i.i.d.-X):*

1. *$\{X_j\}_{j=1}^N$ are **independent** if given $(i_1, ..., i_m) \subset (1, 2, ..., N)$, then for all $x = (x_{i_1}, ..., x_{i_m}) \in \mathbb{R}^m$:*

$$F(x_{i_1}, ..., x_{i_m}) = \prod_{j=1}^m F_j(x_{i_j}), \tag{4.3}$$

where $F(x_{i_1}, ..., x_{i_m})$ is the joint distribution function of $\{X_{i_j}\}_{j=1}^m$.

2. X_j are **identically distributed with** X *if for any j and all x:*

$$F_j(x) = F(x). \tag{4.4}$$

Below we will assume that $N = \infty$ to simplify notation and references to Chapter I.9. The reader can think through the similar construction for finite $N = n$ using the framework of Chapter I.7, or simply observe that any finite n-tuple subset from an infinite sample $\{X_j\}_{j=1}^\infty$ is by definition a random n-sample.

4.1.1 First Construction: $(\mathcal{S}_F', \mathcal{E}_F', \mu_F')$

Let the probability space $(\mathcal{S}, \mathcal{E}, \mu)$ and random variable $X : \mathcal{S} \longrightarrow \mathbb{R}$ be given, and let $F(x)$ denote the distribution function of X. In this section we provide the first construction of a probability space $(\mathcal{S}_F', \mathcal{E}_F', \mu_F')$ and i.i.d. $\{X_j\}_{j=1}^N$. The notation for the sample space is to connote that this space is constructed using the Borel measure induced by the distribution function $F(x)$.

To begin, F is increasing and right continuous as noted in Notation 3.5. By Proposition I.5.23, this function induces a measure μ_F defined on $\mathcal{B}(\mathbb{R})$, the Borel sigma algebra on $\mathbb{R}$, such that for all right semi-closed intervals:

$$\mu_F(a, b] = F(b) - F(a).$$

See Section 6.1.1 for a more detailed review of this development.

By Proposition 3.4:

$$F(b) - F(a) = \mu\left(X^{-1}(a, b]\right),$$

and thus for all right semi-closed intervals:

$$\mu_F(a, b] = \mu\left(X^{-1}(a, b]\right).$$

In other words, the set functions μ_F and $\mu\left(X^{-1}\right)$ agree on the **semi-algebra** $\mathcal{A}'$ of right semi-closed intervals on $\mathbb{R}$. Then by Proposition I.5.11, the definitional extension of these set functions will also agree on the associated algebra $\mathcal{A}$ of finite disjoint unions of $\mathcal{A}'$-sets. Since μ_F is a measure on this algebra, so too is $\mu\left(X^{-1}\right)$. The uniqueness theorem of Proposition I.6.14 now applies to assure that these measures agree on $\sigma(\mathcal{A})$, the smallest sigma algebra that contains $\mathcal{A}$. Noting that $\sigma(\mathcal{A}) = \sigma(\mathcal{A}')$ by definition, Proposition I.8.1 asserts that $\sigma(\mathcal{A}) = \mathcal{B}(\mathbb{R})$.

In summary, the measure μ_F of Proposition I.5.23 satisfies for all $A \in \mathcal{B}(\mathbb{R})$:

$$\mu_F(A) = \mu\left(X^{-1}(A)\right). \tag{4.5}$$

Thus $\mu_F(\mathbb{R}) = 1$ and $(\mathbb{R}, \mathcal{B}(\mathbb{R}), \mu_F)$ is a **probability space**. Specifically, $(\mathbb{R}, \mathcal{B}(\mathbb{R}), \mu_F)$ is the **probability space on $\mathbb{R}$ induced by X and $(\mathcal{S}, \mathcal{E}, \mu)$.**

For completeness we note that there is also a **complete probability space** given by this proposition, $(\mathbb{R}, \mathcal{M}_F(\mathbb{R}), \mu_F)$ with $\mathcal{B}(\mathbb{R}) \subset \mathcal{M}_F(\mathbb{R})$. While random variables are only defined to be Borel measurable, meaning $X^{-1}(A) \in \mathcal{E}$ only for Borel sets A, (4.5) will extend to this complete probability space by Proposition I.6.24 if we use the Proposition I.6.20 completion $\left[\mu\left(X^{-1}\right)\right]^C$ in place of the measure $\mu\left(X^{-1}\right)$.

Now consider $\{(\mathbb{R}_i, \mathcal{B}(\mathbb{R}_i), \mu_{F_i})\}_{i=1}^\infty$, a countable collection of **identical copies** of this space that are **indexed only for notational purposes**. By Proposition I.9.20, we can construct the infinite dimensional complete probability space:

$$(\mathbb{R}^\mathbb{N}, \sigma(\mathbb{R}^\mathbb{N}), \mu_\mathbb{N}),$$

where:

$$\mathbb{R}^{\mathbb{N}} \equiv \{(x_1, x_2, ...)|x_i \in \mathbb{R}_i\}.$$

The measure $\mu_{\mathbb{N}}$ is uniquely defined on the complete sigma algebra $\sigma(\mathbb{R}^{\mathbb{N}})$, which contains the sigma algebra $\sigma(\mathcal{A}^+)$ generated by the algebra $\mathcal{A}^+$. That is, $\sigma(\mathcal{A}^+) \subset \sigma(\mathbb{R}^{\mathbb{N}})$.

The algebra $\mathcal{A}^+$ is the collection of **general finite dimensional measurable rectangles** or **general cylinder sets** in $\mathbb{R}^{\mathbb{N}}$. In detail, we say that $H \in \mathcal{A}^+$ if with $x \equiv (x_1, x_2, ...)$:

$$H = \{x|(x_{j(1)}, x_{j(2)}, ...x_{j(n)}) \in A\},$$

for some n-tuple of positive integers $J = (j(1), j(2), ..., j(n))$ and $A \in \mathcal{B}(\mathbb{R}^n)$, the Borel sigma algebra on $\mathbb{R}^n$.

On $\mathcal{A}^+$, $\mu_{\mathbb{N}}$ is defined by:

$$\mu_{\mathbb{N}}(H) = \mu_J(A),$$

where μ_J is the finite dimensional product probability measure on $(\mathbb{R}^n, \mathcal{B}(\mathbb{R}^n))$ induced by $\{\mu_{F_{j(i)}}\}_{i=1}^n$, and derived in Proposition I.7.20. In the special case where A is a measurable rectangle, $A = \prod_{j=1}^n A_j$ for $A_j \in \mathcal{B}(\mathbb{R})$, the measure $\mu_J(A)$ is representable in terms of μ_F since $\mu_{F_{j(i)}} = \mu_F$ for all i:

$$\mu_{\mathbb{N}}(H) = \prod_{j=1}^n \mu_F(A_j). \tag{4.6}$$

We now state the first construction of a sample space and sample for a random variable X defined on $(\mathcal{S}, \mathcal{E}, \mu)$. Note that X_j defined in (4.8) is the **projection mapping** on $\mathbb{R}^{\mathbb{N}}$ to the jth coordinate, which in Notation I.9.4 is denoted π_j.

Proposition 4.4 (I. Probability space $(\mathcal{S}'_F, \mathcal{E}'_F, \mu'_F)$ and i.i.d. $\{X_j\}_{j=1}^\infty$) *Let the probability space $(\mathcal{S}, \mathcal{E}, \mu)$ and random variable $X : \mathcal{S} \longrightarrow \mathbb{R}$ with distribution function $F(x)$ be given. Define:*

$$(\mathcal{S}'_F, \mathcal{E}'_F, \mu'_F) \equiv (\mathbb{R}^{\mathbb{N}}, \sigma(\mathbb{R}^{\mathbb{N}}), \mu_{\mathbb{N}}), \tag{4.7}$$

where this countable probability space is the product space defined with identical spaces $\{(\mathbb{R}_i, \mathcal{B}(\mathbb{R}_i), \mu_{F_i})\}_{i=1}^\infty$.

Define $X_j : \mathbb{R}^{\mathbb{N}} \to \mathbb{R}$ on this space by:

$$X_j : (x_1, x_2, ...) = x_j. \tag{4.8}$$

Then $\{X_j\}_{j=1}^\infty$ is a sample of X.

Proof. *First, each X_j is measurable and thus a random variable on $(\mathbb{R}^{\mathbb{N}}, \sigma(\mathbb{R}^{\mathbb{N}}), \mu_{\mathbb{N}})$ since (4.8) yields for $A \in \mathcal{B}(\mathbb{R})$:*

$$X_j^{-1}(A) = \{x \in \mathbb{R}^{\mathbb{N}}|x_j \in A\}.$$

Thus $X_j^{-1}(A) \in \mathcal{A}^+$, a general cylinder set, and measurability follows since $\mathcal{A}^+ \subset \sigma(\mathbb{R}^{\mathbb{N}})$.

Further, from (4.6) with $H \equiv X_j^{-1}(A)$ and then (4.5):

$$\mu_{\mathbb{N}}(X_j^{-1}(A)) = \mu_F(A) \equiv \mu(X^{-1}(A)). \tag{1}$$

This proves (4.2), and thus $\{X_j\}_{j=1}^\infty$ are identically distributed with X.

For independence, let $(i_1, ..., i_m) \subset (1, 2, ...)$ and $\{A_j\}_{j=1}^m \subset \mathcal{B}(\mathbb{R})$ be given. Then by (4.8):

$$\bigcap_{j=1}^m X_{i_j}^{-1}(A_j) = \left\{x|(x_{i_1}, x_{i_2}, ...x_{i_m}) \in \prod_{j=1}^m A_j\right\}.$$

Thus by (4.6) with $H = \bigcap_{j=1}^m X_{i_j}^{-1}(A_j)$, and then (4.5) and (1):

$$\mu_{\mathbb{N}}\left(\bigcap_{j=1}^m X_{i_j}^{-1}(A_j)\right) = \prod_{j=1}^m \mu_F(A_j) \equiv \prod_{j=1}^m \mu_{\mathbb{N}}(X_{i_j}^{-1}(A_j)).$$

This proves (4.1), and thus $\{X_j\}_{j=1}^\infty$ are independent. ∎

A second construction of a probability space and i.i.d. random variables will be seen in Proposition 4.13, and a third construction for discrete random variables in Proposition 4.17.

In Chapter IV.5, this construction will be generalized somewhat to produce a sample space for independent, but not identically distributed random variables. The reader is encouraged to contemplate this generalization now as an exercise.

4.2 Simulation of Random Variables - Theory

In this section we derive results which provide the necessary theoretical framework for stochastic simulations, which require mathematically generating random samples of random variables. In Chapter IV.5 we will return to this topic and investigate examples. The general result here is that if one can generate samples of U, a random variable with a continuous uniform distribution on $[0,1]$ as in (1.19), then by a relatively simple numerical procedure, such samples can be transformed into samples of the random variable X of interest.

For stochastic simulations, generating random samples of such U is now very easy as many mathematical software programs have a built-in function which does exactly this. For example, in Microsoft Excel this function is **RAND()**, while in MathWorks MATLAB it is called **rand**.

After deriving distributional and independence results, we return to a second construction of $(\mathcal{S}', \mathcal{E}', \mu')$ and i.i.d. $\{X_j\}_{j=1}^N$. Again we assume $N = \infty$ for notational consistency.

4.2.1 Distributional Results

We begin with the simplest and strongest result, which applies in the case when the distribution function $F(x)$ for X is **continuous and strictly increasing** on $\mathbb{R}$. Part 1 of this proposition is of interest in some applications, but it is really part 2 that is key to the problem of generating random samples. Part 2 states that if $F(x)$ is both continuous and strictly increasing, and U is a random variable with a continuous uniform distribution on $[0,1]$, then $F^{-1}(U)$ has distribution function $F(x)$, the same as the random variable X.

Part 2 is expressed with U defined on a general probability $(\mathcal{S}', \mathcal{E}', \mu')$. This space is quite arbitrary, and can equal the original space $(\mathcal{S}, \mathcal{E}, \mu)$. For specificity, we can set:

$$(\mathcal{S}', \mathcal{E}', \mu') = ((0,1), \mathcal{B}((0,1)), m),$$

as defined in 1 of Exercise 1.29, using $(\mathbb{R}, \mathcal{B}(\mathbb{R}), m)$ and Lebesgue measure m, and then let U be the identity function, $U(s) = s$.

For the proof of this result, note that if F is continuous and strictly increasing on $\mathbb{R}$, then the range of $F(x)$ satisfies $Rng F = (0,1)$. There F^{-1} is well defined pointwise by $F^{-1}(y) = x$ if $F(x) = y$. It is then also the case that F^{-1} is continuous and strictly increasing on $(0,1)$.

Proposition 4.5 (Continuous, strictly increasing $F(x)$) *Let $(\mathcal{S}, \mathcal{E}, \mu)$ be given and $X : (\mathcal{S}, \mathcal{E}, \mu) \to (\mathbb{R}, \mathcal{B}(\mathbb{R}), m)$ a random variable with distribution function $F(x)$ that is both continuous and strictly increasing. Then:*

1. *$F(X) : (\mathcal{S}, \mathcal{E}, \mu) \to ((0,1), \mathcal{B}((0,1)), m)$ defined by:*

$$F(X)(s) = F(X(s)),$$

is a random variable on $\mathcal{S}$ which has a continuous uniform distribution on $(0,1)$.

2. If $U : (\mathcal{S}', \mathcal{E}', \mu') \to ((0,1), \mathcal{B}((0,1)), m)$ has a continuous uniform distribution, then $F^{-1}(U) : (\mathcal{S}', \mathcal{E}', \mu') \to (\mathbb{R}, \mathcal{B}(\mathbb{R}), m)$ defined by:

$$F^{-1}(U)(s') = F^{-1}(U(s')),$$

is a random variable on $\mathcal{S}'$ which has distribution function $F(x)$.

Proof. For notational clarity we put random variable subscripts on distribution functions *other* than $F(x)$, the distribution function for X.

1. As defined, $F(X)$ is a composite function defined on $(\mathcal{S}, \mathcal{E}, \mu)$ with inverse defined on $(0,1)$ by $F(X)^{-1} \equiv X^{-1}F^{-1}$. Since $F : (\mathbb{R}, \mathcal{B}(\mathbb{R}), m) \to ((0,1), \mathcal{B}((0,1)), m)$ is continuous, $F^{-1}(A) \in \mathcal{B}(\mathbb{R})$ if $A \in \mathcal{B}((0,1))$ by Proposition I.3.12, and hence $X^{-1}F^{-1}(A) \in \mathcal{E}$. Thus $F(X)$ is a random variable on $\mathcal{S}$.

   The uniform distribution function on $[0,1]$ is given by $F_U(y) = y$, so we seek to show that $F_{F(X)}(y) = y$. Let $y \in (0,1)$ be given, and assume that $y = F(x)$. It follows that $x = F^{-1}(y)$ and thus since F^{-1} is continuous and strictly increasing on $(0,1)$:

$$
\begin{aligned}
F_{F(X)}(y) &\equiv \mu\left[\{F(X) \leq y\}\right] \\
&= \mu\left[\{X \leq x\}\right] \\
&\equiv F(x) \\
&= y.
\end{aligned}
$$

   Thus the distribution function of $F(X)$ is the continuous uniform distribution function on $(0,1)$.

2. Since $F^{-1} : ((0,1), \mathcal{B}((0,1)), m) \to (\mathbb{R}, \mathcal{B}(\mathbb{R}), m)$ is continuous, it is Borel measurable as above and hence $F^{-1}(U)$ is a random variable on $(\mathcal{S}', \mathcal{E}', \mu')$. For $x \in \mathbb{R}$, we need to show that $F_{F^{-1}(U)}(x) = F(x)$. To this end, since U has a uniform distribution and F is increasing:

$$
\begin{aligned}
F_{F^{-1}(U)}(x) &\equiv \mu'\left[\{F^{-1}(U) \leq x\}\right] \\
&= \mu'\left[\{U \leq F(x)\}\right] \\
&= F(x).
\end{aligned}
$$

∎

Example 4.6 (X normally distributed) Let $F(x)$ be the distribution function associated with the standard normal density function in (1.23):

$$F(x) = \frac{1}{\sqrt{2\pi}} \int_{-\infty}^{x} \exp\left(-y^2/2\right) dy.$$

Then $F(x)$ satisfies the hypotheses of Proposition 4.5, being continuous and strictly increasing.

   As is often the case, F^{-1} here does lend itself to a tractable analytic representation, so it is necessary to numerically estimate this function. Using symmetry of the integrand, it is only necessary to estimate for $x > 0$:

$$F_0(x) = \frac{1}{\sqrt{2\pi}} \int_{0}^{x} \exp\left(-y^2/2\right) dy.$$

This can be done by various numerical integration routines to provide a table of values $(x, F_0(x))$. *This table can be extended by symmetry to* $F(x)$:

$$F(x) = \begin{cases} F_0(x) + 1/2, & x > 0, \\ 1/2 - F_0(|x|), & x < 0. \end{cases}$$

Finally, interpolation can be used to estimate $F^{-1}(y)$ *on* $(0, 1)$.

By Proposition 4.5, $F^{-1}(U)$ *has the standard normal distribution when* U *has the continuous uniform distribution on* $(0, 1)$. *As noted in the introduction, we can set* $(\mathcal{S}', \mathcal{E}', \mu') = ((0, 1), \mathcal{B}((0, 1)), m)$ *as defined in Exercise 1.29, and then let* $U : ((0, 1), \mathcal{B}((0, 1)), m) \to ((0, 1), \mathcal{B}((0, 1)), m)$ *be the identity function,* $U(s) = s$.

*For an application to **random sampling** or **stochastic simulation**, imagine that we have generated a **random sample** $\{u_i\}_{i=1}^n$ from this uniform distribution. This is a numerical sample, which by definition, is generated by a mathematical routine which reflects values from i.i.d. uniform variates* $\{U_i\}_{i=1}^n$, *meaning* $u_i = U_i(s_i)$, *for some collection* $\{s_i\}_{i=1}^n$. *The objective of this process is for the random sample* $\{u_i\}_{i=1}^n$ *to be "approximately" independent and uniformly distributed. Intuitively, the process is such that these approximate statements are intended to become increasingly exact as* $n \to \infty$.

Thus for any $(a, b) \subset (0, 1)$, *to be uniformly distributed requires that:*

$$\Pr\{u_i \in (a, b)\} \approx b - a,$$

where this probability is simply m/n *with* m *the count of* $\{u_i \in (a, b)\}$. *Independence of these simulated variates implies that for any subset* $(i_1, ..., i_m) \subset (1, ..., n)$ *and collection* $\{(a_j, b_j)\}_{j=1}^m$ *with* $(a_j, b_j) \subset (0, 1)$:

$$\Pr\{u_{i_j} \in (a_j, b_j) \text{ all } j\} \approx \prod_{j=1}^m (b_j - a_j).$$

Proposition 4.5 now states that $\{F^{-1}(u_i)\}_{i=1}^n$ *are approximately normally distributed, and we will see below that this collection is in fact independent. Thus* $\{F^{-1}(u_i)\}_{i=1}^n$ *are i.i.d. and is a random sample of the standard normal distribution.*

Besides numerical integration, there are also various polynomial and rational function-based formulas for $F_0(x)$ *for* $x > 0$ *with maximum errors which can be estimated. The literature is vast here because of the historical usefulness of these results. Much was published prior to the mid-1960s, before extensive numerical computations on computers were feasible, when analytically tractable formulas were then very desirable. With continuous improvements in what is computationally feasible, these formulaic results are less necessary today, and numerical integration is often used instead.*

Such integration routines are often built into the mathematical software to generate random normal variates directly. For example, in Microsoft Excel this function is NORM.S.INV.

While Proposition 4.5 is often applicable, many distribution functions are neither continuous nor strictly increasing, so we next generalize this result. In order to do so it will be necessary to replace F^{-1} by the left continuous inverse F^*. While F^{-1} need not exist, F^* always exists by Proposition 3.16, and $F^* = F^{-1}$ when F is continuous and strictly increasing by Proposition 3.22.

What is clear with a moment of thought is that part 1 of Proposition 4.5 has little hope of generalization.

Example 4.7 (On Example 3.15) *Looking back to Example 3.15:*

$$F(x) = \begin{cases} 0, & x < 0, \\ 0.5x, & 0 \le x < 1, \\ 0.5, & 1 \le x < 2, \\ 1, & x \ge 2. \end{cases}$$

For part 1 of Proposition 4.5, it is clear that the random variable $F(X)$ cannot possibly be continuous and uniformly distributed on $(0,1)$, since the range of $F(X)$ excludes $(0.5, 1.0)$. The problem with F of course, is the discontinuity at $x = 2$.

That said, $F_{F(X)}(y)$ can be evaluated for this example for $y \in [0,1]$ by:

$$F_{F(X)}(y) = m\{F(X) \le y\},$$

with m Lebesgue measure. It then follows that:

$$F_{F(X)}(y) = \begin{cases} y, & 0 \le y \le 0.5, \\ 0.5, & 0.5 \le y < 1.0, \\ 1, & y = 1. \end{cases}$$

Thus, even though not uniformly distributed:

$$F_{F(X)}(y) \le y,$$

for all y.

On the other hand, there is hope that part 2 of Proposition 4.5 remains true in this case, since for this example,

$$F^*(y) = \begin{cases} 2y, & 0 < y \le 0.5, \\ 2.0, & 0.5 < y \le 1.0. \end{cases}$$

If U has a continuous uniform distribution, it seems intuitively plausible that $F^*(U)$ will replicate $F(x)$ when $0 < y \le 0.5$. And since the event $0.5 < y \le 1.0$ has probability measure 0.5, it appears feasible that $F^*(U)$ will replicate $F(x)$ when $0.5 \le y \le 1.0$, as desired.

We now make these observations precise, and also address what can be said about the random variable $F(X)$.

Proposition 4.8 (General $F(x)$) *Let $(\mathcal{S}, \mathcal{E}, \mu)$ be given and $X : (\mathcal{S}, \mathcal{E}, \mu) \to (\mathbb{R}, \mathcal{B}(\mathbb{R}), m)$ a random variable with distribution function $F(x)$ and left continuous inverse $F^*(y)$ defined on $(0,1)$. Then:*

1. $F(X) : (\mathcal{S}, \mathcal{E}, \mu) \to ((0,1), \mathcal{B}((0,1)), m)$ *defined by:*

$$F(X)(s) = F(X(s)),$$

   is a random variable on $\mathcal{S}$ with distribution function $F_{F(X)}(y)$ satisfying:

$$F_{F(X)}(y) \le y.$$

Further,

$$F_{F(X)}(y) = y,$$

if and only if $F(x)$ is continuous.

2. *If $U : (\mathcal{S}', \mathcal{E}', \mu') \to ((0,1), \mathcal{B}((0,1)), m)$ has a continuous uniform distribution, then $F^*(U) : (\mathcal{S}', \mathcal{E}', \mu') \to (\mathbb{R}, \mathcal{B}(\mathbb{R}), m)$ defined by:*

$$F^*(U)(s') = F^*(U(s')),$$

is a random variable on $(0,1)$, which has distribution function $F(x)$.

Proof. *We prove part 2 first, since this result is needed for the proof of part 1. For notational clarity we put random variable subscripts on distribution functions **other** than $F(x)$, the distribution function for X.*

1. *Recall that $F^* : ((0,1), \mathcal{B}((0,1)), m) \to (\mathbb{R}, \mathcal{B}(\mathbb{R}), m)$ is Borel measurable by Proposition 3.16, and thus $(F^*)^{-1}(A) \in \mathcal{B}((0,1))$ for $A \in \mathcal{B}(\mathbb{R})$. It then follows that $F^*(U)$ is a random variable on $(\mathcal{S}', \mathcal{E}', \mu')$ since for $A \in \mathcal{B}(\mathbb{R})$:*

$$[F^*(U)]^{-1}(A) = U^{-1}\left[(F^*)^{-1}(A)\right] \in \mathcal{E}'.$$

To show that $F_{F^*(U)}(x) = F(x)$ for $x \in \mathbb{R}$, recall that $F^*(y) \le x$ if and only if $y \le F(x)$ by (3.8). Thus since U has a continuous uniform distribution:

$$
\begin{aligned}
F_{F^*(U)}(x) &= \mu'\left[\{F^*(U) \le x\}\right] \\
&= \mu'\left[U \le F(x)\right] \\
&= F(x).
\end{aligned}
$$

2. *As noted in Notation 3.5, $F : (\mathbb{R}, \mathcal{B}(\mathbb{R}), m) \to ((0,1), \mathcal{B}((0,1)), m)$ is Borel measurable and hence $F(X)$ is a random variable on $(\mathcal{S}, \mathcal{E}, \mu)$ as in part 2 above. Also, the random variables $X : \mathcal{S} \to \mathbb{R}$ and $F^*(U) : \mathcal{S} \to \mathbb{R}$ have the same distribution function by part 2, so for $y \in (0,1)$:*

$$
\begin{aligned}
F_{F(X)}(y) &= \mu\left[\{F(X) \le y\}\right] \\
&= \mu\left[\{F(F^*(U)) \le y\}\right].
\end{aligned}
$$

Now by (3.10), $y \le F\left(F^*(y)\right)$ for all y, and so by monotonicity of F on $\mathbb{R}$ and F^* on $(0,1)$:

$$
\begin{aligned}
F_{F(X)}(y) &\le \mu\left[\{F(F^*(U)) \le F\left(F^*(y)\right)\}\right] \\
&= \mu\left[\{F^*(U) \le F^*(y)\}\right] \\
&= \mu\left[\{U \le y\}\right] \\
&= y.
\end{aligned}
$$

If F is continuous, then $y = F\left(F^*(y)\right)$ in the above derivation by Proposition 3.22, and it then follows that $F_{F(X)}(y) = y$.

If F is discontinuous at some $x \in \mathbb{R}$, then of necessity it is left discontinuous at this point and so $F(x^-) < F(x)$, where $F(x^-)$ denotes the left limit. Choose y so that:

$$F(x^-) < y < F(x).$$

Then $F(X) \le y$ if and only if $X < x$, and so:

$$
\begin{aligned}
F_{F(X)}(y) &= \mu\left[\{F(X) \le y\}\right] \\
&= \mu\left[\{X < x\}\right] \\
&= F(x^-) \\
&< y.
\end{aligned}
$$

■

4.2.2   Independence Results

The prior section demonstrated that it is easy, at least in theory, to generate a random variable X with given arbitrary distribution function $F(x)$. Starting with a continuous, uniformly distributed random variable U on $(0, 1)$ for example, we simply define $X = F^*(U)$. In all cases, the distribution function of X will be the given distribution function $F(x)$. An application of this was seen in Proposition 3.6. If $F(x)$ is continuous, we can conversely generate a continuous, uniformly distributed random variable U on $(0, 1)$ from any such X by defining $U = F(X)$.

These results also have an application to random sampling, or the stochastic simulation of a random variable. As noted in the introductory comments to this section and in Example 4.6, the generation of random samples of random variables requires that the variates be correctly distributed, and that they also be independent.

We prove next that independence of the continuous uniformly distributed variates $\{U_j\}_{j=1}^n$ assures independence of the resultant variates $\{X_j\}_{j=1}^n$ defined as $\{F^*(U_j)\}_{j=1}^n$. Similarly, independence of identically distributed variates $\{X_j\}_{j=1}^n$ assures independence of the variates $\{U_j\}_{j=1}^n$ defined as $\{F(X_j)\}_{j=1}^n$, though these variates will be continuous uniformly distributed only when $F(x)$ is continuous by Proposition 4.8.

Both independence results are in fact a corollary of Proposition 3.56 since $F(x)$ is Borel measurable as noted in Notation 3.5, and $F^*(y)$ is Borel measurable by Proposition 3.16. However we provide a direct proof which may be of independent interest. The proof presented requires continuity of $F(x)$ for part 2. Though this is not needed for independence by Proposition 3.56, continuity **is** required for the distributional result on U which is the primary case of interest.

Note that the general independence result based on Proposition 3.56, as well as the result below, are true for $n = \infty$. This follows because independence of an infinite collection is defined in terms of independence of all finite subsets of variables by Definition 3.47.

Proposition 4.9 (Independence; F, F^*) *Let $(\mathcal{S}, \mathcal{E}, \mu)$ be given and $X : (\mathcal{S}, \mathcal{E}, \mu) \to (\mathbb{R}, \mathcal{B}(\mathbb{R}), m)$ a random variable with distribution function $F(x)$ and left continuous inverse $F^*(y)$.*

1. *If $\{U_j\}_{j=1}^n$ are independent, continuous, uniformly distributed random variables on $(0, 1)$, then:*
$$\{X_j\}_{j=1}^n \equiv \{F^*(U_j)\}_{j=1}^n,$$
are independent random variables with distribution function $F(x)$.

2. *If $\{X_j\}_{j=1}^n$ are independent random variables with continuous distribution function $F(x)$, then:*
$$\{U_j\}_{j=1}^n \equiv \{F(X_j)\}_{j=1}^n,$$
are independent random variables with a continuous uniform distribution.

Proof.

1. *Each X_j has distribution function $F(x)$ by Proposition 4.8.*

   Independence of $\{U_j\}_{j=1}^n$ implies that there exists a probability space $(\mathcal{S}', \mathcal{E}', \mu')$ with $U_j : \mathcal{S}' \to ((0, 1), \mathcal{B}(0, 1), m)$. Since each U_j has a continuous uniform distribution, it follows from Proposition 3.53 that the joint distribution function of $U \equiv (U_1, U_2, ..., U_n)$

is given by:

$$
\begin{aligned}
F_U(y_1, y_2, ..., y_n) &\equiv \mu' \left[\bigcap_{j=1}^{n} \{U_j \leq y_j\} \right] \\
&= \prod_{j=1}^{n} F_{U_j}(y_j) \\
&= \prod_{j=1}^{n} y_j.
\end{aligned}
$$

As above, $X_j \equiv F^*(U_j)$ is a random variable on $(\mathcal{S}', \mathcal{E}', \mu')$ since F^* is Borel measurable. For $X \equiv (X_1, X_2, ..., X_n)$, the joint distribution is defined:

$$
F_X(x_1, x_2, ..., x_n) \equiv \mu' \left[\bigcap_{j=1}^{n} \{X_j \leq x_j\} \right].
$$

For each j, $X_j \leq x_j$ and $F^*(U_j) \leq x_j$ have the same probability by Proposition 4.8, and by (3.8) this last condition is equivalent to $U_j \leq F(x_j)$. Hence:

$$
\begin{aligned}
F_X(x_1, x_2, ..., x_n) &\equiv \mu' \left[\bigcap_{j=1}^{n} \{U_j \leq F(x_j)\} \right] \\
&= \prod_{j=1}^{n} F_{U_j}(F(x_j)) \\
&= \prod_{j=1}^{n} F(x_j).
\end{aligned}
$$

Thus $\{X_j\}_{j=1}^{n}$ are independent by Proposition 3.53.

2. *The distributional result is again from Proposition 4.8 since $F(x)$ is assumed continuous. If $\{X_j\}_{j=1}^{n}$ are independent random variables with distribution function $F(x)$, then there exists a probability space $(\mathcal{S}', \mathcal{E}', \mu')$ with $X_j : \mathcal{S}' \rightarrow (\mathbb{R}, \mathcal{B}(\mathbb{R}), m)$. For $X \equiv (X_1, X_2, ..., X_n)$, the joint distribution is given by Proposition 3.53:*

$$
F_X(x_1, x_2, ..., x_n) \equiv \mu' \left[\bigcap_{j=1}^{n} \{X_j \leq x_j\} \right] = \prod_{j=1}^{n} F(x_j).
$$

Now $U_j = F(X_j)$ is a random variable on $(\mathcal{S}', \mathcal{E}', \mu')$ since F is Borel measurable. Also $F(F^*(y_j)) = y_j$ by continuity and Proposition 3.22, and since F is increasing it follows that for $U \equiv (U_1, U_2, ..., U_n)$:

$$
\begin{aligned}
F_U(y_1, y_2, ..., y_n) &\equiv \mu' \left[\bigcap_{j=1}^{n} \{U_j(s) \leq y_j\} \right] \\
&= \mu' \left[\bigcap_{j=1}^{n} \{F(X_j) \leq F(F^*(y_j))\} \right] \\
&= \mu' \left[\bigcap_{j=1}^{n} \{X_j \leq F^*(y_j)\} \right] \\
&= \prod_{j=1}^{n} F(F^*(y_j)) \\
&= \prod_{j=1}^{n} y_j.
\end{aligned}
$$

Thus $\{U_j\}_{j=1}^{n}$ are independent by Proposition 3.53, since $y_j = F_{U_j}(y_j)$ is the distribution function for U_j.

∎

4.2.3 Second Construction: $(\mathcal{S}'_U, \mathcal{E}'_U, \mu'_U)$

In this section we provide another construction of a probability space $(\mathcal{S}'_U, \mathcal{E}'_U, \mu'_U)$ and i.i.d. $\{X_j\}_{j=1}^{\infty}$ using the results developed above. The notation here for this space is to connote that this space is constructed using the Borel measure induced by the uniform distribution function on $(0, 1)$. Much of the initial setup is the same as that for Section 4.1.1, but we include some of the details here to clarify the modification we then introduce.

Let the probability space $(\mathcal{S}, \mathcal{E}, \mu)$ and random variable $U : \mathcal{S} \longrightarrow \mathbb{R}$ be given, with continuous, uniform distribution function:

$$F_U(y) = \begin{cases} 0, & y \le 0, \\ y, & 0 < y \le 1, \\ 1, & 1 < y. \end{cases}$$

By Proposition I.5.23, this distribution gives rise to a measure μ_{F_U} defined on $\mathcal{B}(\mathbb{R})$ such that for all right semi-closed intervals:

$$\mu_{F_U}(a, b] = F_U(b) - F_U(a). \tag{4.9}$$

See Section 6.1.1 for a more detailed review of this development.

Proposition 3.4 obtains that for all such intervals:

$$F_U(b) - F_U(a) = \mu\left(U^{-1}(a, b]\right),$$

and thus the set functions μ_{F_U} and $\mu\left(U^{-1}\right)$ agree on the semi-algebra $\mathcal{A}'$ of right semi-closed intervals on $\mathbb{R}$:

$$\mu_{F_U}(a, b] = \mu\left(U^{-1}(a, b]\right).$$

Repeating the derivation in Section 4.1.1 for (4.5), we obtain that for all $A \in \mathcal{B}(\mathbb{R})$:

$$\mu_{F_U}(A) = \mu\left(U^{-1}(A)\right). \tag{4.10}$$

Thus $\mu_{F_U}(\mathbb{R}) = 1$ and $(\mathbb{R}, \mathcal{B}(\mathbb{R}), \mu_{F_U})$ is a **probability space**, and specifically, the **probability space on $\mathbb{R}$ induced by U and $(\mathcal{S}, \mathcal{E}, \mu)$**.

Letting $\{(\mathbb{R}_i, \mathcal{B}(\mathbb{R}_i), \mu_{F_{U_i}})\}_{i=1}^{\infty}$ denote a countable collection of identical copies of this space, again indexed only for notational purposes, we can, by Proposition I.9.20, construct the infinite dimensional complete probability space:

$$(\mathbb{R}^{\mathbb{N}}, \sigma(\mathbb{R}^{\mathbb{N}}), \mu_{\mathbb{N}}).$$

As noted in Section 4.1.1, $\mu_{\mathbb{N}}$ is uniquely defined on $\sigma(\mathbb{R}^{\mathbb{N}})$, and this complete sigma algebra contains $\sigma(\mathcal{A}^+)$, the sigma algebra generated by the algebra $\mathcal{A}^+$ of general cylinder sets in $\mathbb{R}^{\mathbb{N}}$.

In detail, if $H \in \mathcal{A}^+$ then:

$$H = \{(x_{j(1)}, x_{j(2)}, \ldots x_{j(n)}) \in A\},$$

for some n-tuple of positive integers $J = (j(1), j(2), \ldots, j(n))$ and $A \in \mathcal{B}(\mathbb{R}^n)$, the Borel sigma algebra on $\mathbb{R}^n$.

On $\mathcal{A}^+$, $\mu_{\mathbb{N}}$ is defined by:

$$\mu_{\mathbb{N}}(H) = \mu_J(A),$$

where μ_J is the finite dimensional product probability measure on $(\mathbb{R}^n, \mathcal{B}(\mathbb{R}^n))$ induced by $\{\mu_{F_{U_{j(i)}}}\}_{i=1}^n$, and derived in Proposition I.7.20. In the special case where A is a measurable rectangle, $A = \prod_{j=1}^{n} A_j$ for $A_j \in \mathcal{B}(\mathbb{R})$, the measure $\mu_J(A)$ is representable in terms of μ_{F_U} as in (4.6), since $\mu_{F_{U_{j(i)}}} = \mu_{F_U}$ for all i:

$$\mu_{\mathbb{N}}(H) = \prod_{j=1}^{n} \mu_{F_U}(A_j). \tag{4.11}$$

While $(\mathbb{R}^{\mathbb{N}}, \sigma(\mathbb{R}^{\mathbb{N}}), \mu_{\mathbb{N}})$ is formally and notationally a probability space on $\mathbb{R}^{\mathbb{N}}$, we claim that $(0,1)^{\mathbb{N}} \subset \mathbb{R}^{\mathbb{N}}$ is an **essential subset,** and that this probability space is **equivalent** to $\left((0,1)^{\mathbb{N}}, \sigma\left((0,1)^{\mathbb{N}}\right), \mu_{\mathbb{N}}\right)$, the probability space of 1 of Exercise 1.29. First, a definition.

Definition 4.10 (Essential subsets; equivalent spaces) *Given a probability space* $(\mathcal{S}, \mathcal{E}, \mu)$, *a subset* $B \subset \mathcal{S}$ *is **essential** if* $B \in \mathcal{E}$ *and* $\mu(B) = 1$.

If B *is essential, we say that the probability spaces* $(\mathcal{S}, \mathcal{E}, \mu)$ *and* $(B, \sigma(B), \mu)$ *are **equivalent probability spaces,** denoted:*

$$(\mathcal{S}, \mathcal{E}, \mu) \sim (B, \sigma(B), \mu),$$

where:

$$\sigma(B) \equiv \left\{ A \bigcap B \mid A \in \mathcal{E} \right\}. \tag{4.12}$$

For any $B \in \mathcal{E}$ and $\sigma(B)$ as defined in (4.12), the triplet $(B, \sigma(B), \mu)$ is a measure space by 1 of Exercise 1.29, and it is apparently a probability space if and only if $\mu(B) = 1$. The motivation for calling $(\mathcal{S}, \mathcal{E}, \mu)$ and $(B, \sigma(B), \mu)$ equivalent probability spaces is the following simple but useful result.

Proposition 4.11 (On equivalent spaces) *If* $(\mathcal{S}, \mathcal{E}, \mu)$ *and* $(B, \sigma(B), \mu)$ *are equivalent probability spaces by Definition 4.10, then:*

1. *For any* $A \in \mathcal{E}$ *there exists* $A' \in \sigma(B)$ *with:*

$$\mu(A) = \mu(A').$$

2. *If* X *is a random variable on* $(\mathcal{S}, \mathcal{E}, \mu)$ *with distribution function* $F(x)$, *then* X_B, *the* **restriction** *of* X *to* B, *is a random variable on* $(B, \sigma(B), \mu)$ *with the same distribution function.*

Proof. *The first observation follows with* $A' \equiv A \bigcap B$, *a set in* $\sigma(B)$ *by definition, since by finite additivity:*

$$\mu(A) = \mu\left(A \bigcap B\right) + \mu\left(A \bigcap \tilde{B}\right).$$

This equals $\mu(A \bigcap B)$ *by monotonicity of* μ *since* $A \bigcap \tilde{B} \subset \tilde{B}$ *and thus* $\mu\left(\tilde{B}\right) = 0$.

For 2, if $C \in \mathcal{B}(\mathbb{R})$, *then* $X^{-1}(C) \in \mathcal{E}$ *since* X *is a random variable on* $\mathcal{S}$. *Then* $X_B^{-1}(C) \in \sigma(B)$ *by definition of restriction and (4.12):*

$$X_B^{-1}(C) \equiv X^{-1}(C) \bigcap B.$$

For the distribution function of X_B, *by an application of part 1:*

$$\begin{aligned} F_{X_B}(x) &\equiv \mu\{s \in B \mid X_B(s) \le x\} \\ &= \mu\left[\{s \in \mathcal{S} \mid X(s) \le x\} \bigcap B\right] \\ &= \mu[\{s \in \mathcal{S} \mid X(s) \le x\}] \\ &= F_X(x). \end{aligned}$$

∎

We now prove that $(0,1)^{\mathbb{N}}$ is essential in $(\mathbb{R}^{\mathbb{N}}, \sigma(\mathbb{R}^{\mathbb{N}}), \mu_{\mathbb{N}})$, where:

$$(0,1)^{\mathbb{N}} \equiv \{x \in \mathbb{R}^{\mathbb{N}} | 0 < x_i < 1 \text{ for all } i\}.$$

Proposition 4.12 ($(0,1)^{\mathbb{N}}$ is essential in $(\mathbb{R}^{\mathbb{N}}, \sigma(\mathbb{R}^{\mathbb{N}}), \mu_{\mathbb{N}})$) *With the notation above, the set $(0,1)^{\mathbb{N}}$ is essential in $(\mathbb{R}^{\mathbb{N}}, \sigma(\mathbb{R}^{\mathbb{N}}), \mu_{\mathbb{N}})$, where this countable product probability space is defined with identical spaces $\{(\mathbb{R}_i, \mathcal{B}(\mathbb{R}_i), \mu_{F_{U_i}})\}_{i=1}^{\infty}$.*

Consequently:

$$(\mathbb{R}^{\mathbb{N}}, \sigma(\mathbb{R}^{\mathbb{N}}), \mu_{\mathbb{N}}) \sim \left((0,1)^{\mathbb{N}}, \sigma\left((0,1)^{\mathbb{N}}\right), \mu_{\mathbb{N}}\right), \tag{4.13}$$

where:

$$\sigma\left((0,1)^{\mathbb{N}}\right) \equiv \left\{A \bigcap (0,1)^{\mathbb{N}} | A \in \sigma(\mathbb{R}^{\mathbb{N}})\right\}. \tag{4.14}$$

Proof. *To prove that $(0,1)^{\mathbb{N}}$ is essential is to show that $(0,1)^{\mathbb{N}}$ is $\sigma(\mathbb{R}^{\mathbb{N}})$-measurable, and that $\mu_{\mathbb{N}}\left((0,1)^{\mathbb{N}}\right) = 1$. Then (4.13) follows by Definition 4.10.*

For measurability, let $H_n \in \mathcal{A}^+$ be given by $J_n \equiv (1, ..., n)$ and $A_n \equiv (0,1)^n \in \mathcal{B}(\mathbb{R}^n)$. Then $\{H_n\}_{n=1}^{\infty}$ is a nested collection of $\mathcal{A}^+$-sets, $H_{n+1} \subset H_n$ for all n, with $\bigcap_{n=1}^{\infty} H_n = (0,1)^{\mathbb{N}}$. Thus $(0,1)^{\mathbb{N}}$ is $\sigma(\mathbb{R}^{\mathbb{N}})$-measurable since $\mathcal{A}^+ \subset \sigma(\mathbb{R}^{\mathbb{N}})$.

To prove that $\mu_{\mathbb{N}}\left((0,1)^{\mathbb{N}}\right) = 1$, continuity from above of $\mu_{\mathbb{N}}$ obtains:

$$\mu_{\mathbb{N}}\left((0,1)^{\mathbb{N}}\right) = \lim_{n \to \infty} \mu_{\mathbb{N}}(H_n) \equiv \lim_{n \to \infty} \mu_{J_n}((0,1)^n),$$

where μ_{J_n} is the finite dimensional product probability measure on $(\mathbb{R}^n, \mathcal{B}(\mathbb{R}^n))$ induced by $\{\mu_{F_{U_{j(i)}}}\}_{i=1}^n$. To complete the proof, we claim that $\mu_{J_n}((0,1)^n) = 1$ for all n.

Let n be given and define $A_k \equiv (0, 1 - 1/k]^n$. Then $\{A_k\}_{k=1}^{\infty}$ is a nested collection of $\mathcal{B}(\mathbb{R}^n)$-sets, $A_k \subset A_{k+1}$ for all k, with $\bigcup_{k=1}^{\infty} A_k = (0,1)^n$. Thus by continuity from below of μ_{J_n}:

$$\mu_{J_n}((0,1)^n) = \lim_{k \to \infty} \mu_{J_n}\left((0, 1 - 1/k]^n\right).$$

Now by (4.11):

$$\mu_{J_n}\left((0, 1 - 1/k]^n\right) = \prod_{j=1}^{n} \mu_{F_U}((0, 1 - 1/k]),$$

to which we apply (4.9) to obtain:

$$\mu_{J_n}\left((0, 1 - 1/k]^n\right) = (1 - 1/k)^n.$$

Thus:

$$\mu_{J_n}((0,1)^n) = \lim_{k \to \infty} (1 - 1/k)^n = 1,$$

completing the proof that $(0,1)^{\mathbb{N}}$ is essential in $(\mathbb{R}^{\mathbb{N}}, \sigma(\mathbb{R}^{\mathbb{N}}), \mu_{\mathbb{N}})$. ∎

Besides the conclusions of Proposition 4.11 on equivalent spaces, this provides a proof of the existence of the probability space $\left((0,1)^{\mathbb{N}}, \sigma\left((0,1)^{\mathbb{N}}\right), \mu_{\mathbb{N}}\right)$, and expands, at least formally, the applicability of Proposition I.9.20.

Proposition 4.13 (II. Probability space $(\mathcal{S}'_U, \mathcal{E}'_U, \mu'_U)$ and i.i.d. $\{X_j\}_{j=1}^{\infty}$) *Let the probability space $(\mathcal{S}, \mathcal{E}, \mu)$ and random variable $U : \mathcal{S} \longrightarrow (0,1)$ with continuous, uniform distribution function $F_U(x)$ be given. Let a random variable X be defined on this or some other probability space, with distribution function $F(x)$.*

With the notation of Proposition 4.12, define:

$$(\mathcal{S}'_U, \mathcal{E}'_U, \mu'_U) \equiv \left((0,1)^{\mathbb{N}}, \sigma\left((0,1)^{\mathbb{N}}\right), \mu_{\mathbb{N}}\right), \tag{4.15}$$

and let $U_j : (0,1)^{\mathbb{N}} \to (0,1)$ *be defined by:*

$$U_j : (y_1, y_2, ...) = y_j. \tag{4.16}$$

With F^* *the left continuous inverse of* F, *define* $X_j : (0,1)^{\mathbb{N}} \to \mathbb{R}$ *by:*

$$X_j = F^*(U_j). \tag{4.17}$$

That is, $X_j : (y_1, y_2, ...) = F^*(y_j)$.

Then $\{X_j\}_{j=1}^{\infty}$ *is a sample of* X.

Proof. *First,* U_j *is measurable and thus a random variable on* $\left((0,1)^{\mathbb{N}}, \sigma\left((0,1)^{\mathbb{N}}\right), \mu_{\mathbb{N}}\right)$ *since for* $A \in \mathcal{B}(0,1)$, *(4.16) yields:*

$$
\begin{aligned}
U_j^{-1}(A) &= \{y \in (0,1)^{\mathbb{N}} | y_j \in A\} \\
&= \{y \in \mathbb{R}^{\mathbb{N}} | y_j \in A\} \bigcap (0,1)^{\mathbb{N}}.
\end{aligned}
$$

Thus $U_j^{-1}(A) \in \sigma\left((0,1)^{\mathbb{N}}\right)$ *by definition (4.14), since* $\{y \in \mathbb{R}^{\mathbb{N}} | y_j \in A\} \in \mathcal{A}^+ \subset \sigma(\mathbb{R}^{\mathbb{N}})$ *as a general cylinder set.*

Applying (4.11) with $H \equiv U_j^{-1}(A)$ *and then (4.10):*

$$\mu_{\mathbb{N}}(U_j^{-1}(A)) = \mu_{F_U}(A) \equiv \mu(U^{-1}(A)), \tag{1}$$

which is (4.2). That is, $\{U_j\}_{j=1}^{\infty}$ *are identically distributed with* U, *and thus all have a continuous, uniform distribution function.*

For independence of $\{U_j\}_{j=1}^{\infty}$, *let* $(i_1, ..., i_m) \subset (1, 2, ...)$ *and* $\{A_j\}_{j=1}^m \subset \mathcal{B}(0,1)$ *be given. Then by (4.16):*

$$\bigcap_{j=1}^m U_{i_j}^{-1}(A_j) = \left\{y \in (0,1)^{\mathbb{N}} | (y_{i_1}, y_{i_2}, ... y_{i_m}) \in \prod_{j=1}^m A_j\right\}.$$

Thus by (4.11), then (4.10) and (1):

$$\mu_{\mathbb{N}}\left(\bigcap_{j=1}^m U_{i_j}^{-1}(A_j)\right) = \prod_{j=1}^m \mu_{F_U}(A_j) \equiv \prod_{j=1}^m \mu_{\mathbb{N}}(U_{i_j}^{-1}(A_j)),$$

which is (4.1).

Hence $\{U_j\}_{j=1}^{\infty}$ *are independent random variables on* $((0,1)^{\mathbb{N}}, \sigma((0,1)^{\mathbb{N}}), \mu_{\mathbb{N}})$, *with continuous, uniform distribution functions. It now follows from Proposition 4.9 that* $\{X_j\}_{j=1}^{\infty}$ *are independent random variables on* $((0,1)^{\mathbb{N}}, \sigma((0,1)^{\mathbb{N}}), \mu_{\mathbb{N}})$, *with common distribution function* $F(x)$.

In other words, $\{X_j\}_{j=1}^{\infty}$ *is a sample of* X. ∎

4.3 An Alternate Construction for Discrete Random Variables

In this section we develop an alternative characterization for Proposition 4.4 when X is a a **discrete random variable.** While this earlier result applies in this case, it can be more simply stated using the framework and notation of **discrete probability theory.** We will not recharacterize Proposition 4.13 in this case, as both in theory and application, that result applies equally well with general or discrete random variables.

Discrete probability theory is simply the application of general notions of probability theory to discrete random variables. If $(\mathcal{S}, \mathcal{E}, \mu)$ is a probability space, we say that $Y : \mathcal{S} \to \mathbb{R}$ is a **discrete random variable** if it has finite $(N < \infty)$ or countable range $(N = \infty)$:

$$Rng(Y) = \{y_j\}_{j=1}^N.$$

If $N = 1$, so $Rng(Y) = \{y_0\}$, we say that Y is a **degenerate random variable**.

In some applications it is more convenient to denote the indexing of the countable case by:

$$Rng(Y) = \{y_j\}_{j=-\infty}^\infty,$$

but we will not make this distinction as any countable set can be indexed either way.

Defining:

$$p_j \equiv \mu\left(Y^{-1}(y_j)\right),$$

the distribution function of Y is then given:

$$F_Y(x) = \sum_{y_j \leq x} p_j.$$

This distribution function induces a Borel measure μ_{F_Y} by Proposition I.5.23, and as in previous sections this measure satisfies for all $A \in \mathcal{B}(\mathbb{R})$:

$$\mu_{F_Y}(A) = \mu\left(Y^{-1}(A)\right). \tag{4.18}$$

Thus $\mu_{F_Y}(\mathbb{R}) = 1$ and $(\mathbb{R}, \mathcal{B}(\mathbb{R}), \mu_{F_Y})$ is the **probability space on $\mathbb{R}$ induced by Y and** $(\mathcal{S}, \mathcal{E}, \mu)$.

But this is not the model typically contemplated for this real probability space in **discrete probability theory**. Instead, the typical representation of this space is as a **discrete probability space**.

A real or general probability space:

$$(Z, \sigma(Z), p),$$

is said to be a discrete probability space if Z is a finite $(N < \infty)$ or countable $(N = \infty)$ collection of points:

$$Z = \{z_j\}_{j=1}^N.$$

The probability measure $p \equiv \{p_j\}_{j=1}^N$, with $\sum_{j=1}^N p_j = 1$ and all $p_j > 0$ for well-definedness, is defined on Z by:

$$p_j \equiv p(z_j),$$

and $\sigma(Z)$ is the power sigma algebra $\sigma(P(Z))$ of all subsets of Z, with the probability of any such set defined additively.

If Y is a discrete random variable on $(\mathcal{S}, \mathcal{E}, \mu)$ as above, and $(Z, \sigma(Z), p)$ represents the real discrete probability space defined on $Z \equiv Rng(Y)$ with associated probabilities, then in the notation of Definition 4.10:

$$(\mathbb{R}, \mathcal{B}(\mathbb{R}), \mu_{F_Y}) \sim (Z, \sigma(Z), p), \tag{4.19}$$

meaning these probability spaces are **equivalent**. For this assertion, we need to verify several statements:

1. $Z \in \mathcal{B}(\mathbb{R})$ and $\mu_{F_Y}(Z) = 1$, so Z is essential in $(\mathbb{R}, \mathcal{B}(\mathbb{R}), \mu_{F_Y})$:

   The Borel sigma algebra contains all open sets, and thus all closed sets including individual points $\{z\}$. Consequently $\mathcal{B}(\mathbb{R})$ contains a finite or countable collection of such points, such as Z. Then $\mu_{F_Y}(Z) = 1$ follows from (4.18) and countable additivity.

2. The sigma algebra $\sigma(Z)$, defined above as the power sigma algebra on Z, equals that given as in (4.12):

$$\sigma(Z) \equiv \left\{ A \bigcap Z \mid A \in \mathcal{B}(\mathbb{R}) \right\}.$$

Certainly $\sigma(Z)$ contains the set on the right by definition of power set. The reverse inclusion follows since $z_j \in \mathcal{B}(\mathbb{R})$ for all j as noted in 1.

3. For any set $B \in \sigma(Z)$:

$$\mu_{F_Y}(B) = p(B).$$

where p is the pointwise measure defined on Z.

By countable additivity it is enough to prove this for $B = \{z_j\}$. Then $\mu_{F_Y}(B) = p_j$ by (4.18), and $p_j \equiv p(z_j)$ on $(Z, \sigma(Z), p)$ by definition.

Thus for the random variables of discrete probability theory, we can model the induced real probability space as the real Borel space $(\mathbb{R}, \mathcal{B}(\mathbb{R}), \mu_{F_Y})$, or as the discrete probability space $(Z, \sigma(Z), p)$, depending on the application in hand.

4.3.1 Third Construction: $(\mathcal{S}'_p, \mathcal{E}'_p, \mu'_p)$

In this section we reframe the Proposition 4.4 construction as the sample space $(\mathcal{S}'_p, \mathcal{E}'_p, \mu'_p)$ for discrete i.i.d. $\{Y_j\}_{j=1}^{\infty}$, by generalizing the results developed above. The notation for the sample space here is to connote that this space is constructed using the Borel measure induced by the distribution function of a discrete random variable Y with associated probability function p. Our primary goal is to generalize the result in (4.19) to the infinite probability spaces:

$$(\mathbb{R}^{\mathbb{N}}, \sigma(\mathbb{R}^{\mathbb{N}}), \mu_{\mathbb{N}}) \sim (Z^{\mathbb{N}}, \sigma(Z^{\mathbb{N}}), \mu_{\mathbb{N}}).$$

The space on the left will be constructed in the usual way using countable copies of $(\mathbb{R}, \mathcal{B}(\mathbb{R}), \mu_{F_Y})$, while the space on the right will generalize $(Z, \sigma(Z), p)$. In particular, $\sigma(Z^{\mathbb{N}})$ as given by Definition 4.10 is again the power sigma algebra on $Z^{\mathbb{N}}$ as above, and $\mu_{\mathbb{N}}$ on $\sigma(Z^{\mathbb{N}})$ will be characterized in terms of a probability function $p_{\mathbb{N}}$ that will extend calculations with $p(z)$ in a natural way. Much of the initial setup is the same as that for Sections 4.1.1 and 4.2.3, but we include some details here to clarify the modification we then introduce.

Consider $\{(\mathbb{R}_i, \mathcal{B}(\mathbb{R}_i), \mu_{Y_i})\}_{i=1}^{\infty}$, a countable collection of identical copies of the Borel space induced by Y and $(\mathcal{S}, \mathcal{E}, \mu)$, indexed only for notational purposes as before. By Proposition I.9.20, construct the infinite dimensional complete probability space:

$$(\mathbb{R}^{\mathbb{N}}, \sigma(\mathbb{R}^{\mathbb{N}}), \mu_{\mathbb{N}}),$$

where:

$$\mathbb{R}^{\mathbb{N}} \equiv \{(x_1, x_2, ...) \mid x_i \in \mathbb{R}_i\}.$$

The measure $\mu_{\mathbb{N}}$ is uniquely defined on the complete sigma algebra $\sigma(\mathbb{R}^{\mathbb{N}})$, which contains the sigma algebra $\sigma(\mathcal{A}^+)$ generated by the algebra $\mathcal{A}^+$. In other words, $\sigma(\mathcal{A}^+) \subset \sigma(\mathbb{R}^{\mathbb{N}})$.

The algebra $\mathcal{A}^+$ is the collection of general cylinder sets in $\mathbb{R}^{\mathbb{N}}$. If $H \in \mathcal{A}^+$:

$$H = \{x \mid (x_{j(1)}, x_{j(2)}, ... x_{j(n)}) \in A\},$$

for some n-tuple of positive integers $J = (j(1), j(2), ..., j(n))$ and $A \in \mathcal{B}(\mathbb{R}^n)$, the Borel sigma algebra on $\mathbb{R}^n$.

On $\mathcal{A}^+$, $\mu_{\mathbb{N}}$ is defined by:

$$\mu_{\mathbb{N}}(H) = \mu_J(A),$$

where μ_J is the finite dimensional product probability measure on $(\mathbb{R}^n, \mathcal{B}(\mathbb{R}^n))$ induced by $\{\mu_{Y_{j(i)}}\}_{i=1}^n$, and derived in Proposition I.7.20. In the special case where $A = \prod_{j=1}^n A_j$ for $A_j \in \mathcal{B}(\mathbb{R})$, the measure $\mu_J(A)$ is representable in terms of μ_Y since $\mu_{Y_{j(i)}} = \mu_Y$ for all i:

$$\mu_{\mathbb{N}}(H) = \prod_{j=1}^n \mu_Y(A_j). \tag{4.20}$$

We now prove that $Z^{\mathbb{N}}$ is essential in $(\mathbb{R}^{\mathbb{N}}, \sigma(\mathbb{R}^{\mathbb{N}}), \mu_{\mathbb{N}})$, where:

$$Z^{\mathbb{N}} \equiv \{(z_1, z_2, \ldots) | z_i \in Z_i\}.$$

Proposition 4.14 ($Z^{\mathbb{N}}$ is essential in $(\mathbb{R}^{\mathbb{N}}, \sigma(\mathbb{R}^{\mathbb{N}}), \mu_{\mathbb{N}})$) *With the notation above, the set $Z^{\mathbb{N}}$ is essential in $(\mathbb{R}^{\mathbb{N}}, \sigma(\mathbb{R}^{\mathbb{N}}), \mu_{\mathbb{N}})$, where this countable product probability space is defined with identical spaces $\{(\mathbb{R}_i, \mathcal{B}(\mathbb{R}_i), \mu_{F_{Y_i}})\}_{i=1}^{\infty}$.*

Consequently:

$$(\mathbb{R}^{\mathbb{N}}, \sigma(\mathbb{R}^{\mathbb{N}}), \mu_{\mathbb{N}}) \sim (Z^{\mathbb{N}}, \sigma(Z^{\mathbb{N}}), \mu_{\mathbb{N}}). \tag{4.21}$$

Proof. *To simplify notation, let $\{y_j\}_{j=1}^{\infty}$ be an enumeration of the countably many points in Z. We leave it as an exercise to verify this proof in the case where Z is finite.*

Define $A^{(n)} = \prod_{k=1}^n A_k$ where $A_k \equiv \{y_j\}_{j=1}^{\infty}$ for all k. Then $A_k \in \mathcal{B}(\mathbb{R})$ by property 1 of the previous section, and thus $A^{(n)} \in \mathcal{B}(\mathbb{R}^n)$. Now define $H_n \in \mathcal{A}^+$ by:

$$H_n \equiv \{x \in \mathbb{R}^{\mathbb{N}} | (x_1, x_2, \ldots x_n) \in A^{(n)}\}.$$

Then:

$$Z^{\mathbb{N}} = \bigcap_{n=1}^{\infty} H_n \in \sigma(\mathcal{A}^+),$$

and thus $Z^{\mathbb{N}} \subset \sigma(\mathbb{R}^{\mathbb{N}})$ is measurable.

Now $\{H_n\}_{n=1}^{\infty}$ is a nested sequence of measurable sets, $H_{n+1} \subset H_n$, and so by continuity from above of $\mu_{\mathbb{N}}$:

$$\mu_{\mathbb{N}}(Z^{\mathbb{N}}) = \lim_{n \to \infty} \mu_{\mathbb{N}}(H_n).$$

It now follows that $\mu_{\mathbb{N}}(Z^{\mathbb{N}}) = 1$ since by (4.20):

$$\mu_{\mathbb{N}}(H_n) = \prod_{k=1}^n \mu_F(A_k) = \left(\sum_{j=1}^{\infty} p_j\right)^n = 1.$$

Thus $Z^{\mathbb{N}}$ is $\sigma(\mathbb{R}^{\mathbb{N}})$-measurable and $\mu_{\mathbb{N}}(Z^{\mathbb{N}}) = 1$. So $Z^{\mathbb{N}}$ is essential in $(\mathbb{R}^{\mathbb{N}}, \sigma(\mathbb{R}^{\mathbb{N}}), \mu_{\mathbb{N}})$ and (4.21) follows by Definition 4.10. ∎

In general, the $\mu_{\mathbb{N}}$ measure of a set cannot be calculated directly from the probability measure $p \equiv \{p(z_j)\}_{j=1}^{\infty}$ defined on $(Z, \sigma(Z), p)$ of (4.19). However, the following result provides a simple pointwise result.

Corollary 4.15 (A pointwise characterization of $\mu_{\mathbb{N}}$ on $(Z^{\mathbb{N}}, \sigma(Z^{\mathbb{N}}), \mu_{\mathbb{N}})$) *If μ_{F_Y} is the Borel measure associated with a degenerate random variable, and thus $\mu_{F_Y}(z_0) = p(z_0) = 1$ for some z_0, then $\mu_{\mathbb{N}}(z) = 1$ for the unique $z \equiv (z_0, z_0, \ldots) \in Z^{\mathbb{N}}$.*

In all other cases, $\mu_{\mathbb{N}}(z) = 0$ for all $z \in Z^{\mathbb{N}}$.

Proof. *If $z \equiv (z_1, z_2, \ldots) \in Z^{\mathbb{N}}$, then $(z_1, z_2, \ldots z_n) \in \mathcal{B}(\mathbb{R}^n)$ for any n and thus $H_n \in \sigma(\mathcal{A}^+)$ if defined:*

$$H_n \equiv \{x \in \mathbb{R}^{\mathbb{N}} | x_j = z_j, \text{ for } 1 \le j \le n\}.$$

Then by (4.20):

$$\mu_{\mathbb{N}}(H_n) = \prod_{j=1}^n p(z_j).$$

Now $\{H_n\}_{n=1}^{\infty}$ *is a nested sequence of measurable sets,* $H_{n+1} \subset H_n$ *for all* n, *and* $z = \bigcap_{n=1}^{\infty} H_n$. *Thus by continuity from above:*

$$\mu_{\mathbb{N}}(z) = \lim_{n \to \infty} \prod_{j=1}^{n} p(z_j).$$

If μ_{F_Y} *is induced by a degenerate random variable with* $\mu_{F_Y}(z_0) = p(z_0) = 1$ *for some* z_0, *then* $Z^{\mathbb{N}} = \{z\}$ *with* $z = (z_0, z_0, ...)$ *and* $\mu_{\mathbb{N}}(z) = 1$. *Otherwise* $\max p(z_j) < 1$ *and so:*

$$\mu_{\mathbb{N}}(z) \leq \lim_{n \to \infty} (\max p(z_j))^n = 0.$$

■

Example 4.16 (Uncountability of $Z^{\mathbb{N}}$; One exception) *Except for the degenerate case, the Corollary 4.15 result that $\mu_{\mathbb{N}}(z) = 0$ for all $z \in Z^{\mathbb{N}}$, and yet $\mu_{\mathbb{N}}(Z^{\mathbb{N}}) = 1$, compels the conclusion that $Z^{\mathbb{N}}$ must be uncountable. That is, if $Z = \{z_j\}_{j=1}^{N}$ with $1 < N \leq \infty$, then $Z^{\mathbb{N}}$ is uncountable. The general proof by contradiction is that if $Z^{\mathbb{N}}$ is countable, then $\mu_{\mathbb{N}}(Z^{\mathbb{N}}) = 0$ by countable additivity of $\mu_{\mathbb{N}}$.*

*A simple example of this conclusion is to let $(\mathcal{S}, \mathcal{E}, \mu)$ be a probability space and $Y : \mathcal{S} \to \mathbb{R}$ a **binomial random variable**:*

$$Rng(Y) = \{0, 1\},$$

with $\mu\left[Y^{-1}(1)\right] = p'$, $\mu\left[Y^{-1}(0)\right] = 1 - p'$, *for* $0 < p' < 1$. *The associated distribution function $F_Y(x)$ is defined on $\mathbb{R}$:*

$$F_Y(x) = \begin{cases} 0, & x < 0. \\ 1 - p', & 0 \leq x < 1, \\ 1, & 1 \leq x. \end{cases}$$

In the notation of the introduction the associated discrete probability space $(Z, \sigma(Z), p)$ has $Z = \{0, 1\}$, the measure p is defined:

$$p(\cdot) \equiv \mu\left[Y^{-1}(\cdot)\right],$$

so $p(0) = 1 - p'$ and $p(1) - p'$. Then, $\sigma(Z)$ is the power sigma algebra:

$$\sigma(Z) = \{\emptyset, 0, 1, \{0, 1\}\}.$$

The probability space $\left(Z^{\mathbb{N}}, \sigma\left(Z^{\mathbb{N}}\right), \mu_{\mathbb{N}}\right)$ then has uncountable $Z^{\mathbb{N}}$, since by definition:

$$Z^{\mathbb{N}} \equiv \{(z_1, z_2, ...) | z_i \in \{0, 1\}\}.$$

Thus every point $z \in Z^{\mathbb{N}}$ is a countable sequence of 0s and 1s. By the identification:

$$(z_1, z_2, ...) \equiv 0.z_1 z_2 ...,$$

*we see that points of $Z^{\mathbb{N}}$ can be put in "one-to-one" correspondence with $x \in [0, 1]$, expressed in binary units. That is, $Z^{\mathbb{N}}$ has the **cardinality** $\mathfrak{c}$ of the continuum $\mathbb{R}$ (see below).*

We put one-to-one in quotes because there is a small ambiguity in binary (or decimal or any base system) expansions when a number x has a finite expansion. For example in binary, let:

$$x = 0.z_1 z_2 ... z_{n-1} 1000000000...$$

This expansion is equivalent to:

$$x = 0.z_1 z_2 ... z_{n-1} 0111111111...,$$

and thus one can have two sequences $z \in Z^{\mathbb{N}}$ that are identified with any such real number.

This implies that $Z^{\mathbb{N}}$ has "more" points than does $[0,1]$, and greater cardinality. But this conclusion is invalid. Since the collection of reals in $[0,1]$ with finite binary expansions is a subset of the rationals, this ambiguity only happens countably many times. In the mathematics of cardinal addition, adding a smaller cardinal (countable) to a larger cardinal (the continuum) obtains the larger cardinal. With this same arithmetic, $[0,1]$ and $(0,1)$ have the same cardinality, while $(0,1)$ and $\mathbb{R}$ have the same cardinality by the one-to-one correspondence $y = \tan(\pi(x+1/2))$.

We now state the third construction of a probability space and sample for a discrete random variable Y defined on $(\mathcal{S}, \mathcal{E}, \mu)$.

Proposition 4.17 *(III. **Probability space** $(\mathcal{S}'_p, \mathcal{E}'_p, \mu'_p)$ **and i.i.d.** $\{Y_j\}_{j=1}^{\infty}$) Let the probability space $(\mathcal{S}, \mathcal{E}, \mu)$ and discrete random variable $Y : \mathcal{S} \longrightarrow \mathbb{R}$ be given, and with the notation of Proposition 4.14, let:*

$$(\mathcal{S}'_p, \mathcal{E}'_p, \mu'_p) \equiv (Z^{\mathbb{N}}, \sigma(Z^{\mathbb{N}}), \mu_{\mathbb{N}}). \tag{4.22}$$

Define $Y_j : Z^{\mathbb{N}} \to \mathbb{R}$ by:

$$Y_j : (z_1, z_2, ...) = z_j. \tag{4.23}$$

Then $\{Y_j\}_{j=1}^{\infty}$ is a sample of Y.

Proof. *First, Y_j is measurable and thus a random variable on $(Z^{\mathbb{N}}, \sigma(Z^{\mathbb{N}}), \mu_{\mathbb{N}})$ since for $A \in \mathcal{B}(\mathbb{R})$, (4.23) yields:*

$$Y_j^{-1}(A) = \{x \in \mathbb{R}^{\mathbb{N}} | x_j \in A\}.$$

Thus $Y_j^{-1}(A) \in \mathcal{A}^+$, a general cylinder set, and measurability follows since $\mathcal{A}^+ \subset \sigma(\mathbb{R}^{\mathbb{N}})$.

Further, from (4.20) with $H \equiv X_j^{-1}(A)$ and (4.18):

$$\mu_{\mathbb{N}}(Y_j^{-1}(A)) = \mu_F(A) \equiv \mu(Y^{-1}(A)), \tag{1}$$

so these variates are identically distributed by (4.2).

For independence, let $(i_1, ..., i_m) \subset (1, 2, ...)$ and $\{A_j\}_{j=1}^{m} \subset \mathcal{B}(\mathbb{R})$ be given. Then by (4.23):

$$\bigcap_{j=1}^{m} Y_{i_j}^{-1}(A_j) = \left\{ x | (x_{i_1}, x_{i_2}, ... x_{i_m}) \in \prod_{j=1}^{m} A_j \right\}.$$

Thus by (4.20) with $H = \bigcap_{j=1}^{m} Y_{i_j}^{-1}(A_j)$, and (4.18) and (1):

$$\mu_{\mathbb{N}} \left(\bigcap_{j=1}^{m} Y_{i_j}^{-1}(A_j) \right) = \prod_{j=1}^{m} \mu_F(A_j) \equiv \prod_{j=1}^{m} \mu_{\mathbb{N}}(Y_{i_j}^{-1}(A_j)),$$

which is (4.1). ∎

We end this chapter with an exercise that will be useful in the next chapter. It will be seen that the measures of the identified sets are consistent with the general binomial measure in (1.8).

Exercise 4.18 (Binomial measure space) *Let $\left(Z^{\mathbb{N}}, \sigma\left(Z^{\mathbb{N}}\right), \mu_{\mathbb{N}}\right)$ be the binomial space of Example 4.16, and $\{Y_j\}_{j=1}^{\infty}$ defined as in (4.23). For any integer $n \geq 0$ and real $a \geq 0$, define:*

$$A_n(a) = \left\{ z | \sum_{k=1}^{n} Y_k(z) \leq a \right\}.$$

Show that:

$$\mu_{\mathbb{N}}[A_n(a)] = \sum_{j=0}^{\lfloor a \rfloor} \binom{n}{j} p^j q^{n-j},$$

*where $q \equiv 1 - p$ and $\lfloor a \rfloor$ denotes the **greatest integer** less than or equal to a.*

Defining:

$$A'_n(a) = \left\{ z \mid \sum_{k=1}^{n} Y_k(z) \geq a \right\},$$

show that:

$$\mu_{\mathbb{N}} \left[A'_n(a) \right] \leq \sum_{j=\lfloor a \rfloor}^{n} \binom{n}{j} p^j q^{n-j}.$$

Note that this last upper bound is an equality when $\lfloor a \rfloor$ is an integer, and otherwise is an equality with lower sum limit of $\lfloor a \rfloor + 1$. Hint: These sets are general cylinder sets given by $J = (1, ..., n)$ and $B_n(a) \in \mathcal{B}(\mathbb{R}^n)$, respectively $B'_n(a) \in \mathcal{B}(\mathbb{R}^n)$. Then $\mu_{\mathbb{N}} \left[A_n(a) \right] = \mu_J \left[B_n(a) \right]$, for example. Identify the points in these Borel sets.

5

Limit Theorems for RV Sequences

In this chapter we initiate the study of limit theorems for random variable sequences. The first major section is dedicated to the classical versions of the weak and strong laws of large numbers, applicable to binomial variable sequences. The sample spaces of Chapter 4 provide the formal structure within which these results can be framed and proved.

The second major section then investigates more general results for random variable sequences under various modes of convergence, two of which echo the ideas underlying the weak and strong laws.

More results of this type will be found in Chapter 8 within the study of weak convergence of distribution functions, and in Chapter 9 in the study of tail events. Book IV will continue the development of this theory and application, and more yet will be found in Book VI.

Perhaps needless to say, limit theorems on random variable sequences and distribution functions play a critical role in probability theory and its applications to finance and elsewhere.

5.1 Two Limit Theorems for Binomial Sequences

In this section we study special cases of two important results applied to the binomial measure space constructible by Proposition 4.17. Specifically, we require an infinite product probability space associated with the probability space $(Y, \sigma(Y), \mu_B)$ of general coin flips, where we represent the probability space $Y = \{H, T\}$ by the numerical space $Y \equiv \{y | y \in \{1, 0\}\}$. The probability measure μ_B is defined on Y as in (1.7):

$$\mu_B(1) = p, \qquad \mu_B(0) = 1 - p,$$

and the sigma algebra $\sigma(Y)$ is the power sigma algebra given by:

$$\sigma(Y) = \{\emptyset, 0, 1, Y\}.$$

Defining

$$B : (Y, \sigma(Y), \mu_B) \to (\mathbb{R}, \mathcal{B}(\mathbb{R}), m),$$

by $B(y) = y$, then B is a random variable with distribution function defined on $\mathbb{R}$ by:

$$F_B(x) = \begin{cases} 0, & x < 0, \\ 1 - p, & 0 \le x < 1, \\ 1, & 1 \le x. \end{cases}$$

As derived in Section 4.3.1, the associated measure μ_{F_B} of Proposition I.5.23 satisfies for all $A \in \mathcal{B}(\mathbb{R})$:

$$\mu_{F_B}(A) = \mu_B\left(B^{-1}(A)\right),$$

and $(\mathbb{R}, \mathcal{B}(\mathbb{R}), \mu_{F_B})$ is the **probability space on $\mathbb{R}$ induced by B and $(Y, \sigma(Y), \mu_B)$**.

Let $(Y^{\mathbb{N}}, \sigma(Y^{\mathbb{N}}), \mu_{\mathbb{N}})$ be the associated probability space of Proposition 4.14, where:

$$Y^{\mathbb{N}} \equiv \{(y_1, y_2, ...)|y_j \in \{0, 1\}\},$$

and define a collection of random variables $\{B_j\}_{j=1}^{\infty}$ as in Proposition 4.17. Thus $B_j :$ $(Y^{\mathbb{N}}, \sigma(Y^{\mathbb{N}}), \mu_{\mathbb{N}}) \to (\mathbb{R}, \mathcal{B}(\mathbb{R}), m)$ is defined by:

$$B_j : (y_1, y_2, ...) = y_j, \tag{5.1}$$

and these random variables are independent and i.i.d.-B.

Each point $y \in Y^{\mathbb{N}}$ can thus be envisioned as an infinite sequence of coin flips, where Hs and Ts are numerically represent by 1s and 0s, respectively. For each j, the random variable B_j identifies the outcome of the jth coin flip.

In this section we study convergence properties as $n \to \infty$ of the **averaging sequence** or **mean sequence** $\{M_n\}_{n=1}^{\infty}$ defined for $n = 1, 2, ...$, by:

$$M_n \equiv \frac{1}{n} \sum\nolimits_{j=1}^{n} B_j. \tag{5.2}$$

Note that $M_n : (Y^{\mathbb{N}}, \sigma(Y^{\mathbb{N}}), \mu_{\mathbb{N}}) \to (\mathbb{R}, \mathcal{B}(\mathbb{R}), m)$ is a random variable for all n since $\sigma(Y^{\mathbb{N}})$ is the power sigma algebra on $Y^{\mathbb{N}}$ by Proposition 4.14. It may also be of interest to explicitly identify $M_n^{-1}(x)$ for $x \in \mathbb{R}$, noting that:

$$Rng(M_n) = \{j/n\}_{j=0}^{n}.$$

5.1.1   The Weak Law of Large Numbers

There are several forms of the so-called **weak law of large numbers**, and the version addressed here and applicable to the binomial product space $(Y^{\mathbb{N}}, \sigma(Y^{\mathbb{N}}), \mu_{\mathbb{N}})$ is known as **Bernoulli's theorem**, named for **Jacob Bernoulli** (1654–1705) who first derived it. This result makes a statement about the limit of the measures of sets $\{A_n(\epsilon)\}_{n=1}^{\infty}$ for $\epsilon > 0$. With p given as above, the set $A_n(\epsilon) \subset Y^{\mathbb{N}}$ is defined:

$$A_n(\epsilon) \equiv \{|M_n - p| \geq \epsilon\}, \tag{5.3}$$

where we notationally suppress $y \in Y^{\mathbb{N}}$ in this definition for simplicity.

The set sequence $\{A_n(\epsilon)\}_{n=1}^{\infty}$ is not nested and it is an exercise to verify that:
· If $y \in A_n(\epsilon)$, then:
    · $y \in A_{n+1}(\epsilon)$ if $y_{n+1} = 1$;
    · y need not be a member of $A_{n+1}(\epsilon)$ if $y_{n+1} = 0$.
· If $y \notin A_n(\epsilon)$, then:
    · $y \notin A_{n+1}(\epsilon)$ if $y_{n+1} = 0$;
    · y need not be outside $A_{n+1}(\epsilon)$ if $y_{n+1} = 1$.
The **weak law states** that for any $\epsilon > 0$:

$$\mu_{\mathbb{N}}[A_n(\epsilon)] \to 0, \text{ as } n \to \infty.$$

Thus for any $\delta > 0$ there is an $N \equiv N(\delta)$ so that $\mu_{\mathbb{N}}[A_n(\epsilon)] < \delta$ for all $n \geq N$. Equivalently, this result can be expressed:

$$\mu_{\mathbb{N}}\left[\widetilde{A}_n(\epsilon)\right] \to 1, \text{ as } n \to \infty.$$

In other words, for any $\delta > 0$, there is an $N \equiv N(\delta)$ so that $\mu_{\mathbb{N}}\left[\tilde{A}_n(\epsilon)\right] > 1 - \delta$ for all $n \geq N$.

Now for $\mu_{\mathbb{N}}[A_n(\epsilon)]$ to be well defined requires that $A_n(\epsilon)$ be measurable, $A_n(\epsilon) \in \sigma(Y^{\mathbb{N}})$, and this follows because $\sigma(Y^{\mathbb{N}})$ is the power sigma algebra on $Y^{\mathbb{N}}$ by Proposition 4.14. In fact, these subsets of $Y^{\mathbb{N}}$ are definable in terms of finitely many indexes, and thus are general cylinder sets in $\sigma(Y^{\mathbb{N}})$.

For example, let $b \equiv (b_1, ..., b_n)$ be any one of the 2^n n-tuples of coin flips, and define the general cylinder set:

$$A_b \equiv \{B_i(y) = b_i \text{ for } i \leq n\}.$$

Then:

$$A_n(\epsilon) = \bigcup_b{}' A_b,$$

where this union is over the finitely many A_b-sets for which $\left|\sum_{j=1}^n b_j/n - p\right| \geq \epsilon$. Thus $A_n(\epsilon)$ is a general cylinder set, defined by $J = (1, ..., n)$ and $C_n \in \mathcal{B}(\mathbb{R}^n)$, where C_n is the collection of identified b-points.

The weak law addresses the limit of the probabilities of a sequence of events, which are related as noted above, and not the probability of any particular event. Put another way, the weak law does not uniquely identify an event in $\sigma(Y^{\mathbb{N}})$ that has measure zero.

That said, the event $\bigcap_{n=1}^{\infty} A_n(\epsilon)$ is related to the weak law:

$$\bigcap_{n=1}^{\infty} A_n(\epsilon) = \{y \in Y^{\mathbb{N}} |\; |M_n(y) - p| \geq \epsilon \text{ for all } n\}.$$

Since $\bigcap_{n=1}^{\infty} A_n(\epsilon) \subset A_n(\epsilon)$ for every n, the **weak law implies** that this intersection event has measure 0:

$$\mu_{\mathbb{N}}\left(\bigcap_{n=1}^{\infty} A_n(\epsilon)\right) = 0.$$

This set is not uniquely identified by the weak law, in that the same result is true for $\bigcap_{n=m}^{\infty} A_n(\epsilon)$ for any m, as well as for $\bigcap_{m=1}^{\infty} A_{n_m}(\epsilon)$, for any sequence $\{n_m\}_{m=1}^{\infty}$ that is unbounded.

Conversely, that the intersection of countably many events has measure 0 does not imply that the associated sequence of event measures converges to zero. This is trivial for disjoint sets, but also true for collections with nonempty finite intersections.

Example 5.1 $(\mu_{\mathbb{N}}(\bigcap_{n=1}^m B_n') > 0$ **all** m and $\mu_{\mathbb{N}}(\bigcap_n B_n') = 0 \nRightarrow \mu_{\mathbb{N}}(B_n') \to 0)$ *With random variables* $\{B_j\}_{j=1}^{\infty}$ *as given above, define the set* $B_n' = B_n^{-1}(1)$:

$$B_n' = \{y \in Y^{\mathbb{N}} | y_n = 1\}.$$

Then B_n' *is a general cylinder set with* $\mu_{\mathbb{N}}(B_n') = p$, *and for every* m:

$$\mu_{\mathbb{N}}\left(\bigcap_{n=1}^m B_n'\right) = p^m.$$

These measure statements should be formally verified as an exercise.

Since $\{\bigcap_{n \leq m} B_n'\}_{m=1}^{\infty}$ *is a nested decreasing sequence of sets, continuity from above of* $\mu_{\mathbb{N}}$ *applies to yield:*

$$\mu_{\mathbb{N}}\left(\bigcap_{n=1}^{\infty} B_n'\right) = 0.$$

However:

$$\mu_{\mathbb{N}}(B_n') = p \nrightarrow 0.$$

The implication of this example is relevant to the proof of the weak law. A proof that $\mu_{\mathbb{N}}(\bigcap_{n=1}^{\infty} A_n(\epsilon)) = 0$ would be a very weak result, and too weak to imply the weak law we seek, that $\mu_{\mathbb{N}}[A_n(\epsilon)] \to 0$.

The **weak law also implies** that:

$$\mu_{\mathbb{N}}[\liminf A_n(\epsilon)] = 0.$$

Recall Definition 2.1, that the event $\liminf A_n(\epsilon)$ is defined:

$$\liminf A_n(\epsilon) \equiv \bigcup_m \bigcap_{n \geq m} A_n(\epsilon).$$

Because $\bigcap_{n=m}^{\infty} A_n(\epsilon) \subset A_j(\epsilon)$ for every $j \geq m$, the weak law implies that $\mu_{\mathbb{N}}[\bigcap_{n=m}^{\infty} A_n(\epsilon)] = 0$ for every m, and hence so too for the union of such sets which is $\liminf A_n(\epsilon)$.

Again, the reverse conclusion is invalid. By Example 5.1:

$$\mu_{\mathbb{N}}[\liminf B'_n] = 0,$$

yet $\mu_{\mathbb{N}}(B'_n) \not\to 0$. Thus a proof that $\mu_{\mathbb{N}}[\liminf A_n(\epsilon)] = 0$ is again too weak to imply the weak law that $\mu_{\mathbb{N}}[A_n(\epsilon)] \to 0$.

Remark 5.2 (On $\limsup A_n(\epsilon)$) *It would be logical to investigate the event $\limsup_n A_n(\epsilon)$, and this we do in the next section. There it will be seen that:*

$$\mu_{\mathbb{N}}(\limsup A_n(\epsilon)) = 0.$$

It turns out that this is a strong result, and one that implies the weak law.

We now establish Bernoulli's version of the weak law of large numbers. To do so, we utilize a bound for $\mu_{\mathbb{N}}[A_n(\epsilon)]$ in (5.5) known as **Bernstein's inequality.** It is named for **Sergei Natanovich Bernstein** (1880–1968), who developed several inequalities of this type. This inequality will play a critical role in the next section as well, and provides a very powerful bound on the rate at which $\mu_{\mathbb{N}}[A_n(\epsilon)] \to 0$ as $n \to \infty$. We will return to this point in Section 9.1.

To simplify notation, we continue to omit the qualifier $y \in Y^{\mathbb{N}}$ in the sets below.

Proposition 5.3 (Bernoulli's theorem) *Given the binomial product probability space $(Y^{\mathbb{N}}, \sigma(Y^{\mathbb{N}}), \mu_{\mathbb{N}})$, random variables $\{M_n\}_{n=1}^{\infty}$ defined in (5.2), and any $\epsilon > 0$:*

$$\mu_{\mathbb{N}}[\{|M_n(y) - p| \geq \epsilon\}] \to 0, \ \text{as } n \to \infty. \tag{5.4}$$

Proof. *By finite additivity:*

$$\begin{aligned} \mu_{\mathbb{N}}[\{|M_n(y) - p| \geq \epsilon\}] &= \mu_{\mathbb{N}}[\{M_n(y) \geq p + \epsilon\}] \\ &\quad + \mu_{\mathbb{N}}[\{M_n(y) \leq p - \epsilon\}]. \end{aligned}$$

The result of (5.4) will follow if we show that the measure of these sets can be made as small as desired by taking n large.

*To this end, recall Exercise 4.18 with $m = \lfloor n(p + \epsilon) \rfloor$, the **greatest integer** less than or equal to $n(p + \epsilon)$. Denoting $q \equiv 1 - p$, we have for any $\lambda > 0$:*

$$\begin{aligned} \mu_{\mathbb{N}}[\{M_n(y) \geq p + \epsilon\}] &\leq \sum_{j=m}^{n} \binom{n}{j} p^j q^{n-j} \\ &\leq \sum_{j=0}^{n} \binom{n}{j} p^j q^{n-j} e^{[j-n(p+\epsilon)]\lambda} \\ &= \left(pe^{\lambda q} + qe^{-\lambda p}\right)^n e^{-\lambda n \epsilon}. \end{aligned}$$

This inequality follows because $e^{[j-n(p+\epsilon)]\lambda} \geq 1$ exactly when $j \geq m$.

Now $e^y \leq y + e^{y^2}$ for all y by an application of the Taylor series in (1.12) with a remainder term:

$$e^y = 1 + y + (\lambda_1 y)^2 / 2, \ \text{with } 0 < \lambda_1 < 1,$$
$$y + e^{y^2} = 1 + y + y^2 + (\lambda_2 y^2)^2 / 2, \ \text{with } 0 < \lambda_2 < 1.$$

Applying this inequality to $e^{\lambda q}$ and $e^{-\lambda p}$ obtains

$$\mu_{\mathbb{N}} [\{M_n(y) \geq p + \epsilon\}] \leq e^{-\lambda n \epsilon} \left(pe^{\lambda^2 q^2} + qe^{\lambda^2 p^2} \right)^n$$
$$\leq e^{-\lambda n \epsilon} e^{\lambda^2 n},$$

since $0 < p^2, q^2 < 1$ and $p + q = 1$. The tightest upper bound is obtained by choosing $\lambda = \epsilon/2$, which obtains:

$$\mu_{\mathbb{N}} [\{M_n(y) \geq p + \epsilon\}] \leq e^{-n\epsilon^2/4}.$$

A similar analysis with Exercise 4.18 and $m = \lfloor n(p - \epsilon) \rfloor$ obtains:

$$\mu_{\mathbb{N}} [\{M_n(y) \leq p - \epsilon\}] = \sum_{j=0}^{m} \binom{n}{j} p^j q^{n-j}$$
$$\leq \sum_{j=0}^{n} \binom{n}{j} p^j q^{n-j} e^{[n(p-\epsilon)-j]\lambda}$$
$$= \left(pe^{-\lambda q} + qe^{\lambda p} \right)^n e^{-\lambda n \epsilon}$$
$$\leq e^{-n\epsilon^2/4}.$$

Combining produces **Bernstein's inequality:**

$$\mu_{\mathbb{N}} [\{|M_n(y) - p| \geq \epsilon\}] \leq 2e^{-n\epsilon^2/4}, \tag{5.5}$$

and the result follows. ■

Corollary 5.4 (Bernoulli's theorem) *On the binomial product space $(Y^{\mathbb{N}}, \sigma(Y^{\mathbb{N}}), \mu_{\mathbb{N}})$, for any $\epsilon > 0$:*

$$\mu_{\mathbb{N}} [\{|M_n(y) - p| < \epsilon\}] \to 1, \ \text{as } n \to \infty. \tag{5.6}$$

Proof. *Immediate by finite additivity:*

$$\mu_{\mathbb{N}} [\{|M_n(y) - p| < \epsilon\}] \geq 1 - 2e^{-n\epsilon^2/4}.$$

■

Notation 5.5 (Convergence in probability) *The result of the weak law is often expressed as stating that $M_n(y) \equiv \sum_{j=1}^{n} B_j/n$ **converges in probability to** p, and denoted $M_n(y) \to_P p$. See Section 5.2 for more on this and related notions.*

5.1.2 The Strong Law of Large Numbers

There are several forms of the so-called **strong law of large numbers.** The version addressed here and applicable to the binomial product space $(Y^{\mathbb{N}}, \sigma(Y^{\mathbb{N}}), \mu_{\mathbb{N}})$ is known as **Borel's theorem,** named for **Émile Borel** (1871–1956) who first derived it. We begin by defining two events which will be shown to be **tail events** as introduced in Definition 2.12:

1. Given $\epsilon > 0$ and $A_n(\epsilon)$ as defined in (5.3):

$$A_n(\epsilon) \equiv \{|M_n(y) - p| \geq \epsilon\},$$

we define the **limit supremum set** $A_S(\epsilon)$ by:

$$A_S(\epsilon) \equiv \limsup A_n(\epsilon). \tag{5.7}$$

Recall Definition 2.1, that the event $\limsup A_n(\epsilon)$ is defined:

$$\limsup A_n(\epsilon) \equiv \bigcap_m \bigcup_{n \geq m} A_n(\epsilon).$$

Thus $A_S(\epsilon)$ is the set on which $|M_n(y) - p| \geq \epsilon$ infinitely often as $n \to \infty$.

2. The **convergence set** C_S is defined as the set on which $M_n(y)$ converges to p as $n \to \infty$:

$$C_S \equiv \left\{ \lim_{n \to \infty} M_n(y) = p \right\}. \tag{5.8}$$

Here we are again using the standard set notation which suppresses the qualifier $y \in Y^{\mathbb{N}}$.

Since $\sigma(Y^{\mathbb{N}})$ is the power sigma algebra by Proposition 4.14, $A_S(\epsilon)$ for any $\epsilon > 0$ and C_S are events. These events are also related as can be derived by considering the complement $\widetilde{C}_S$. Then $y \in \widetilde{C}_S$ if and only if there exists $\epsilon_0 > 0$ so that $|M_n(y) - p| \geq \epsilon_0$ infinitely often, which is to say:

$$y \in \limsup A_n(\epsilon_0).$$

Consequently:

$$\widetilde{C}_S = \bigcup_{\epsilon > 0} A_S(\epsilon), \tag{5.9}$$

where this union is taken over all rational ϵ, or a sequence $\epsilon_k \to 0$. These countability restrictions on ϵ are allowable because $\{A_S(\epsilon)\}$ is a **nested collection of sets** with respect to ϵ:

$$A_S(\epsilon) \subset A_S(\epsilon') \text{ if } \epsilon' \leq \epsilon. \tag{5.10}$$

We next seek to show that all $A_S(\epsilon)$ and C_S are tail events, and moreover, tail events that will allow the application of Kolmogorov's zero-one law of Proposition 2.15. Thus we need to identify a sequence of independent sets with respect to which we will define the tail sigma algebra $\mathcal{T} \subset \sigma(Y^{\mathbb{N}})$ using (2.7).

To this end, as in Example 5.1, let:

$$B_n' = \{y_n = 1\}.$$

Then B_n' is a cylinder set and $\mu_{\mathbb{N}}(B_n') = p$. Also, $\{B_n'\}_{n=1}^\infty$ is a collection of **mutually independent sets**. For $n \neq m$:

$$\mu_{\mathbb{N}}(B_n' \cap B_m') = p^2 = \mu_{\mathbb{N}}(B_n')\mu_{\mathbb{N}}(B_m'),$$

and this result extends to any finite collection of distinct sets.

We now show that $A_S(\epsilon) \in \mathcal{T}$ and $C_S \in \mathcal{T}$ for the tail sigma algebra defined by:

$$\mathcal{T} \equiv \mathcal{T}\left(\{B_n'\}_{n=1}^\infty\right).$$

To set the stage, the definition of $\mathcal{T}$ in (2.7) requires the sigma algebra $\sigma(B_n', B_{n+1}', B_{n+2}', ...)$ for each n. This sigma algebra includes all sets B_j' for $j \geq n$ and their complements, as well as all finite and countable unions and intersections of these sets. In particular, this sigma algebra contains every $A \in \sigma(Y^{\mathbb{N}})$ which is definable by specifying the values of any finite or infinite collection of y_j-values for $j \geq n$. Indeed, if $J = (j_1, j_2, \ldots)$ is a finite or infinite increasing index set with $j_1 \geq n$, and A is a set defined by specifying values for the y_j-values for $j \in J$, then:

$$A = \bigcap_{j \in J} D_j,$$

where $D_j = B'_j$ if $y_j = 1$ and $D_j = \widetilde{B'}_j$ if $y_j = 0$.

The intuition underlying the following proposition is that for the sets defined in (5.7) and (5.8), the values of y_j for $j \leq m$ are irrelevant for any m.

Proposition 5.6 $(A_S(\epsilon), C_S \in \mathcal{T})$ *With the definitions in (5.7) and (5.8), and* $\mathcal{T} \equiv \mathcal{T}$ $(\{B'_n\}_{n=1}^{\infty})$:

    1. $A_S(\epsilon) \in \mathcal{T}$ for every $\epsilon > 0$.

    2. $C_S \in \mathcal{T}$.

Proof. *For statement 1, note that for any fixed m and $N(m) \geq = 0$:*

$$\begin{aligned}
A_S(\epsilon) &\equiv \bigcap_{n=1}^{\infty} \bigcup_{k=n}^{\infty} \{|M_k(y) - p| \geq \epsilon\} \\
&= \bigcap_{n=m+N(m)}^{\infty} \bigcup_{k=n}^{\infty} \{|M_k(y) - p| \geq \epsilon\}.
\end{aligned} \tag{1}$$

The last set contains $A_S(\epsilon)$ by construction, and we show equality by proof by contradiction. Assume there exists y so that:

$$\begin{aligned}
y &\in \bigcap_{n=m+N(m)}^{\infty} \bigcup_{k \geq n}^{\infty} \{|M_k(y) - p| \geq \epsilon\}, \\
y &\notin \bigcap_{n=1}^{\infty} \bigcup_{k \geq n}^{\infty} \{|M_k(y) - p| \geq \epsilon\}.
\end{aligned}$$

This implies that $y \notin \bigcup_{k \geq n}^{\infty} \{|M_k(y) - p| \geq \epsilon\}$ for some n with $1 \leq n < m + N(m)$. So for all $k \geq n$:

$$|M_k(y) - p| < \epsilon,$$

contradicting that y is in the first set. Thus we can define $A_S(\epsilon)$ by (1).

If $y \in A_S(\epsilon)$, then for every $n > m$ there exists $k \geq n$ with:

$$|M_k(y) - p| \geq \epsilon.$$

Thus for every m there exists an unbounded sequence of such ks. Denoting $B_j(y) \equiv y_j$, then for any such $k > m$:

$$M_k(y) = mM_m(y)/k + \sum_{j=m+1}^{k} y_j/k.$$

But $mM_m(y)/k \leq m/k$ implies that for any such k:

$$\left| \sum_{j=m+1}^{k} y_j/k - p \right| \geq \epsilon - m/k.$$

Choose $N(m)$ so that for some k in this sequence, $m/(m + N(m)) = m/k < \epsilon/2$. Then:

$$\begin{aligned}
A_S(\epsilon) &= \bigcap_{n=m+N(m)}^{\infty} \bigcup_{k \geq n}^{\infty} \{|M_k(y) - p| \geq \epsilon\} \\
&\subset \bigcap_{n=m+N(m)}^{\infty} \bigcup_{k \geq n}^{\infty} \left\{ \left| \sum_{j=m+1}^{k} y_j/k - p \right| \geq \epsilon/2 \right\},
\end{aligned}$$

and this last set is an element of $\sigma(B'_{m+1}, B'_{m+2}, B'_{m+3}, \ldots)$. *That this is true for every* m *obtains that* $A_S(\epsilon) \in \mathcal{T}$, *which proves statement 1.*

For statement 2, let ϵ_k *be a sequence of rational numbers with* $\epsilon_k \to 0$ *and let* m *be arbitrary. Applying De Morgan's law to the identity in (5.9) obtains* $C_S = \bigcap_{\epsilon_k} \tilde{A}_S(\epsilon_k)$. *Then using the identity for* $A_S(\epsilon)$ *from (1) with* $N(m) = 0$ *obtains:*

$$C_S = \bigcap_{\epsilon_k} \bigcup_{n \geq m} \bigcap_{j \geq n} \{|M_j(y) - p| < \epsilon_k\}.$$

Hence, C_S *is definable by countably many unions and intersections of sets from* $\sigma(B'_{m+1}, B'_{m+2}, B'_{m+3}, \ldots)$, *and this is true for every* m, *so* $C_S \in \mathcal{T}$. $\blacksquare$

Corollary 5.7 (Kolmogorov's Zero-One Law) *With the definitions in (5.7) and (5.8):*

    1. *For every* $\epsilon > 0$,

$$\mu_{\mathbb{N}}[A_S(\epsilon)] \equiv \mu_{\mathbb{N}}[\limsup \{|M_n(y) - p| \geq \epsilon\}] \in \{0, 1\}. \qquad (5.11)$$

    2. *For the convergence set:*

$$\mu_{\mathbb{N}}[C_S] \equiv \mu_{\mathbb{N}}\left[\left\{\lim_{n \to \infty} M_n(y) = p\right\}\right] \in \{0, 1\}. \qquad (5.12)$$

Proof. *This immediately follows from Kolmogorov's Zero-One Law of Proposition 2.15 by the above result that* $A_S(\epsilon) \in \mathcal{T}$ *and* $C_S \in \mathcal{T}$. $\blacksquare$

Given (5.9), the values of $\mu_{\mathbb{N}}[A_S(\epsilon)]$ and $\mu_{\mathbb{N}}[C_S]$ are complementary, thus reducing the possibilities from Corollary 5.7.

Proposition 5.8 (Kolmogorov's Zero-One Law) *With the notation above:*

$$\mu_{\mathbb{N}}[C_S] = 1 \text{ if and only if } \mu_{\mathbb{N}}[A_S(\epsilon)] = 0 \text{ for all } \epsilon > 0. \qquad (5.13)$$

If $\mu_{\mathbb{N}}[C_S] = 0$, *then there exists* $\epsilon_0 > 0$ *so that* $\mu_{\mathbb{N}}[A_S(\epsilon)] = 1$ *for all* $\epsilon \leq \epsilon_0$.
Proof. *From the last lines of the proof of Proposition 5.6 with rational* $\epsilon_k \to 0$:

$$\tilde{C}_S = \bigcup_{\epsilon_k} \limsup A_n(\epsilon_k).$$

If $\mu_{\mathbb{N}}[C_S] = 1$, *then* $\mu_{\mathbb{N}}[\limsup A_n(\epsilon_k)] = 0$ *for all such* ϵ_k *and hence for all* $\epsilon > 0$ *since* $\{\limsup A_n(\epsilon)\}$ *is nested with respect to* ϵ *by (5.10).*

Conversely, by (5.10) and continuity from below of $\mu_{\mathbb{N}}$:

$$\begin{aligned} \mu_{\mathbb{N}}\left[\tilde{C}_S\right] &= \lim_k \mu_{\mathbb{N}}[\limsup A_n(\epsilon_k)] \\ &= \lim_k \mu_{\mathbb{N}}[A_S(\epsilon_k)]. \end{aligned}$$

So if $\mu_{\mathbb{N}}[A_S(\epsilon_k)] = 0$ *for all* k, *then* $\mu_{\mathbb{N}}[C_S] = 1$.

Now if $\mu_{\mathbb{N}}[C_S] = 0$ *and thus* $\mu_{\mathbb{N}}\left[\tilde{C}_S\right] = 1$, *then by (5.9):*

$$\tilde{C}_S = \bigcup_{\epsilon > 0} A_S(\epsilon),$$

and thus $\mu_{\mathbb{N}}[A_S(\epsilon_0)] = 1$ *for at least one* $\epsilon_0 > 0$. *Then by (5.10),* $\mu_{\mathbb{N}}[A_S(\epsilon)] = 1$ *for all* $\epsilon \leq \epsilon_0$. $\blacksquare$

The **strong law of large numbers is** stated next, and known as **Borel's theorem** when restricted to the binomial measure case. This result is typically stated as $\mu_\mathbb{N}[C_S] = 1$, but by the above proposition, can equally well be stated in terms of $\mu_\mathbb{N}[A_S(\epsilon)] = 0$ for all $\epsilon > 0$.

Proposition 5.9 (Borel's theorem) *For the binomial product space* $(Y^\mathbb{N}, \sigma(Y^\mathbb{N}), \mu_\mathbb{N})$, *and* $\{M_n\}_{n=1}^\infty$ *defined in (5.2):*

$$\mu_\mathbb{N}\left[\left\{\lim_{n\to\infty} M_n(y) = p\right\}\right] = 1. \tag{5.14}$$

Equivalently, for all $\epsilon > 0$:

$$\mu_\mathbb{N}[\limsup\{|M_n(y) - p| \geq \epsilon\}] = 0. \tag{5.15}$$

Proof. *We prove (5.15) and then (5.14) follows from the prior proposition. Bernstein's inequality in (5.5) provides:*

$$\mu_\mathbb{N}[\{|M_n(y) - p| \geq \epsilon\}] \leq 2e^{-n\epsilon^2/4}.$$

This implies that for all $\epsilon > 0$,

$$\sum_{n=1}^\infty \mu_\mathbb{N}[\{|M_n(y) - p| \geq \epsilon\}] < \infty,$$

and hence by the Borel-Cantelli lemma of Proposition 2.6:

$$\mu_\mathbb{N}[\limsup\{|M_n(y) - p| \geq \epsilon\}] = 0 \text{ for all } \epsilon > 0.$$

∎

Notation 5.10 (Strong law convergence) *The result of the strong law in (5.14) is often notationally expressed in one of several ways. With* $M_n(y)$ *defined in (5.2):*

1. $M_n(y)$ **converges almost everywhere to** p, *denoted:*

$$M_n(y) \to_{a.e.} p;$$

2. $M_n(y)$ **converges almost surely to** p, *denoted:*

$$M_n(y) \to_{a.s.} p;$$

3. $M_n(y)$ **converges to** p **with probability one,** *denoted:*

$$M_n(y) \to_1 p.$$

In general, we will use the terminology of "almost surely" or "with probability 1," as these are the probability space versions of the measure space convergence terminology of "almost everywhere."

See Section 5.2 below for more on this notion.

5.1.3   Strong Laws versus Weak Laws

Both the weak law and strong law are satisfied in the binomial product space $(Y^{\mathbb{N}}, \sigma(Y^{\mathbb{N}}), \mu_{\mathbb{N}})$, so it is only natural to wonder if these laws are in fact equivalent. It turns out that the strong law is indeed stronger than the weak law in the sense that the strong law always implies the weak law, but the converse is not generally true.

Abstractly, the weak law and strong law begin with a sequence of measurable sets $\{A_n(\epsilon)\}$ and state the following:

1. **Weak Law:**
$$\mu_{\mathbb{N}}[A_n(\epsilon)] \to 0 \text{ as } n \to \infty.$$

2. **Strong Law:**
$$\mu_{\mathbb{N}}[\limsup_n [A_n(\epsilon)]] = 0 .$$

For the binomial space investigation above, the set $A_n(\epsilon)$ was defined in (5.3).

From (2.5), any such strong law implies:
$$\limsup \mu_{\mathbb{N}}[A_n(\epsilon)] = \liminf \mu_{\mathbb{N}}[A_n(\epsilon)] = 0 ,$$

and hence by definition:
$$\lim_n \mu_{\mathbb{N}}[A_n(\epsilon)] = 0,$$

which is the corresponding weak law.

On the other hand, the Borel-Cantelli lemma of Proposition 2.6 states that if the series $\sum_{n=1}^{\infty} \mu_{\mathbb{N}}[A_n(\epsilon)]$ converges, then $\mu_{\mathbb{N}}[\limsup[A_n(\epsilon)]] = 0$. In general, a weak law does not assure such convergence. But in the particular case of the binomial measure, the Bernstein inequality proved the weak law, and also showed that $\mu_{\mathbb{N}}[A_n(\epsilon)] \to 0$ fast enough to guarantee convergence of $\sum_{n=1}^{\infty} \mu_{\mathbb{N}}[A_n(\epsilon)]$. Thus the associated strong law was assured in this case by the Borel-Cantelli lemma.

In summary, we can anticipate that in any application for which a weak law holds, of critical importance will be the rate of convergence of $\mu_{\mathbb{N}}[A_n(\epsilon)] \to 0$. When this convergence is fast enough to assure convergence of $\sum_{n=1}^{\infty} \mu_{\mathbb{N}}[A_n(\epsilon)]$, a corresponding strong law will then also hold by the Borel-Cantelli theorem. In the absence of convergence, a definitive result is implied by the Borel-Cantelli lemma only in the case of independent $A_n(\epsilon)$-sets. Then the strong law will fail, and indeed fail decisively with $\mu_{\mathbb{N}}[\limsup[A_n(\epsilon)]] = 1$.

5.2   Convergence of Random Variables 1

Given a probability space $(\mathcal{S}, \mathcal{E}, \mu)$ and a sequence of random variables $\{X_n\}_{n=1}^{\infty}$ defined on $\mathcal{S}$, we are interested in the study of questions of convergence, $X_n \to X$, to some random variable X defined on $\mathcal{S}$. There are many definitions of convergence and in this section we focus on the two introduced in the prior section, plus a new mode of convergence. As it will generally not be the case that convergence occurs everywhere or nowhere on $\mathcal{S}$, it will be important to ensure that the set of points on which convergence occurs is in fact measurable, so that statements can be made about the probability of such convergence.

Generalizing the investigation in the prior section, in this section we investigate the following modes of convergence:

- $X_n \to_P X$: **convergence in probability**;

- $X_n \to_{a.s.} X$: **almost sure convergence**, or equivalently, $X_n \to_1 X$: **convergence with probability** 1;

- $X_n \to_d X$: **convergence in law** or **convergence in distribution**.

5.2.1 Notions of Convergence

Let $\{X_n\}_{n=1}^\infty$ and X be random variables defined on $(\mathcal{S}, \mathcal{E}, \mu)$. Generalizing (5.3), define:

$$A_n(\epsilon) \equiv \{|X_n - X| \geq \epsilon\}, \tag{5.16}$$

which is shorthand for $\{s \in \mathcal{S} | |X_n(s) - X(s)| \geq \epsilon\}$. Note that for any $\epsilon > 0$ and n, $A_n(\epsilon) \in \mathcal{E}$.

To see this, let $Y \equiv X_n - X$, a random variable on $(\mathcal{S}, \mathcal{E}, \mu)$. Then $A_n(\epsilon)$ is the pre-image of a Borel set:

$$A_n(\epsilon) = Y^{-1}\left\{(-\infty, \epsilon] \bigcup [\epsilon, \infty)\right\},$$

and consequently $\mu(A_n(\epsilon))$ is well defined for all $\epsilon > 0$ and n.

Definition 5.11 (Convergence in Probability) *Given a probability space $(\mathcal{S}, \mathcal{E}, \mu)$ and random variables $\{X_n\}_{n=1}^\infty$ and X, we say that X_n **converges to X in probability,** denoted $X_n \to_P X$, if $\lim_{n \to \infty} \mu(A_n(\epsilon)) = 0$ for every $\epsilon > 0$. In other words, for every $\epsilon > 0$:*

$$\lim_{n \to \infty} \mu(\{|X_n - X| \geq \epsilon\}) = 0. \tag{5.17}$$

If $\{X_n\}_{n=1}^\infty$ and X are random vectors defined on $\mathcal{S}$ with range in $\mathbb{R}^m$, then $|X_n - X|$ in (5.17) is interpreted in terms of the standard norm on $\mathbb{R}^m$. Specifically, if $X = (X^{(1)}, ..., X^{(m)})$ and similarly for X_n:

$$|X_n - X|^2 \equiv \sum_{j=1}^m \left(X_n^{(j)} - X^{(j)}\right)^2.$$

Remark 5.12 (On general measure spaces) *Convergence in probability has a counterpart in the more general measure space context. Then random variables are replaced by measurable functions, and one uses the terminology that f_n **converges to f in measure** if for all $\epsilon > 0$:*

$$\lim_{n \to \infty} \mu(\{|f_n - f| \geq \epsilon\}) = 0.$$

Exercise 5.13 (On "for all $\epsilon > 0$") *While the above definition states that (5.17) is satisfied for all $\epsilon > 0$, it is enough to know that this statement is true for all $\epsilon \in (0, \delta)$ for arbitrary $\delta > 0$, or for a sequence $\epsilon_j \to 0$. Prove this.*

Example 5.14 (Binomial probability space) *In the prior section, the probability space $(\mathcal{S}, \mathcal{E}, \mu)$ was an infinite product probability space $(Y^{\mathbb{N}}, \sigma(Y^{\mathbb{N}}), \mu_{\mathbb{N}})$ associated with the binomial space $(Y, \sigma(Y), \mu_B)$, where $Y = \{0, 1\}$ and μ_B is defined on Y as in (1.7):*

$$\mu_B(1) = p, \ \mu_B(0) = 1 - p.$$

Hence if $y \in Y^{\mathbb{N}}$, then y_j denotes the jth component which equals 0 or 1. The sequence of random variables $\{X_n\}_{n=1}^\infty$ on $Y^{\mathbb{N}}$ was defined in (5.2) and denoted M_n, and X was defined by the constant random variable, $X(y) = p$.

*In **Bernoulli's theorem** of Proposition 5.3, it was then shown that M_n **converges to p in probability,** so $X_n \to_P X$.*

In addition to convergence in probability, we are also interested in pointwise convergence, defined by the following set which generalizes (5.8):

$$C_S \equiv \left\{ \lim_{n \to \infty} X_n = X \right\}. \tag{5.18}$$

But before we can speak of the probability of this pointwise convergence set, it must be verified that $C_S \in \mathcal{E}$.

To this end, let ϵ_k be a countable sequence of real numbers with $\epsilon_k \to 0$, and let m be arbitrary. Then with m arbitrary:

$$
\begin{aligned}
C_S &= \bigcap_{\epsilon_k} \bigcup_{N \geq m} \bigcap_{n \geq N} \{|X_n - X| < \epsilon_k\} \\
&= \bigcap_{\epsilon_k} \bigcup_{N \geq m} \bigcap_{n \geq N} \widetilde{A}_n(\epsilon),
\end{aligned}
$$

and hence $C_S \in \mathcal{E}$ since $A_n(\epsilon) \in \mathcal{E}$. Thus we can investigate the probability of this event.

Definition 5.15 (Convergence with Probability 1) *Given a probability space $(\mathcal{S}, \mathcal{E}, \mu)$ and random variables $\{X_n\}_{n=1}^{\infty}$ and X, we say that X_n **converges to X with probability** 1, or, X_n **converges to X almost surely,** denoted $X_n \to_1 X$, or, $X_n \to_{a.s.} X$, if $\mu(C_S) = 1$. In other words:*

$$\mu \left(\left\{ \lim_{n \to \infty} X_n = X \right\} \right) = 1. \tag{5.19}$$

If $\{X_n\}_{n=1}^{\infty}$ and X are random vectors defined on $\mathcal{S}$ with range in $\mathbb{R}^m$, then $\{\lim_{n \to \infty} X_n = X\}$ in (5.19) is interpreted in terms of the standard norm on $\mathbb{R}^m$.

Remark 5.16 (On general measure spaces) *Convergence with probability 1 also has a counterpart in the more general measure space context. One then uses the terminology that f_n **converges to f almost everywhere,** or, f_n **converges to f except on a set of measure** 0, if:*

$$\mu \left(\left\{ \lim_{n \to \infty} f_n = f \right\} \right) = 1.$$

Example 5.17 (Binomial probability space) *Continuing with the notation of Example 5.14, it was shown in **Borel's theorem** of Proposition 5.9 that $X_n \equiv M_n$ **converges to** $X = p$ **with probability** 1, so $X_n \to_{a.s.} p$.*

We have the counterpart to Proposition 5.8 next.

Proposition 5.18 (Characterization of $X_n \to_1 X$) *With the notation above, $X_n \to_1 X$ if and only if for every $\epsilon > 0$:*

$$\mu \left(\limsup_n A_n(\epsilon) \right) = 0.$$

Proof. *This proof is identical to the proof of Proposition 5.8 where $\limsup_n A_n(\epsilon)$ was denoted $A_S(\epsilon)$, and is left as an exercise.* ∎

The final notion of convergence in this section requires no investigation into the measurability of a defining set, and thus in one sense is the simplest notion. For this definition, recall that by Proposition I.3.60 and Remark I.3.61, a distribution function F is increasing, right continuous, has left limits, and is thus continuous except for at most countably many points.

Definition 5.19 (Convergence in Distribution) *Given a probability space* $(\mathcal{S}, \mathcal{E}, \mu)$ *and random variables* $\{X_n\}_{n=1}^\infty$ *and* X *with associated distribution functions* $\{F_n\}_{n=1}^\infty$ *and* F, *we say that* X_n **converges in distribution** *to* X, *or*, **converges in law**, *denoted* $X_n \to_d X$, *or*, $X_n \Rightarrow X$, *if* $F_n(x) \to F(x)$ *at every continuity point of* F.

This definition applies directly if $\{X_n\}_{n=1}^\infty$ *and* X *are random vectors with range in* $\mathbb{R}^m$ *with associated joint distribution functions* $\{F_n\}_{n=1}^\infty$ *and* F.

This definition also applies in the general case where $\{X_n\}_{n=1}^\infty$ *and* X *are defined on different probability spaces,* $\{(\mathcal{S}_n, \mathcal{E}_n, \mu_n)\}_{n=1}^\infty$ *and* $(\mathcal{S}, \mathcal{E}, \mu)$.

Example 5.20 (On convergence in distribution) *Convergence in distribution is a very "weak" form of convergence, and not only because it is related to the concept of weak convergence of distribution functions of Chapter 8.*

The notions that $X_n \to_{a.s.} X$ *or* $X_n \to_P X$ *yield statements about* $X(s)$ *based on the values of* $X_n(s)$. *In the first case of convergence almost surely,* $X_n(s) \to X(s)$ *pointwise except (at most) on a set of probability zero. In the second case of convergence in probability,* $\mu\{|X_n - X| < \epsilon\} \to 1$ *for all* $\epsilon > 0$. *In other words, given any error tolerance* ϵ, X_n *will be within* ϵ *of* X *on a set with measure that approaches 1 as* $n \to \infty$.

The present notion, that $X_n \to_d X$, *only provides information on the behavior of the distribution functions, and no other "predictive" information about* X *based on the values of* X_n.

For example, let $\{X_n\}$ *be a sample of random variables defined on some* $(\mathcal{S}, \mathcal{E}, \mu)$ *with a given distribution function. In other words, these variates are independent and identically distributed as studied in Chapter 4. Then by construction,* $X_n \to_d X_j$ *for any fixed* j. *Thus, every sample of random variables converges in distribution to an arbitrary variate* X_j *in the sequence. But by definition, independence assures that there can be no "predictive" information about* X_j *based on these values of* X_n.

5.2.2 Convergence Relationships

The first result connects these modes of convergence in terms of "strength," meaning which modes imply other modes.

Proposition 5.21 ($X_n \to_{a.s.} X \Rightarrow X_n \to_P X \Rightarrow X_n \to_d X$) *Let* $\{X_n\}_{n=1}^\infty$, X *be random variables on* $(\mathcal{S}, \mathcal{E}, \mu)$.

1. *If* X_n *converges to* X *with probability 1, then* X_n *converges to* X *in probability. In other words:*

$$X_n \to_{a.s.} X \;\Rightarrow\; X_n \to_P X.$$

2. *If* X_n *converges to* X *in probability, then* X_n *converges to* X *in distribution. In other words:*

$$X_n \to_P X \;\Rightarrow\; X_n \to_d X.$$

Proof. *For property 1, (2.5) obtains that for any* $\epsilon > 0$:

$$\mu\left(\limsup_n A_n(\epsilon)\right) \ge \limsup_n \mu(A_n(\epsilon)).$$

Thus Proposition 5.18 assures that $\lim_n \mu(A_n(\epsilon)) = 0$, *which is convergence in probability by (5.17).*

For property 2, convergence $X_n \to_P X$ *means that for every* $\epsilon > 0$:

$$\lim_{n \to \infty} \mu(\{|X_n - X| \ge \epsilon\}) = 0. \tag{1}$$

As events in $\mathcal{S}$, for any x and $\epsilon > 0$:

$$\{X_n \le x\} \subset \{X \le x + \epsilon\} \bigcup \{|X_n - X| \ge \epsilon\},$$

and

$$\{X \le x - \epsilon\} \subset \{X_n \le x\} \bigcup \{|X_n - X| \ge \epsilon\}.$$

These are left as an exercise, and are justified by considering complements, that $A \subset B$ if and only if $\tilde{B} \subset \tilde{A}$, and applying de Morgan's laws.

By monotonicity and subadditivity of μ, this implies that:

$$F(x - \epsilon) - \mu(\{|X_n - X| \ge \epsilon\}) \le F_n(x) \le F(x + \epsilon) + \mu(\{|X_n - X| \ge \epsilon\}).$$

Letting $n \to \infty$, we cannot yet assume that $\lim F_n(x)$ exists, and thus we take a limit inferior and limit superior of the values in these inequalities (Definition I.3.42). Applying (1) and I.(3.14) obtains that for any $\epsilon > 0$:

$$F(x - \epsilon) \le \liminf F_n(x) \le \limsup F_n(x) \le F(x + \epsilon).$$

If x is a continuity point of $F(x)$, then both $F(x - \epsilon)$ and $F(x + \epsilon)$ converge to $F(x)$ as $\epsilon \to 0$, and thus $\lim F_n(x) = F(x)$ and $X_n \to_d X$. ∎

Example 5.22 (On Bernoulli's theorem: $X_n \to_P X \Rightarrow X_n \to_d X$) *Recall Example 5.14, that for any $\epsilon > 0$:*

$$\mu_\mathbb{N} [\{|X_n - p| \ge \epsilon\}] \to 0 \text{ as } n \to \infty,$$

where $X_n(y) \equiv \sum_{j=1}^n B_j(y)/n$. Thus $X_n \to_P p$.

*Then property 2 of the above proposition states that $X_n \to_d p$, where $X \equiv p$ is defined as the constant or "degenerate" random variable. The distribution function of X is given by $F(x) = \chi_{[p,\infty)}(x)$, the **characteristic function** of $[p, \infty)$, defined to be 1 on this set, and 0 elsewhere. Thus F has one discontinuity at $x = p$.*

Hence if F_n denotes the distribution function of X_n, then for $x \ne p$:

$$F_n(x) \to \chi_{[p,\infty)}(x).$$

That is, $F_n(x) \to 0$ for $x < p$ and $F_n(x) \to 1$ for $x > p$.

Regarding property 1 of Proposition 5.21, the average of n independent binomial random variables converges to p both with probability 1 by Borel's theorem of Proposition 5.9, and converges in probability by Bernoulli's theorem of Proposition 5.3. In general, however, convergence with probability 1 is stronger than convergence in probability.

Example 5.23 ($X_n \to_{a.s.} X \Rightarrow X_n \to_P X$, but not conversely) *Recall 1 of Exercise 1.29 to define the probability space $[0, 1]$ with Lebesgue measure m. On this space, define random variables $X_n = \chi_{I_n}(x)$, the characteristic functions of a collection of intervals I_n, so $\chi_{I_n}(x) = 1$ if $x \in I_n$ and $\chi_{I_n}(x) = 0$ otherwise. Define the intervals I_n as follows. Let $s_m = 2^{m+1} - 1$, so $s_0 = 1$ and $s_{m+1} - s_m = 2^{m+1}$. Define $I_1 = (0, 1]$, and for n with $s_m < n \le s_{m+1}$, define 2^{m+1} intervals by:*

$$I_{s_m+j} = \left(\frac{j-1}{2^{m+1}}, \frac{j}{2^{m+1}} \right], \quad j = 1, 2, ..., 2^{m+1}.$$

Then $X_n \to_P 0$, because for any ϵ with $0 < \epsilon < 1$:

$$\{|X_{s_m+j}(x)| \ge \epsilon\} = I_{s_m+j},$$

and $m(I_{s_m+j}) = 1/2^{m+1} \to 0$.

On the other hand, $X_n \not\to_{a.e.} 0$, since the convergence set:

$$C_S \equiv \{\lim_{n \to \infty} X_n(x) = 0\} = \{0\}.$$

To see this, let $x \neq 0$ be given. Then for any m there is a j so that $x \in I_{s_m+j}$, and hence $X_{s_m+j}(x) = 1$ infinitely often.

In summary, not only is $m(C_S) \neq 1$, but here $m(C_S) = 0$. Thus we have convergence in probability, and convergence with probability 0.

Regarding property 2 of Proposition 5.21, first $X_n \to_d X$ need not imply $X_n \to_P X$ simply by definition, since $X_n \to_d X$ does not even require that these random variables be defined on the same probability space. Convergence for $X_n \to_P X$ does require all variables be defined on the same space to make $\mu(\{|X_n - X| \geq \epsilon\})$ well defined.

But even without this definitional problem, $X_n \to_P X$ is a stronger form of convergence than $X_n \to_d X$.

Example 5.24 ($X_n \to_P X \Rightarrow X_n \to_d X$, but not conversely)

Let $\{X_n\}_{n=1}^{\infty}$ be a sequence of binomial random variables defined on some space $(\mathcal{S}, \mathcal{E}, \mu)$, with $\mu(X_n^{-1}(0)) = 1 - p$, $\mu(X_n^{-1}(1)) = p$. Let X be another binomial random variable with the same parameter p, independent of $\{X_n\}_{n=1}^{\infty}$. This collection is assured to exist by Proposition 4.17 with all independent binomials by defining $(\mathcal{S}, \mathcal{E}, \mu) = (Z^{\mathbb{N}}, \sigma(Z^{\mathbb{N}}), \mu_{\mathbb{N}})$, $X = Y_1$ and $X_n = Y_{n+1}$ for all n. Alternatively, for $\{X_n\}_{n=1}^{\infty}$ not independent, we can also define on this space $X = Y_1$ and $X_n = Y_2$ for all n.

Then $X_n \to_d X$ because $F_n(x) = F(x)$ for all x. But $X_n \not\to_P X$ because for any $0 < \epsilon < 1$:

$$\mu(\{|X_n - X| \geq \epsilon\}) = 2p(1 - p),$$

for all n by independence.

While Example 5.23 makes it clear that $X_n \to_P X$ does not in general imply that $X_n \to_{a.e.} X$, a closer look at this example reveals that there are many subsequences of random variables $\{X_{n_m}\}$, so that $X_{n_m} \to_{a.e.} X$. For example, define X_{n_m} with $n_m = s_m \equiv 2^{m+1} - 1$ for $m \geq 1$. Then $X_{n_m} = \chi_{I_{n_m}}(x)$ equals 1 on $(0, 1/2^{m+1}]$ and is 0 elsewhere, and thus $X_{n_m} \to_{a.e.} 0$. In fact here, convergence is **everywhere**.

The following result generalizes this example with the aid of the **Borel-Cantelli lemma**.

Proposition 5.25 ($X_n \to_P X \Rightarrow X_{n_m} \to_{a.e} X$) Let $\{X_n\}_{n=1}^{\infty}$, X be random variables on $(\mathcal{S}, \mathcal{E}, \mu)$.

If X_n converges to X in probability, then there exists a subsequence $\{X_{n_m}\}_{m=1}^{\infty}$ that converges to X with probability 1. In other words:

$$X_n \to_P X \Rightarrow X_{n_m} \to_{a.e} X ,$$

for some subsequence $\{X_{n_m}\}_{m=1}^{\infty}$.

Proof. If $X_n \to_P X$, then $\mu\{|X_n - X| \geq \epsilon\} \to 0$ as $n \to \infty$ for any $\epsilon > 0$. Letting $\epsilon = 2^{-m}$, choose n_m so that $\mu\{|X_n - X| \geq 2^{-m}\} \leq 2^{-m}$ for $n \geq n_m$. We can assume that $\{n_m\}_{m=1}^{\infty}$ is increasing by using $n'_m \equiv \max_{k \leq m}\{n_m\}$. Define $A_m = \{|X_{n_m} - X| \geq 2^{-m}\}$, and note that $\sum_{m=1}^{\infty} \mu(A_m) < \infty$.

By the Borel-Cantelli lemma, $\mu(\limsup A_m) = 0$, or equivalently with $B^c \equiv \widetilde{B}$ the complement of B:

$$\mu[(\limsup A_m)^c] = 1.$$

Applying de Morgan's laws to (2.1):

$$(\limsup A_m)^c = \bigcup_{m=1}^{\infty} \bigcap_{k \geq m}^{\infty} \tilde{A}_k$$
$$= \bigcup_{m=1}^{\infty} \bigcap_{k \geq m}^{\infty} \{|X_{n_k} - X| < 2^{-k}\}.$$

Thus if $s \in (\limsup A_m)^c$, it follows that $s \in \bigcap_{k \geq m}^{\infty} \{|X_{n_k} - X| < 2^{-k}\}$ for some m, and thus $X_{n_k}(s) \to X(s)$.

In other words, $(\limsup A_m)^c$ is the convergence set of X_{n_m} to X, and this set has probability 1, and so $X_{n_m} \to_{a.e} X$. ∎

Remark 5.26 ($X_n \to_d X \nRightarrow X_{n_m} \to_P X$) *Example 5.24 demonstrates that not only does $X_n \to_d X$ not imply that $X_n \to_P X$, but there is no subsequence of X_n for which such convergence in probability holds. Thus there is no counterpart to Proposition 5.25 in this case.*

The next result shows that for a constant (also called "degenerate") random variable X, convergence in distribution to X and convergence in probability to X are equivalent. In other words, if $X \equiv a$ is a constant random variable, meaning $F(x) = \chi_{[a,\infty)}(x)$, then $X_n \to_P X$ if and only if $X_n \to_d X$.

This seems surprising initially, since the notion that $X_n \to_P X$ requires all random variables be defined on a common probability space, whereas $X_n \to_d X$ does not require this. But note that when $X = a$ that $X_n \to_P a$ makes sense even when the random variables $\{X_n\}$ are not defined on a common probability space. Indeed,

$$\{|X_n - a| \geq \epsilon\} = \{X_n \geq a + \epsilon\} \bigcup \{X_n \leq a - \epsilon\},$$

and so the limit $\lim_{n \to \infty} \mu_n(\{|X_n - a| \geq \epsilon\})$ is well defined even with different probability spaces $\{(\mathcal{S}_n, \mathcal{E}_n, \mu_n)\}_{n=1}^{\infty}$.

Proposition 5.27 ($X_n \to_P a$ **if and only if** $X_n \to_d a$) *If $\{X_n\}_{n=1}^{\infty}$ is a sequence of random variables on $(\mathcal{S}, \mathcal{E}, \mu)$, then $X_n \to_P a$ if and only if $X_n \to_d a$:*

$$X_n \to_P a \iff X_n \to_d a.$$

Proof. *By Proposition 5.21, there is only left to prove that $X_n \to_d a$ implies $X_n \to_P a$. As noted above, the event $\{|X_n - a| \geq \epsilon\}$ is the union of $\{X_n \geq a + \epsilon\}$ and $\{X_n \leq a - \epsilon\}$, and since $\{X_n \geq a + \epsilon\} \subset \{X_n > a + \epsilon/2\}$, finite subadditivity obtains:*

$$\mu_n(\{|X_n - a| \geq \epsilon\}) \leq \mu_n(\{X_n \leq a - \epsilon\}) + \mu_n(X_n > a + \epsilon/2)$$
$$= F_n(a - \epsilon) + 1 - F_n(a + \epsilon/2).$$

But $X_n \to_d a$ implies that for all $x \neq a$, which is the only discontinuity point of $F(x) = \chi_{[a,\infty)}(x)$, that $F_n(x) \to 0$ for $x < a$ and $F_n(x) \to 1$ for $x > a$. Thus $\mu_n(|X_n(s) - a| \geq \epsilon\}) \to 0$ as $n \to \infty$, and so $X_n \to_P a$. ∎

The last result of this section is a version of the **continuous mapping theorem,** and more generally known as the **mapping theorem,** which will be more fully developed in Section 8.5 using the machinery provided by **Skorokhod's representation theorem.** This result will again be generalized in Book VI as an application of the **portmanteau theorem** on weak convergence of measures in $\mathbb{R}^n$.

Proposition 5.28 (A continuous mapping theorem) *If $\{X_n\}_{n=1}^\infty$ is a sequence of random variables on $(\mathcal{S}, \mathcal{E}, \mu)$, and g is a continuous function on $\mathbb{R}$, then:*

1. *$X_n \to_P a$ implies that $g(X_n) \to_P g(a)$;*
2. *$X_n \to_d a$ implies that $g(X_n) \to_d g(a)$.*

Proof. *First note that by continuity of g, $\{g(X_n)\}_{n=1}^\infty$ is a sequence of random variables on $(\mathcal{S}, \mathcal{E}, \mu)$. To see this, recall that if $A \in \mathcal{B}(\mathbb{R})$, then $g^{-1}(A) \in \mathcal{B}(\mathbb{R})$ by Proposition I.3.12, and hence:*

$$g(X_n)^{-1}(A) = X_n^{-1}\left(g^{-1}(A)\right) \in \mathcal{E}.$$

Thus by Proposition 5.27, only one of these statements requires proof.

To prove part 1, it follows from continuity of g at a that given $\epsilon > 0$, there is a δ so that $|g(X_n) - g(a)| < \epsilon$ if $|X_n - a| < \delta$. As a statement on the probability space $(\mathcal{S}, \mathcal{E}, \mu)$, this implies:

$$\{|X_n - a| < \delta\} \subset \{|g(X_n) - g(a)| < \epsilon\},$$

and hence by monotonicity of μ:

$$\mu\{|X_n - a| < \delta\} \leq \mu\{|g(X_n) - g(a)| < \epsilon\}.$$

Thus if $X_n \to_P a$, then $\mu\{|X_n - a| < \delta\} \to 1$, and then also $\mu\{|g(X_n) - g(a)| < \epsilon\} \to 1$. In other words, $g(X_n) \to_P g(a)$. ∎

5.2.3 Slutsky's Theorem

Slutsky's theorem, named for **Evgeny "Eugen" Slutsky** (1880–1948), addresses the following question. If $X_n \to_d X$ and $Y_n \to Y$, where "$\to$" denotes convergence in some manner, does $X_n + Y_n \to_d X + Y$ or $X_n Y_n \to_d XY$? Asked differently, is there a mode of convergence: $Y_n \to Y$, for which convergence in distribution of $X_n \to_d X$ is preserved in these other sequences?

The following version of Slutsky's theorem provides affirmative results when $Y_n \to_P a$, or equivalently by Proposition 5.27, $Y_n \to_d a$. But this latter "weaker" mode of convergence is a special case of this result that does not generalize. See Example 5.31 and Remark 5.32.

Thus the following statement emphasizes the assumption $Y_n \to_P a$.

Proposition 5.29 (Slutsky's theorem) *Given sequences of random variables $\{X_n\}_{n=1}^\infty$ and $\{Y_n\}_{n=1}^\infty$ defined on $(\mathcal{S}, \mathcal{E}, \mu)$ with $X_n \to_d X$ and $Y_n \to_P a$, then:*

1. *$X_n + Y_n \to_d X + a$;*
2. *$X_n Y_n \to_d aX$;*
3. *$X_n/Y_n \to_d X/a$, if $a \neq 0$.*

By Proposition 5.27, these conclusions also follow if $Y_n \to_d a$.
Proof. *For statement 1, note that for any $\epsilon > 0$:*

$$
\begin{aligned}
F_{X_n + Y_n}(x) &\leq \mu\left(\{X_n + Y_n \leq x\} \bigcap \{|Y_n - a| \leq \epsilon\}\right) + \mu\{|Y_n - a| > \epsilon\} \\
&\leq \mu\{X_n \leq x - a + \epsilon\} + \mu\{|Y_n - a| > \epsilon\}.
\end{aligned}
$$

These inequalities and those below are justified by set inclusions and monotonicity of measures, and left as an exercise. Since $Y_n \to_P a$ and $X_n \to_d X$, there exists $\epsilon_j \to 0$ so that:

$$\limsup F_{X_n + Y_n}(x) \leq \limsup F_{X_n}(x - a + \epsilon_j) = F_X(x - a + \epsilon_j).$$

The last step uses an equality by choosing $x - a + \epsilon_j$ to be a continuity point of F_X. These continuity points are dense in $\mathbb{R}$ as noted in Notation 3.5, and thus for any x we can choose $\epsilon_j \to 0$ as noted.

Similarly,

$$
\begin{aligned}
1 - F_{X_n + Y_n}(x) &\leq \mu\left(\{X_n + Y_n > x\} \bigcap \{|Y_n - a| \leq \epsilon\}\right) + \mu\left(\{|Y_n - a| > \epsilon\}\right) \\
&\leq \mu\{X_n > x - a - \epsilon\} + \mu\{|Y_n - a| > \epsilon\}.
\end{aligned}
$$

With the same justifications, we choose $\epsilon'_j \to 0$ depending on x so that:

$$
\begin{aligned}
\limsup\left(1 - F_{X_n + Y_n}(x)\right) &\leq \limsup \mu\{X_n > x - a - \epsilon'_j\} \\
&= 1 - F_X(x - a - \epsilon'_j),
\end{aligned}
$$

and so:

$$
\limsup\left(-F_{X_n + Y_n}(x)\right) \leq -F_X(x - a - \epsilon'_j).
$$

Now $\limsup[-f_n] = -\liminf[f_n]$ by Remark I.3.44, and this obtains:

$$
\liminf F_{X_n + Y_n}(x) \geq F_X(x - a - \epsilon'_j).
$$

Combining yields:

$$
\begin{aligned}
F_X(x - a - \epsilon'_j) &\leq \liminf F_{X_n + Y_n}(x) \\
&\leq \limsup F_{X_n + Y_n}(x) \leq F_X(x - a + \epsilon_j).
\end{aligned}
$$

Since $F_X(y - a) = F_{X+a}(y)$, this yields:

$$
\begin{aligned}
F_{X+a}(x - \epsilon'_j) &\leq \liminf F_{X_n + Y_n}(x) \\
&\leq \limsup F_{X_n + Y_n}(x) \leq F_{X+a}(x + \epsilon_j).
\end{aligned}
$$

If x is a continuity point of F_{X+a}, then letting the associated $\epsilon_j, \epsilon'_j \to 0$ obtains that $\lim F_{X_n + Y_n}(x) = F_{X+a}(x)$.

For statement 2, write $X_n Y_n = aX_n + (Y_n - a)X_n$. Now $aX_n \to_d aX$, which is left as an exercise. This result will then follow from part 1 once it is proved that $(Y_n - a)X_n \to_P 0$.

To this end, let $\epsilon > 0$ and $\delta > 0$. Then by finite additivity and monotonicity:

$$
\begin{aligned}
\mu\{(Y_n - a)X_n > \epsilon/2\} &= \mu\left(\{|(Y_n - a)X_n| > \epsilon/2\} \bigcap \{|Y_n - a| \leq \delta\}\right) \\
&\quad + \mu\left(\{|(Y_n - a)X_n| > \epsilon/2\} \bigcap \{|Y_n - a| > \delta\}\right) \\
&\leq \mu\left(\{|(Y_n - a)X_n| > \epsilon/2\} \bigcap \{|Y_n - a| \leq \delta\}\right) + \mu\{|Y_n - a| > \delta\} \\
&\leq \mu\left(\{|X_n| > \epsilon/2\delta\}\right) + \mu\{|Y_n - a| > \delta\}.
\end{aligned}
$$

The last step reflects that for $x, y \geq 0$:

$$
\{yx > \epsilon/2\} \bigcap \{y \leq \delta\} \subset \{x > \epsilon/2\delta\}.
$$

Because $Y_n \to_P a$, the second term above converges to zero as $n \to \infty$ for any fixed δ. If δ is chosen so that $\epsilon/2\delta$ is a continuity point of F_X, then the above inequality obtains:

$$
\limsup \mu\{(Y_n - a)X_n > \epsilon/2\} \leq 1 - F_X(\epsilon/2\delta).
$$

Because such continuity points $\epsilon/2\delta$ are dense, this upper bound can be made arbitrarily small by choosing $\epsilon/2\delta$ large, and thus:

$$\lim \mu\{(Y_n - a)X_n \geq \epsilon/2\} = 0.$$

This proves that $(Y_n - a)X_n \to_P 0$ as claimed.

The result in statement 3 follows from 2 if it can be shown that $1/Y_n \to_P 1/a$ when $a \neq 0$. First:

$$\{|1/Y_n - 1/a| \geq \epsilon\} = \{1/Y_n \geq 1/a + \epsilon\} \bigcup \{1/Y_n \leq 1/a - \epsilon\}.$$

Using a Taylor series analysis, then for $0 < \xi < \epsilon$:

$$\begin{aligned} a/(1+a\epsilon) &= a - a^2\epsilon + a^3\epsilon^2 - a^4\epsilon^3(1+a\xi)^{-4} \\ &< a - \left(a^2\epsilon - a^3\epsilon^2\right). \end{aligned}$$

Thus:

$$\{1/Y_n \geq 1/a + \epsilon\} = \{Y_n \leq a/(1+a\epsilon)\} \subset \{Y_n \leq a - \left(a^2\epsilon - a^3\epsilon^2\right)\}.$$

Analogously:

$$\{1/Y_n \leq 1/a - \epsilon\} \subset \{Y_n \geq a + \left(a^2\epsilon - a^3\epsilon^2\right)\},$$

and so:

$$\{|1/Y_n - 1/a| \geq \epsilon\} \subset \{|Y_n - a| \geq a^2\epsilon - a^3\epsilon^2\}.$$

If $a < 0$, then $a^2\epsilon - a^3\epsilon^2 > 0$ for all $\epsilon > 0$ and thus $\mu\{|1/Y_n - 1/a| \geq \epsilon\} \to 0$ because $Y_n \to_P a$. For $a > 0$, $a^2\epsilon - a^3\epsilon^2 > 0$ for all $\epsilon < 1/a$, and again $\mu\{|1/Y_n - 1/a| \geq \epsilon\} \to 0$ for all such ϵ. By Exercise 5.13, this is enough to prove that $Y_n \to_P a$. ∎

Exercise 5.30 (A generalization) *Given sequences of random variables $\{X_n\}_{n=1}^{\infty}$, $\{Y_n\}_{n=1}^{\infty}$ and $\{Z_n\}_{n=1}^{\infty}$ defined on $(\mathcal{S}, \mathcal{E}, \mu)$, with $X_n \to_d X$, $Y_n \to_P a$ and $Z_n \to_P b$, prove that:*

$$X_n Y_n + Z_n \to_d aX + b.$$

Example 5.31 (On $Y_n \to_d Y$) *If $X_n \to_d X$ and $Y_n \to_d Y$, then it need not be the case that $X_n + Y_n \to_d X + Y$ when Y is not constant.*

For example, let X and Y be independent, non-degenerate random variables on $(\mathcal{S}, \mathcal{E}, \mu)$ with the same distribution function. Define $X_n = X$ and $Y_n = -X$. Then $X_n \to_d X$ and $Y_n \to_d -Y$, simply because all variates have the same distribution function. Then $X_n + Y_n \equiv 0$, and so $F_{X_n+Y_n}(x) = \chi_{[0,\infty)}(x)$, the characteristic function of $[0,\infty)$. Since independent, $X - Y$ cannot have this distribution function.

Similarly, if $X_n \to_d X$ and $Y_n \to_d Y$, it need not be the case that $X_n Y_n \to_d XY$ when Y is not constant. Again, let X and Y be independent, non-degenerate random variables with the same distribution, and also $X(\mathcal{S}) \subset (0,\infty)$, say. Then with $X_n = X$ and $Y_n = 1/X$, it again follows that $X_n \to_d X$, $Y_n \to_d 1/Y$, and $X_n Y_n \equiv 1$. Thus $F_{X_n+Y_n}(x) = \chi_{[1,\infty)}(x)$, while by independence XY cannot have this distribution function.

Remark 5.32 (Book VI generalization) *The above statement of Slutsky's theorem is the most common version, but it can be significantly generalized to statements about random vectors, and more generally stated in a form reminiscent of the continuous mapping theorem. For example, if $X_n \to_d X$, $Y_n \to_P a$, and $g(x,y)$ is a continuous function, then $g(X_n, Y_n) \to_d g(X, a)$. These versions will be addressed in Book VI using the **Portmanteau theorem** on weak convergence of measures in $\mathbb{R}^n$. This theorem will provide an alternative characterization of convergence in distribution using the integration theory of Book V.*

5.2.4   Kolmogorov's Zero-One Law

Kolmogorov's zero-one law is named for **Andrey Kolmogorov** (1903–1987) as noted in Section 2.3. It states that the probability of every **tail event** is 0 or 1, when the tail sigma algebra $\mathcal{T} \equiv \mathcal{T}(\{A_n\}_{n=1}^{\infty})$ is defined in terms of independent events $\{A_n\}_{n=1}^{\infty}$. The generalization pursued in this section is to associate the convergence of a sequence of random variables with a tail event, and then conclude that such convergence occurs with probability 0 or 1.

In order to demonstrate this result, the first step is to restate Proposition 1.27 to produce a result on independent random variables. Recall Definition 3.43 for the sigma algebra $\sigma(X_{j1}, X_{j2}, X_{j3}, ..., X_{jM_j})$, and Definition 1.15 for the notion of independent sigma algebras.

Proposition 5.33 (Independent RVs and sigma algebras) *Given a probability space* $(\mathcal{S}, \mathcal{E}, \mu)$, *let* $\{X_{jk}\}$ *be independent random variables on* $\mathcal{S}$ *with* $1 \le j \le N$ *and* $1 \le k \le M_j$, *where* N *and* M_j *are finite or infinite (countable or uncountable).*

Then $\{\sigma(X_{j1}, X_{j2}, X_{j3}, ..., X_{jM_j})\}_{j=1}^{N}$ *are independent sigma algebras.*

Proof. *Define* A_j *to be the class of finite intersections of sets of the form* $X_{jk}^{-1}(B)$ *with* $B \in \mathcal{B}(\mathbb{R})$ *and* $1 \le k \le M_j$. *That is:*

$$\mathsf{A}_j \equiv \left\{ \bigcap_K X_{jk_i}^{-1}(B_i) \right\},$$

with $K = (k_1, ..., k_n) \subset (1, 2, ..., M_j)$, *where* $\{k_i\}_{i=1}^{n}$ *are distinct, and* $\{B_i\}_{i=1}^{n} \subset \mathcal{B}(\mathbb{R})$. *Then each* A_j *is closed under finite intersections by definition, and* $\{\mathsf{A}_j\}$ *is a collection of independent classes because the original random variables are independent.*

For example, given sets in A_{j_1} *and* A_{j_2}, *since* $\{X_{jk}\}$ *are independent random variables:*

$$\mu\left(\left(\bigcap_{K_1} X_{j_1 k_i}^{-1}(B_i) \right) \bigcap \left(\bigcap_{K_2} X_{j_2 k_i}^{-1}(B_i') \right) \right)$$
$$= \prod_{K_1} \mu\left(X_{j_1 k_i}^{-1}(B_i) \right) \prod_{K_2} \mu\left(X_{j_1 k_i}^{-1}(B_i') \right)$$
$$= \mu\left(\bigcap_{K_1} X_{j_1 k_i}^{-1}(B_i) \right) \mu\left(\bigcap_{K_2} X_{j_1 k_i}^{-1}(B_i') \right).$$

This extends to any finite collection of sets in $\{\mathsf{A}_k\}_{k=1}^{N}$ *Hence by Proposition 1.27 as generalized in Remark 1.28,* $\{\sigma(\mathsf{A}_j)\}_{j=1}^{N}$ *are mutually independent.*

Now $\sigma(X_{jk}) \subset \sigma(\mathsf{A}_j)$ *for all* k *by definition, and so* $\sigma\left(\{\sigma(X_{jk})\}_{k=1}^{M_j} \right) \subset \sigma(\mathsf{A}_j)$. *The reverse inclusion follows from the definition of* A_j, *and thus:*

$$\sigma(\mathsf{A}_j) = \sigma\left(\{\sigma(X_{jk})\}_{k=1}^{M_j} \right).$$

Finally, recalling (3.24):

$$\sigma\left(\{\sigma(X_{jk})\}_{k=1}^{M_j} \right) = \sigma(X_{j1}, X_{j2}, X_{j3}, ..., X_{jM_j}),$$

and the result follows. ∎

To state Kolmogorov's zero-one law, the notion of "tail event" in Definition 2.12 needs to be adapted to the current context of random variables.

Definition 5.34 ($\mathcal{T} \equiv \mathcal{T}(\{X_n\}_{n=1}^{\infty})$ **and tail events)** *Let* $\{X_n\}_{n=1}^{\infty}$ *be a collection of random variables on* $(\mathcal{S}, \mathcal{E}, \mu)$. *The* **tail sigma algebra associated with** $\{X_n\}_{n=1}^{\infty}$, *denoted* $\mathcal{T} \equiv \mathcal{T}\left(\{X_n\}_{n=1}^{\infty}\right)$, *is defined by:*

$$\mathcal{T} = \bigcap_{n=1}^{\infty} \sigma(X_n, X_{n+1}, X_{n+2}, ...), \tag{5.20}$$

where as in Definition 3.43, $\sigma(X_n, X_{n+1}, X_{n+2}, ...)$ is the smallest sigma algebra with respect to which each X_j is measurable for $j \geq n$.

*A **tail event** is any set $A \in \mathcal{T}$.*

Example 5.35 (Random variable convergence and tail events) *Perhaps surprisingly, given a collection of random variables $\{X_n\}_{n=1}^{\infty}$ defined on $(\mathcal{S}, \mathcal{E}, \mu)$, many but not all sets related to the convergence of this sequence are tail events.*

1. ***Convergence of a Sum to a Constant:***

   Given $r \in \mathbb{R}$ define:

   $$A_r \equiv \left\{ \lim_{n \to \infty} \sum_{k=1}^{n} X_k(s) = r \right\}.$$

   Then A_r can be characterized:

   $$A_r = \bigcap_{\epsilon_k} \bigcup_{N \geq 1} \bigcap_{n \geq N} \left\{ \left| \sum_{k=1}^{n} X_k(s) - r \right| < \epsilon_k \right\},$$

   where ϵ_k is a sequence with $\epsilon_k \to 0$. Now for every ϵ_k, it follows by definition that:

   $$\bigcup_{N \geq 1} \bigcap_{n \geq N} \left\{ \left| \sum_{k=1}^{n} X_k(s) - r \right| < \epsilon_k \right\} \in \sigma(X_1, X_2, X_3, ...).$$

   Further, it cannot in general be the case that for given $j > 1$:

   $$\bigcup_{N \geq 1} \bigcap_{n \geq N} \left\{ \left| \sum_{k=1}^{n} X_k(s) - r \right| < \epsilon_k \right\} \in \sigma(X_j, X_{j+1}, X_{j+2}, ...).$$

   Otherwise, this would imply that $\lim_{n \to \infty} \sum_{k=1}^{n} X_k(s) = r$ if and only if $\lim_{n \to \infty} \sum_{k=j}^{n} X_k(s) = r$ for such j, and this is possible only when $X_k(s) \equiv 0$ for $k < j$. Thus in general, $A_r \notin \mathcal{T}$.

2. ***Convergence of an Average to a Constant:***

   Assume that $\{X_n\}_{n=1}^{\infty}$ defined on $(\mathcal{S}, \mathcal{E}, \mu)$ are uniformly bounded pointwise, meaning $|X_n(s)| \leq K(s) < \infty$ for all n. Given $r \in \mathbb{R}$ and $\epsilon_k \to 0$ as above, define:

   $$
   \begin{aligned}
   M_r &= \left\{ \lim_{n \to \infty} \left[\sum_{k=1}^{n} X_k(s)/n \right] = r \right\} \\
   &= \bigcap_{\epsilon_k} \bigcup_{N \geq 1} \bigcap_{n \geq N} \left\{ \left| \sum_{k=1}^{n} X_k(s)/n - r \right| < \epsilon_k \right\}.
   \end{aligned}
   $$

   We claim that for given s, $\sum_{k=1}^{n} X_k(s)/n \to r$ if and only if for any fixed m, $\sum_{k=m}^{n} X_k(s)/n \to r$. Certainly the latter implies the former statement, so assume the former. Then for any $\epsilon > 0$ there exists N so that for $n \geq N$:

   $$\left| \sum_{k=1}^{n} X_k(s)/n - r \right| < \epsilon.$$

But $|a - b| \geq |a| - |b|$, *and so for any* m:

$$\left| r - \sum\nolimits_{k=m}^{n} X_k(s)/n \right| - \left| \sum\nolimits_{k=1}^{m-1} X_k(s)/n \right| \leq \left| r - \sum\nolimits_{k=1}^{n} X_k(s)/n \right|.$$

Uniform boundedness assures that the middle term can be made arbitrarily small for n *large. Thus:*

$$\left\{ \left| \sum\nolimits_{k=m}^{n} X_k(s)/n - r \right| < 2\epsilon \right\},$$

and so $\sum_{k=m}^{n} X_k(s)/n \to r$. *Hence for any* m:

$$M_r = \bigcap_{\epsilon_k} \bigcup_{N \geq m} \bigcap_{n \geq N} \left\{ \left| \sum\nolimits_{k=m}^{n} X_k(s)/n - r \right| < \epsilon_k \right\},$$

and since

$$\bigcup_{N \geq m} \bigcap_{n \geq N} \left\{ \left| \sum\nolimits_{k=m}^{n} X_k(s)/n - r \right| < \epsilon_k \right\} \in \sigma(X_m, X_{m+1}, X_{m+2}, ...),$$

it follows that $M_r \in \mathcal{T}$ *for pointwise uniformly bounded* X_k.

3. **Convergence of a Sum**

While convergence of $\sum_{k=1}^{n} X_k(s)$ *to a given constant* r *is not in general a tail event, the set on which this series converges to any value is a tail event. Specifically, define:*

$$A = \left\{ \lim_{n \to \infty} \sum\nolimits_{k=1}^{n} X_k(s) \text{ exists and is finite} \right\}.$$

By the Cauchy criterion, a series $\lim_{n \to \infty} \sum_{k=1}^{n} a_k$ *converges if and only if for any* $\epsilon > 0$ *there is an* N *so that for all* $n > m > N$:

$$\left| \sum\nolimits_{k=m}^{n} a_k \right| < \epsilon.$$

Hence, with $\epsilon_k \to 0$ *as above and any* m:

$$A = \bigcap_{\epsilon_k} \bigcup_{N \geq m} \bigcap_{n \geq N} \left\{ \left| \sum\nolimits_{k=m}^{n} X_k(s) \right| < \epsilon_k \right\},$$

and $A \in \mathcal{T}$.

We now state and prove **Kolmogorov's zero-one law.**

Proposition 5.36 (Kolmogorov's zero-one law) *Given a probability space* $(\mathcal{S}, \mathcal{E}, \mu)$ *and independent random variables* $\{X_j\}_{j=1}^{\infty}$, *then for any* $A \in \mathcal{T} \equiv \mathcal{T}(\{X_n\}_{n=1}^{\infty})$:

$$\mu(A) \in \{0, 1\}. \tag{5.21}$$

Proof. *The result will follow from Proposition 1.18 if it is proved that every such* A *is independent of itself. From Proposition 5.33:*

$$\sigma(X_1), \ \sigma(X_2), ..., \ \sigma(X_{n-1}), \ \sigma(X_n, X_{n+1}, X_{n+2}, ...),$$

are independent classes for any n. *Since* $A \in \sigma(X_n, X_{n+1}, X_{n+2}, ...)$, *it follows that* $A, \sigma(X_1), \sigma(X_2), ..., \sigma(X_{n-1})$ *is an independent collection for any* n. *Hence* $\{A, \{\sigma(X_j)\}_{j=1}^{\infty}\}$ *is an independent collection by definition of independence for infinite collections.*

Another application of Proposition 5.33 yields that $\sigma(A)$ *and* $\sigma(\{\sigma(X_j)\}_{j=1}^{\infty}) = \sigma(X_1, X_2, ...)$ *are independent classes. But since* $A \in \mathcal{T} \subset \sigma(X_1, X_2, ...)$, *and by definition* $A \in \sigma(A)$, *we conclude that* A *is independent of itself and the result follows.* ∎

mark 5.37 (On Proposition 2.15) *This version of Kolmogorov's zero-one law is in* *t a generalization of that given in Proposition 2.15. To see this, given a probability space* $, \mathcal{E}, \mu)$ *and mutually independent events* $\{A_j\}_{j=1}^{\infty}$, *define random variables* $\{X_j\}_{j=1}^{\infty}$ *by:*

$$X_j(s) = \begin{cases} 1, & s \in A_j, \\ 0, & s \notin A_j. \end{cases}$$

e random variable $X_j(s)$ *has been seen before. It is the **indicator function** or **charac-** **ristic function** of the set* A_j, *and denoted* $1_{A_j}(s)$, $I_{A_j}(s)$ *or* $\chi_{A_j}(s)$. *As before, each* (s) *is a random variable since* $A_j \in \mathcal{E}$.
Then for any j:

$$\sigma(X_j) = \{\emptyset, A_j, \widetilde{A}_j, \mathcal{S}\}.$$

us $\{\sigma(X_j)\}_{j=1}^{\infty}$ *are independent sigma algebras by Proposition 1.19 and thus by definition,* $_j\}_{j=1}^{\infty}$ *are independent random variables.*
The tail sigma algebras defined by (2.7) and (5.20) are equivalent because of (3.24). In *er words, because* $\sigma(X_j) = \sigma(A_j)$, *(3.24) assures that:*

$$\bigcap_{n=1}^{\infty} \sigma(A_n, A_{n+1}, A_{n+2}, ...) = \bigcap_{n=1}^{\infty} \sigma(X_n, X_{n+1}, X_{n+2}, ...).$$

nce,

$$\mathcal{T}(\{A_n\}_{n=1}^{\infty}) = \mathcal{T}(\{X_n\}_{n=1}^{\infty}).$$

In summary, the earlier statement of Kolmogorov's zero-one law can be interpreted as a *cial case of this last result, when the random variables are defined as indicator functions* *independent sets.*

While Kolmogorov's conclusion is very powerful in limiting the potential values of $\mu(A)$ $A \in \mathcal{T}$, it provides no insight in a given application as to which outcome is correct, nor *w* one can determine this outcome.

ample 5.38 (Convergence of an average; convergence set) *Let* M_r *be defined as* *part 2 of Example 5.35 above:*

$$M_r \equiv \left\{ \lim_{n \to \infty} \left[\sum_{k=1}^{n} X_k(s)/n \right] = r \right\}.$$

en M_r *is a tail event for uniformly bounded* X_k, *so by Kolmogorov's zero-one law we* *clude that for any* r, $\mu(M_r) \in \{0, 1\}$. *Clearly, there can be at most one* r *for which* $M_r) = 1$, *so Kolmogorov's zero-one law assures that either* $\mu(M_r) = 0$ *for all* r, *or there* *a unique* r *for which* $\mu(M_r) = 1$. *Further, if* $|X_k| \leq K$, *then this unique* r, *when it exists,* *o satisfies* $|r| \leq K$.
In the case of the binomial probability space in Section 5.1, the unique value of r *was* *nd in Borel's theorem of Proposition 5.9 to be* p, *where* $p \equiv \mu(X^{-1}(1))$.
*In Book IV we will see that this example generalizes in the **weak law of large numbers*** *identify the unique value of* r *when it exists. In the special case where the random variables* *e independent and have the same distribution function* $F(x)$, *we will see that the unique* *ue of* r *is given as the **expectation of** X *when this exists. The expectation of* X *will be* *roduced in Book IV once integrals are developed in Book III. This definition will not be* *npletely formalized until Book VI, using the more general integration theory of Book V.*
As another example, with A *defined in part 3 of Example 5.35 to be the tail event for* *ich* $\lim_{n \to \infty} \sum_{k=1}^{n} X_k(s)$ *exists, we again conclude by Kolmogorov's zero-one law that* $4) \in \{0, 1\}$.

Distribution Functions and Borel Measures

this chapter we show that all the distribution functions introduced in Chapter 3 induce rel probability measures in the respective range spaces $\mathbb{R}$ or $\mathbb{R}^n$. Specifically, we address tribution functions that are:

1. univariate (Definition 3.1);

2. joint (Definition 3.28);

3. marginal (Definition 3.34);

4. conditional (Definition 3.39).

We also seek to confirm that given a function F on $\mathbb{R}$ or $\mathbb{R}^n$ with the characteristic operties of such a distribution function, there exists a probability space $(\mathcal{S}, \mathcal{E}, \mu)$ and a idom variable/vector X defined on this space with distribution function F.

Before beginning, it is worthwhile to identify how the term **distribution function** has peared in this Book II and in Book I in various contexts. While initially looking quite tinct, these notions are closely related. We summarize the results here, with more details low, and also identify what needs to be done in this chapter that is new.

Book I Distribution Functions:

1. Distribution functions were introduced in Definition I.3.56 as **functions associated with general measurable functions defined on a measure space** $(X, \sigma(X), \mu)$. Such distribution functions were shown to have several characteristic properties in Proposition I.3.60, which later proved to be very important.

2. Distribution functions were again introduced in Definition I.5.4 and in I.(5.3) as **functions associated with general or finite Borel measures on** $\mathbb{R}$. In Proposition I.5.7, these distribution functions were shown to share key properties with those addressed in Proposition I.3.60 noted above. Thus while having very different starting points, these distribution functions were seen to share the key **characteristic properties** of being **increasing** and **right continuous**.

   (a) In Proposition I.5.23 it was then proved that starting with a function with these characteristic properties, that a Borel measure on $\mathbb{R}$ could be induced.

   (b) Consistency questions between Borel measures and such distribution functions were investigated in Section I.5.3.

3. **For Borel measures on** $\mathbb{R}^n$, the associated distribution functions were defined in I.(8.2) for finite Borel measures, and in I.(8.12) for general Borel measures. In Propositions I.8.10 and I.8.12, respectively, these distribution functions were seen to share the key **characteristic properties** of being n-**increasing** and **continuous from above.**

   (a) In Proposition I.8.15 it was proved that starting with a function with these characteristic properties, a Borel measure on $\mathbb{R}^n$ could be induced.

(b) The consistence question between measures and such functions was addressed in Corollary I.8.17.

Summary: The focus of Book I was on studying characteristic properties of distribution functions associated with Borel measures on $\mathbb{R}$ or $\mathbb{R}^n$, and on the construction of Borel measures on these spaces induced by functions which possess these characteristic properties. In the case of $\mathbb{R}$, distribution functions associated with general measurable functions were seen to share these characteristic properties, and thus also induce Borel measures on $\mathbb{R}$.

Diagrammatically, with f denoting a measurable function on $(X, \sigma(X), \mu)$ and F denoting a "distribution" function with the identified characteristic properties:

$$f \text{ on } (X, \sigma(X), \mu) \Longrightarrow F \text{ on } \mathbb{R} \iff \mu \text{ on } \mathcal{B}(\mathbb{R}),$$

$$F \text{ on } \mathbb{R}^n \iff \mu \text{ on } \mathcal{B}(\mathbb{R}^n).$$

The **characteristic properties** of F of **increasing** or n–**increasing** assure that the respectively defined set functions μ_F on $\mathcal{B}(\mathbb{R})/\mathcal{B}(\mathbb{R}^n)$ assign nonnegative measures to all sets. The characteristic properties of F of **right continuity** and **continuity from above** assure that these set functions are countably additive and hence measures. In fact, these properties assure that these measures are continuous from above. That countable additivity assures continuity from above is Proposition I.2.45, while the reverse conclusion is Proposition I.6.19.

Book II Distribution Functions:

1. Distribution functions were introduced in Definition 3.1 as **functions associated with general random variables defined on a probability space** $(\mathcal{S}, \mathcal{E}, \mu)$. This definition is identical with that in Book I, since a probability space is a measure space and a random variable is a measurable function. Consequently the characteristic properties of distribution functions of random variables were established in the Book I results.

    (a) By Proposition I.5.23, distribution functions of random variables then induce Borel probability measures on $\mathbb{R}$.

2. Distribution functions were again introduced in Definition 3.28 for **random vectors defined on** $(\mathcal{S}, \mathcal{E}, \mu)$. Since there was no explicit counterpart in Book I, we prove below that these distribution functions share the characteristic properties of distribution functions associated with Borel measures on $\mathbb{R}^n$ as developed in Book I.

    (a) With this result settled, it will follow by Proposition I.8.15 that distribution functions of random vectors induce Borel probability measures on $\mathbb{R}^n$.

Summary: The focus of this book's development is in part to apply Book I results, but also to generalize the role of f above, to more fully reflect the prominent role of random variables or vectors. Diagrammatically, with X denoting a random variable/vector on $(\mathcal{S}, \mathcal{E}, \mu)$ and F the associated distribution function, we have from Book I given the needed detail noted in item 2:

$$X \text{ on } (\mathcal{S}, \mathcal{E}, \mu) \Longrightarrow F \text{ on } \mathbb{R}/\mathbb{R}^n \iff \mu \text{ on } \mathcal{B}(\mathbb{R})/\mathcal{B}(\mathbb{R}^n).$$

This diagram also highlights another goal of this chapter, and that is to confirm that with F now denoting a distribution function with the characteristic properties identified in Book I:

$$X \text{ on } (\mathcal{S}, \mathcal{E}, \mu) \Longleftarrow F \text{ on } \mathbb{R}/\mathbb{R}^n \iff \mu \text{ on } \mathcal{B}(\mathbb{R})/\mathcal{B}(\mathbb{R}^n).$$

other words, we want to confirm that given such a function F, there exists a probability ace $(\mathcal{S}, \mathcal{E}, \mu)$ and random variable/vector X with distribution function F.

Density functions associated with distribution functions on $\mathbb{R}$ will be studied in Book using the Lebesgue integration and differentiation theory of Book III. The case of mul-ariate density functions is deferred to Book VI utilizing the general integration theory of ok V.

---

1   Distribution Functions on $\mathbb{R}$

: begin by restating Proposition 3.60 of Book I in the current context of the distribution iction of a random variable defined on a probability space.

oposition 6.1 (Properties of a d.f. $F(x)$) *Given a random variable X on a proba-ity space $(\mathcal{S}, \mathcal{E}, \mu)$, the distribution function $F(x)$ associated with X has the following operties:*

1. *$F(x)$ is a nonnegative, increasing function on $\mathbb{R}$ which is Borel, and hence, Lebesgue measurable.*

2. *For all x:*
$$\lim_{y \to x+} F(y) = F(x), \tag{6.1}$$

$$\lim_{y \to x-} F(y) = F(x) - \mu(\{X(s) = x\}). \tag{6.2}$$

3. *$F(x)$ has at most countably many discontinuities.*

4. *The limits of $F(x)$ exists as $x \to \pm\infty$:*
$$\lim_{y \to -\infty} F(y) = 0. \tag{6.3}$$

$$\lim_{y \to \infty} F(y) = 1. \tag{6.4}$$

oof. Items 1 through 3 are restatements of the associated Book I results in the notation probability theory. For 2 we also eliminate the requirement that $F(x_0) < \infty$ for some x_0 ce for all x:
$$F(x) \leq \mu(\mathcal{S}) = 1.$$

ially, the conclusion in 4 follows from X being real valued by definition, so $\mu(\{X(s) = \infty\}) = 0$ and $\mu(\mathcal{S}) = 1$. ∎

mark 6.2 (Converse of Proposition 6.1) *Proposition 3.6 and its corollary provide converse for the above result. It was proved there that given a function $F(x)$ satisfying operties 1, (6.1) of 2, and 4 of the above proposition, there exists a probability space $\mathcal{E}, \mu)$ and a random variable X defined on $\mathcal{S}$, so that $F(x)$ is the distribution function X that also satisfies (6.2) of 2.*

See Proposition 6.6 for an alternative construction for this result.

The condition in (6.1) states that $F(y)$ is **right continuous at every** x, while the condition in (6.2) states that $F(y)$ has **left limits at every** x. It follows that $F(y)$ will be **left continuous** at x if and only if $\mu(\{X(s) = x\}) = 0$. In general, however, there can be a "jump" of size $\mu(\{X(s) = x\})$ in the graph of $F(x)$. However, there can be at most countably many such jumps because every such jump contains a rational number, and since $F(x)$ is increasing, each jump contains a different rational number.

In summary, the distribution function of every random variable is an increasing right continuous function, with left limits, and at most countably many discontinuities. Further, a distribution function is continuous if and only if $\mu(\{X(s) = x\}) = 0$ for all x. Functions which are "continuous on the right and with left limits" are sometimes referred to as **càdlàg,** from the French "continu à droite, limite à gauche," though this terminology is primarily applied in the context of stochastic processes, the subject of Books VII–X.

Thus all distribution functions of random variables are càdlàg. As noted in Book I, càdlàg functions are in general not increasing, and this is what makes "with left limits" meaningful in the general case. For right continuous increasing functions, there is always a left limit at every point (Proposition I.5.8).

In the first section of Book IV, we will develop a more detailed characterization of distribution functions which provides a basis for understanding the role of the continuous and discrete probability theories, as well as an insight into cases where these theories are inadequate and more general models are needed.

6.1.1   Probability Measures from Distribution Functions

In this section we apply the development of Section I.5.2, to induce a unique measure μ_F defined on $(\mathbb{R}, \mathcal{B}(\mathbb{R}))$ from a distribution function F defined on $\mathbb{R}$. The construction logic is nearly identical in the higher dimensional spaces in the following sections, though of necessity the details vary. The primary difference in these developments is the starting point of how one defines the measure of simple sets. In one dimension these are the right semi-closed intervals, and in higher dimensions are the right semi-closed rectangles.

Given a probability space $(\mathcal{S}, \mathcal{E}, \mu)$ and random variable X defined on $\mathcal{S}$ with distribution function $F : \mathbb{R} \to \mathbb{R}$, define the generalized "F-length" of a right semi-closed interval $(a, b]$ by:

$$|(a, b]|_F = F(b) - F(a), \tag{6.5}$$

and note that by the definition of $F(x)$ that:

$$|(a, b]|_F = \mu[X^{-1}(a, b]].$$

The collection of such intervals $\mathcal{A}'$ is a semi-algebra (Definition I.6.8), and this F-length definition is then extended to a well-defined set function $\mu_{\mathcal{A}}$ on the algebra $\mathcal{A}$ of all finite disjoint unions of $\mathcal{A}'$-sets (Proposition I.5.11). This extension is defined on disjoint unions additively, and since an $\mathcal{A}$-set can be represented many ways in terms of $\mathcal{A}'$-sets, it must be checked that this definition is well defined. It is then proved that $\mu_{\mathcal{A}}$ is in fact a measure on the algebra $\mathcal{A}$ (Proposition I.5.13). This development required that the function F satisfy the **two characteristic properties** noted in the introduction, of being **increasing** and **right continuous**.

A set function $\mu_{\mathcal{A}}^*$ is then defined on arbitrary $A \subset \mathbb{R}$ by:

$$\mu_{\mathcal{A}}^*(A) = \inf \left\{ \sum_n \mu_{\mathcal{A}}((a_n, b_n]) \mid A \subset \bigcup_n (a_n, b_n] \right\},$$

or equivalently:

$$\mu_{\mathcal{A}}^*(A) = \inf \left\{ \sum_n \mu[X^{-1}(a_n, b_n]] \mid A \subset \bigcup_n (a_n, b_n] \right\}. \tag{6.6}$$

The first representation for $\mu_{\mathcal{A}}^*(A)$ highlights the distribution function F explicitly, since

$$\mu_{\mathcal{A}}((a_n, b_n]) = |(a, b]|_F \equiv F(b_n) - F(a_n),$$

while the second equivalent representation highlights the original probability measure μ and random variable X. The set function $\mu_{\mathcal{A}}^*$ is the **outer measure induced by** $\mu_{\mathcal{A}}$ (Proposition I.6.4).

We now restate the results of Propositions I.5.20 and I.5.23, and also note the uniqueness result from Example I.6.15. For this statement, a set $A \subset \mathbb{R}$ is said to be $\mu_{\mathcal{A}}^*$-**measurable**, also called **Carathéodory measurable with respect to** $\mu_{\mathcal{A}}^*$, if for any set $E \subset \mathbb{R}$:

$$\mu_{\mathcal{A}}^*(E) = \mu_{\mathcal{A}}^*(A \cap E) + \mu_{\mathcal{A}}^*(\widetilde{A} \cap E). \tag{6.7}$$

We change the language below from measure space to probability space since in the current application, $F(-\infty) = 0$ and $F(\infty) = 1$ defined as limits assure that $\mu_F(\mathbb{R}) = 1$.

Proposition 6.3 (The Borel measure μ_F induced by d.f. $F : \mathbb{R} \to \mathbb{R}$)
Given a probability space $(\mathcal{S}, \mathcal{E}, \mu)$ and random variable $X : \mathcal{S} \longrightarrow \mathbb{R}$, the associated distribution function F defined on $\mathbb{R}$ by:

$$F(x) = \mu[X^{-1}(-\infty, x]],$$

induces a unique probability measure μ_F on the Borel sigma algebra $\mathcal{B}(\mathbb{R})$. This measure is defined on right semi-closed intervals $(a, b]$ by:

$$\mu_F[(a, b]] = F(b) - F(a). \tag{6.8}$$

*In detail, with $\mathcal{M}_F(\mathbb{R})$ defined as the **collection of Carathéodory measurable sets**:*

1. *$\mathcal{A} \subset \mathcal{M}_F(\mathbb{R})$, and for all $A \in \mathcal{A}$:*

$$\mu_{\mathcal{A}}^*(A) = \mu_{\mathcal{A}}(A).$$

2. *$\mathcal{M}_F(\mathbb{R})$ is a sigma algebra and contains every set $A \subset \mathbb{R}$ with $\mu_{\mathcal{A}}^*(A) = 0$.*

3. *$\mathcal{M}_F(\mathbb{R})$ contains the Borel sigma algebra, $\mathcal{B}(\mathbb{R}) \subset \mathcal{M}_F(\mathbb{R})$.*

4. *If μ_F denotes the restriction of $\mu_{\mathcal{A}}^*$ to $\mathcal{M}_F(\mathbb{R})$, then μ_F is a probability measure and hence $(\mathbb{R}, \mathcal{M}_F(\mathbb{R}), \mu_F)$ is a complete probability space.*

5. *The probability measure μ_F is the unique extension of $\mu_{\mathcal{A}}$ from $\mathcal{A}$ to the smallest sigma algebra generated by $\mathcal{A}$, which is $\mathcal{B}(\mathbb{R})$ by Proposition I.8.1.*

6. *For all $A \in \mathcal{B}(\mathbb{R})$:*

$$\mu_F(A) = \mu\left[X^{-1}(A)\right]. \tag{6.9}$$

Proof. *The results in 1 - 5 are a restatement of Proposition I.5.23. This earlier result applies because as required, F here is increasing and right continuous by Proposition 6.1.*

Result 6 follows by uniqueness. This identity is true for all $(a, b] \in \mathcal{A}'$ by the definition of $\mu_F[(a, b]]$ in (6.8), and this then extends to $\mathcal{A}$ additively. Thus $\mu_F(A) = \mu\left[X^{-1}(A)\right]$ for all $A \in \mathcal{A}$. As both μ_F and $\mu\left[X^{-1}(\cdot)\right]$ are well-defined measures on $\mathcal{B}(\mathbb{R})$, the uniqueness result in 5 assures that this identity extends to $\mathcal{B}(\mathbb{R})$. ∎

For the μ_F-measure of various types of intervals, from (6.8):

$$\mu_F\left((a,b]\right) = F(b) - F(a), \tag{1}$$

and:

$$\{b\} = \bigcap_{n=1}^{\infty}(b - 1/n, b],$$

it follows by continuity from above of μ:

$$\mu_F\left(\{b\}\right) = F(b) - F(b^-). \tag{2}$$

Here $F(b^-)$ denotes the left limit of F at b. Thus $\mu_F\left(\{b\}\right) = 0$ if $F(x)$ is continuous at b.
   Then by finite additivity:

$$\mu_F\left((a,b)\right) = F(b^-) - F(a), \tag{3}$$

$$\mu_F\left([a,b)\right) = F(b^-) - F(a^-),$$

and:

$$\mu_F\left([a,b]\right) = F(b) - F(a^-). \tag{4}$$

Remark 6.4 ($(\mathbb{R}, \mathcal{B}(\mathbb{R}), \mu_F)$ and $(\mathcal{S}, \mathcal{E}, \mu)$) *Note that the induced probability space $(\mathbb{R}, \mathcal{B}(\mathbb{R}), \mu_F)$ provides a summarized yet typically incomplete view of the original probability space $(\mathcal{S}, \mathcal{E}, \mu)$. In other words, it identifies the probabilities of some, but typically not all events in $\mathcal{E}$.*
   Of course, by measurability of X it is true that for all $A \in \mathcal{B}(\mathbb{R})$:

$$X^{-1}(A) \in \mathcal{E},$$

and thus we can evaluate the μ-probability of this event because it is revealed by μ_F:

$$\mu\left(X^{-1}(A)\right) \equiv \mu_F\left(A\right).$$

The information about $(\mathcal{S}, \mathcal{E}, \mu)$ so revealed by X and $F(x)$ is typically incomplete. While

$$\sigma(X) \equiv \{X^{-1}(A) | A \in \mathcal{B}(\mathbb{R})\}$$

is a sigma algebra and hence a sigma subalgebra of $\mathcal{E}$, it is typically the case that

$$\sigma(X) \subsetneqq \mathcal{E}.$$

In other words, the induced sigma algebra $\sigma(X)$ provides probability information for some, but not all events $B \in \mathcal{E}$.

Example 6.5 (Binomial $(Y^{\mathbb{N}}, \sigma(Y^{\mathbb{N}}), \mu_{\mathbb{N}})$) *Consider the infinite product probability space $(Y^{\mathbb{N}}, \sigma(Y^{\mathbb{N}}), \mu_{\mathbb{N}})$ of general coin flips in Example 4.16. We represent $Y = \{H, T\}$ by the numerical space $Y = \{1, 0\}$, and $\mu_{\mathbb{N}}$ is the measure induced by the probability measure μ_B on Y in (1.7):*

$$\mu_B(1) = p, \ \mu_B(0) = 1 - p.$$

Each point $y \in Y^{\mathbb{N}}$ can be envisioned as an infinite sequence of coin flips where Hs and Ts are represented by $1s$ and $0s$, respectively.
   Define X_j as the projection onto the jth coordinate:

$$X_j(y) = y_j.$$

Then $X_j : Y^{\mathbb{N}} \to \mathbb{R}$ *is a random variable, and the induced Borel space* $(\mathbb{R}, \mathcal{B}(\mathbb{R}), \mu_{F_j})$ *provides a very crude summary of the events in* $\sigma(Y^{\mathbb{N}})$. *This is because:*

$$\sigma(X_j) = \{\emptyset, A_j, \widetilde{A}_j, \mathcal{S}\},$$

where $A_j = \{y | y_j = 1\}$. *This random variable only reveals the* $\mu_{\mathbb{N}}$*-probabilities of these 4 events as* $\{0, p, 1 - p, 1\}$, *respectively.*

More informatively, we can define a random variable X'_j *as the sum of the first* j *coordinates of* y:

$$X'_j(y) = \textstyle\sum_{k=1}^{j} y_k.$$

Then $X'_j : Y^{\mathbb{N}} \to \mathbb{R}$, *and* $\sigma(X'_j)$ *is generated by the* $j + 1$ *sets* A'_{ji}, *defined for* $0 \le i \le j$ *by:*

$$A'_{ji} = \{y | \textstyle\sum_{k=1}^{j} y_k = i\}.$$

The $\mu_{\mathbb{N}}$*-probabilities of these* $j + 1$ *events are given as in Exercise 4.18 by:*

$$\mu_{\mathbb{N}}(A'_{ji}) = \binom{j}{i} p^i (1 - p)^{j-i}.$$

For any given j, *the random variable* X'_j *induces a more descriptive Borel space* $(\mathbb{R}, \mathcal{B}(\mathbb{R}), \mu_{F_j})$ *than does* X_j, *but* $\sigma(X'_j)$ *is still quite crude. For example, it cannot provide* $\mu_{\mathbb{N}}$*-probabilities for an event such as* $B_i = \{y | \sum_{k=1}^{i} y_k = i\}$ *for* $i \ne j$, *because* B_i *is not an event in* $\sigma(X'_j)$.

Even more informatively, let $\{r_n\} = \{2, 3, 5, 7, ...\}$ *be an enumeration of the prime numbers and define* $X''_j : Y^{\mathbb{N}} \to \mathbb{R}$ *by:*

$$X''_j(y) = \prod_{k=1}^{j} (r_k)^{y_k}.$$

Then, by the fundamental theorem of arithmetic, $X''_j(y)$ *is uniquely defined given* $(y_1, y_2, ..., y_j)$. *Hence* $\sigma(X''_j)$ *is generated by the* 2^j *sets* A''_{ji}, *defined by the* 2^j *possible values of* $\prod_{k=1}^{j} (r_k)^{y_k}$. *Hence,* $\sigma(X''_j)$ *is finer yet than* $\sigma(X'_j)$, *and contains every event definable in terms of the first* j *coordinates of* y.

6.1.2 Random Variables from Distribution Functions

As noted in the introduction to this chapter, Book I has addressed the connections between Borel measures and their associated distribution functions. In the language of probability theory, every probability measure μ on $(\mathbb{R}, \mathcal{B}(\mathbb{R}))$ gives rise to a distribution function F_μ that is increasing and right continuous, and this function is unique if we specify that $F_\mu(-\infty) = 0$ and $F_\mu(\infty) = 1$ defined as limits (Proposition I.5.7). Conversely, every such function F gives rise to a unique probability measure μ_F on $(\mathbb{R}, \mathcal{B}(\mathbb{R}))$ (Proposition I.5.23 and Example I.6.15).

Now by Proposition 6.1, given a probability space $(\mathcal{S}, \mathcal{E}, \mu)$ and random variable $X : \mathcal{S} \longrightarrow \mathbb{R}$, the associated distribution function F defined on $\mathbb{R}$ by:

$$F(x) \equiv \mu[X^{-1}(-\infty, x]],$$

is increasing and right continuous, with $F(-\infty) = 0$ and $F(\infty) = 1$ defined as limits. Thus such distribution functions also induce probability measures by Proposition 6.3.

The question investigated in this section is as follows. Given an increasing and right continuous function $F : \mathbb{R} \to \mathbb{R}$ with $F(-\infty) = 0$ and $F(\infty) = 1$ defined as limits, does there exist a probability space $(\mathcal{S}, \mathcal{E}, \mu)$ and a random variable X so that F is the distribution function of X? Equivalently, given a probability measure μ' on $(\mathbb{R}, \mathcal{B}(\mathbb{R}))$, does there exist a probability space $(\mathcal{S}, \mathcal{E}, \mu)$ and a random variable X so that μ' is the Borel measure induced by the distribution function of X? In other words, with $\mu' = \mu_F$?

The reader is encouraged to verify the details of the equivalence of these questions, that if either is answered affirmatively, so too is the other.

Proposition 3.6 provides an explicit construction and affirmative answer to the first question, that from such F we can construct $(\mathcal{S}, \mathcal{E}, \mu)$ and X. The following result provides another construction which will be generalized to joint distribution functions in the next section.

Proposition 6.6 ($(\mathcal{S}, \mathcal{E}, \mu)$ and X induced by d.f. $F : \mathbb{R} \to \mathbb{R}$) *Given a function* $F :$ $\mathbb{R} \to \mathbb{R}$ *that is increasing and right continuous, with* $F(-\infty) = 0$ *and* $F(\infty) = 1$ *defined as limits, there exists a probability space* $(\mathcal{S}, \mathcal{E}, \mu)$ *and a random variable* X *so that* F *is the distribution function of* X.

Proof. *Let* $(\mathcal{S}, \mathcal{E}, \mu) \equiv (\mathbb{R}, \mathcal{B}(\mathbb{R}), \mu_F)$, *the probability space of Proposition 6.3. With* m *denoting Lebesgue measure, define:*

$$X : (\mathbb{R}, \mathcal{B}(\mathbb{R}), \mu_F) \to (\mathbb{R}, \mathcal{B}(\mathbb{R}), m),$$

by $X(y) = y$. *Then* X *is measurable, since* $X^{-1}(A) = A$ *for all* $A \in \mathcal{B}(\mathbb{R})$.

The distribution function of X *is defined:*

$$F_X(y) \equiv \mu_F[X^{-1}(-\infty, y]] = \mu_F[(-\infty, y]].$$

Now:

$$(-\infty, y] = \bigcup_{n=1}^{\infty} (y - n, y],$$

and thus by continuity from below of μ_F, *then (6.8), and finally* $F(-\infty) = 0$:

$$
\begin{aligned}
\mu_F[(-\infty, y]] &= \lim_{n \to \infty} \mu_F[(y - n, y]] \\
&= \lim_{n \to \infty} (F(y) - F(y - n)) \\
&= F(y).
\end{aligned}
$$

∎

6.2 Distribution Functions on $\mathbb{R}^n$

In this section we generalize the above results to distribution functions of random vectors, and thus defined on $\mathbb{R}^n$. In the first section we show that these distribution functions induce unique probability measures on $(\mathbb{R}^n, \mathcal{B}(\mathbb{R}^n))$. Unlike distribution functions defined on $\mathbb{R}$, we have no counterpart to Proposition I.3.60 in the vector valued case, and thus we will first need to prove that joint distribution functions satisfy the necessary characteristic properties to justify the application of Book I results.

The next section will show that given a vector valued function F which satisfies these characteristic properties, there exists a probability space $(\mathcal{S}, \mathcal{E}, \mu)$ and a random vector X so that F is the joint distribution function of X.

We then end the chapter with a discussion of the applicability of these results to marginal and conditional distribution functions.

6.2.1 Probability Measures from Distribution Functions

Recall the joint distribution function introduced in Definition 3.28. If $\{X_j\}_{j=1}^n$ are random variables on $(\mathcal{S}, \mathcal{E}, \mu)$, so each $X_j : \mathcal{S} \longrightarrow \mathbb{R}$, the **random vector** $X = (X_1, X_2, ..., X_n)$ is defined as the transformation:

$$X : \mathcal{S} \longrightarrow \mathbb{R}^n,$$

with value on $s \in \mathcal{S}$ given by:

$$X(s) = (X_1(s), X_2(s), ..., X_n(s)).$$

The **joint distribution function (d.f.)**, or **joint cumulative distribution function (c.d.f.)** associated with X, denoted F or F_X, is defined on $(x_1, x_2, ..., x_n) \in \mathbb{R}^n$ by (3.15):

$$F(x_1, x_2, ..., x_n) = \mu \left[\bigcap_{j=1}^n X_j^{-1}(-\infty, x_j] \right].$$

The goal of this section is to prove that a joint distribution function induces a unique probability measure μ_F on the Borel sigma algebra $\mathcal{B}(\mathbb{R}^n)$. This will be an immediate application of Proposition I.8.15 once it is demonstrated that such distribution functions satisfy the requirements of this Book I result.

In order to state this result we require two definitions from Book I:

Definition 6.7 (Continuous from above; n-increasing) *A function $F : \mathbb{R}^n \to \mathbb{R}$ is said to be **continuous from above** if given $x = (x_1, ..., x_n)$ and a sequence $x^{(m)} = (x_1^{(m)}, ..., x_n^{(m)})$ with $x_i^{(m)} \geq x_i$ for all i and m, and $x^{(m)} \to x$ as $m \to \infty$, then:*

$$F(x) = \lim_{m \to \infty} F(x^{(m)}). \tag{6.10}$$

*Further, F is said to be n-**increasing** or to satisfy the n-**increasing condition** if for any bounded right semi-closed rectangle $\prod_{i=1}^n (a_i, b_i]$:*

$$\sum_x sgn(x) F(x) \geq 0. \tag{6.11}$$

Each $x = (x_1, ..., x_n)$ in the summation is one of the 2^n vertices of this rectangle, so $x_i = a_i$ or $x_i = b_i$, and $sgn(x)$ is defined as -1 if the number of a_i-components of x is odd, and $+1$ otherwise.

While the definition of continuous from above is intuitive, the formula in (6.11) may appear mysterious in the absence of the Book I development. But by Proposition I.8.9, given any finite Borel measure μ on $\mathbb{R}^n$ and associated distribution function F defined by:

$$F(x_1, x_2, ..., x_n) = \mu \left[\prod_{j=1}^n (-\infty, x_j] \right],$$

the μ-measure of any bounded right semi-closed rectangle $\prod_{i=1}^n (a_i, b_i]$ is given by:

$$\mu \left(\prod_{i=1}^n (a_i, b_i] \right) = \sum_x sgn(x) F(x), \tag{6.12}$$

where the summation is over the rectangle's vertices as noted above.

Thus since the measure of all such rectangles must be nonnegative, if F is a given function it must satisfy (6.11) for there to be any chance that it arises as the distribution function of a Borel measure.

The Book I development required that the function F satisfy the **two characteristic properties** noted in the introduction, of being n-**increasing** and **continuous from**

above. To create or induce a measure on $\mathcal{B}(\mathbb{R}^n)$ from such a function, we largely followed the template summarized in Section 6.1.1.

Generalizing (6.5), a set function μ_0 is defined on the collection of **bounded** right semi-closed rectangles by (6.12). This collection of sets, denoted $\mathcal{A}'_B$ in Book I, is **not** a semi-algebra (Definition I.6.8), since complements of $\mathcal{A}'_B$-sets cannot be expressed as **finite** disjoint unions of $\mathcal{A}'_B$-sets. However, this collection is big enough to generate $\mathcal{B}(\mathbb{R}^n)$ by Proposition I.8.1.

We next introduce an outer measure $\mu_F^*(A)$ for an arbitrary set $A \subset \mathbb{R}^n$ as in (6.6), but using collections of $\mathcal{A}'_B$-sets. Finally, define a set $A \subset \mathbb{R}^n$ to be μ_F^*-**measurable**, also called **Carathéodory measurable with respect to** μ_F^*, if (6.7) is satisfied for any set $E \subset \mathbb{R}^n$.

With $\mathcal{M}_F(\mathbb{R}^n)$ defined as the **collection of Carathéodory-measurable sets**, and μ_F defined as the restriction of μ_F^* to $\mathcal{M}_F(\mathbb{R}^n)$, then we obtain in Propositions I.8.15 and I.8.16:

1. $(\mathbb{R}^n, \mathcal{M}_F(\mathbb{R}^n), \mu_F)$ is a complete measure space.

2. $\mathcal{B}(\mathbb{R}^n) \subset \mathcal{M}_F(\mathbb{R}^n)$ and so $(\mathbb{R}^n, \mathcal{B}(\mathbb{R}^n), \mu_F)$ is a measure space.

3. $\mu_F = \mu_0$ on $\mathcal{A}'_B \subset \mathcal{B}(\mathbb{R}^n)$, and so μ_F satisfies (6.12).

4. μ_F is the unique extension of μ_0 from $\mathcal{A}'_B$ to $\mathcal{B}(\mathbb{R}^n)$ and $\mathcal{M}_F(\mathbb{R}^n)$.

The Book I proof of (6.12), and importantly the proof below, requires the **inclusion-exclusion formula** of Proposition I.8.8, so it is simply stated here. It applies in general measure spaces and thus also in probability spaces. For disjoint unions, this result reduces to the finite additivity of measures.

Proposition 6.8 (Inclusion-exclusion formula) *Given measurable sets* $\{A_j\}_{j=1}^n$:

$$
\begin{aligned}
\mu\left[\bigcup_{j=1}^n A_j\right] = {} & \sum_{j=1}^n \mu[A_j] - \sum_{i<j} \mu\left[A_i \bigcap A_j\right] \\
& + \sum_{i<j<k} \mu\left[A_i \bigcap A_j \bigcap A_k\right] - \\
& \cdots + (-1)^{n+1} \mu\left[\bigcap_{j=1}^n A_j\right].
\end{aligned}
\tag{6.13}
$$

Proof. See Proposition I.8.8. ∎

We are now ready for the main result. In order to apply the Book I results noted above, we must prove that joint distribution functions satisfy the necessary characteristic properties of n-increasing and continuity from above. We then change the language below from measure space to probability space since in the current application, $F(-\infty) = 0$ and $F(\infty) = 1$ defined as limits, which assures that $\mu_F(\mathbb{R}^n) = 1$.

By **defined as limits** is meant that for any $\epsilon > 0$, there exists N so that $F(x_1, x_2, ..., x_n) \geq 1 - \epsilon$ if all $x_i \geq N$, and $F(x_1, x_2, ..., x_n) \leq \epsilon$ if all $x_i \leq -N$. The reader is invited to prove that joint distribution functions satisfy these limits.

Proposition 6.9 (The Borel measure μ_F induced by d.f. $F : \mathbb{R}^n \to \mathbb{R}$) *Let* $X_j : \mathcal{S} \longrightarrow \mathbb{R}$ *be random variables on* $(\mathcal{S}, \mathcal{E}, \mu)$, $j = 1, 2, ..., n$, *and* $F: \mathbb{R}^n \to \mathbb{R}$ *the joint distribution function defined in (3.15).*

Then F is continuous from above and satisfies the n-increasing condition.

Thus by Propositions I.8.15 and I.8.16, F induces a unique probability measure μ_F on the Borel sigma algebra $\mathcal{B}(\mathbb{R}^n)$, which satisfies (6.12) for all bounded right semi-closed rectangles in $\mathbb{R}^n$.

In addition, for all $A \in \mathcal{B}(\mathbb{R}^n)$:

$$\mu_F(A) = \mu\left[X^{-1}(A)\right]. \tag{6.14}$$

Proof.

1. **Continuity from above:** Let $x = (x_1, ..., x_n)$ and the sequence $x^{(m)} = (x_1^{(m)}, ..., x_n^{(m)})$ be given as in Definition 6.7. Define $A^{(m)} \in \mathcal{E}$, $A \in \mathcal{E}$ by:

$$A^{(m)} = \bigcap_{j=1}^{n} X_j^{-1}\left(-\infty, x_j^{(m)}\right], \qquad A = \bigcap_{j=1}^{n} X_j^{-1}(-\infty, x_j],$$

and note that $A \subset A^{(m)}$ for all m since $x_j^{(m)} \geq x_j$ for each j.

Now define:

$$B^{(m)} = \bigcap_{k=1}^{m} A^{(k)}.$$

Then $\{B^{(m)}\}_{m=1}^{\infty}$ is a nested sequence, $B^{(m+1)} \subset B^{(m)}$ for all m, and $\bigcap_{m=1}^{\infty} B^{(m)} = \bigcap_{m=1}^{\infty} A^{(m)}$. We claim $A = \bigcap_{m=1}^{\infty} A^{(m)}$. Certainly $A \subset \bigcap_{m=1}^{\infty} A^{(m)}$ since $A \subset A^{(m)}$ for all m. If not equal and $s \in \bigcap_{m=1}^{\infty} A^{(m)} - A$, then $X_j(s) \leq x_j^{(m)}$ for all m and j, and $X_k(s) > x_k$ for some k, contradicting convergence $x^{(m)} \to x$.

Thus $A = \bigcap_{m=1}^{\infty} B^{(m)}$ and by continuity from above of the measure μ:

$$\mu[A] = \lim_{m \to \infty} \mu\left[B^{(m)}\right].$$

Since $x^{(m)} \to x$, for any m there exists $N \geq m$ so that $A^{(m')} \subset B^{(m)}$ for all $m' \geq N$. Hence for all such $m' \geq N$:

$$A \subset A^{(m')} \subset B^{(m)}.$$

By monotonicity it then follows that:

$$\mu(A) \leq \mu\left(A^{(m')}\right) \leq \mu\left(B^{(m)}\right),$$

and so:

$$\mu[A] = \lim_{m \to \infty} \mu\left[A^{(m)}\right].$$

This is continuity from above of F since $F(x_1, x_2, ..., x_n) \equiv \mu[A]$ and similarly $F(x_1^{(m)}, x_2^{(m)}, ..., x_n^{(m)}) \equiv \mu\left[A^{(m)}\right]$.

2. **The n-increasing condition:** Let $\prod_{i=1}^{n}(a_i, b_i] \subset \mathbb{R}^n$ be a bounded right semi-closed rectangle. To simplify notation, for any point $x = (x_1, ..., x_n)$, let A_x or $A_{(x_1, ..., x_n)}$ denote the rectangle:

$$A_x \equiv \prod_{j=1}^{n}(-\infty, x_j].$$

We claim that:

$$A_{(b_1, ..., b_n)} = \prod_{i=1}^{n}(a_i, b_i] \cup \bigcup_{j=1}^{n} A_{x^{(j)}},$$

where for each j in the union, $x_i^{(j)} = b_i$ for $i \neq j$ and $x_j^{(j)} = a_j$.

To prove this, note that if $x \in A_{(b_1, ..., b_n)}$, then $x_j \leq b_j$ for all j. Thus if $x \notin \prod_{i=1}^{n}(a_i, b_i]$, then $x_j \leq a_j$ for at least one j and so $x \in \bigcup_{j=1}^{n} A_{x^{(j)}}$.

As $A_{(b_1,...,b_n)}$ is a disjoint union of these two sets, $X^{-1}\left[A_{(b_1,...,b_n)}\right]$ is again a disjoint union:

$$X^{-1}\left[A_{(b_1,...,b_n)}\right] = X^{-1}\left[\prod_{i=1}^n (a_i, b_i]\right] \bigcup X^{-1}\left[\bigcup_{j=1}^n A_{x^{(j)}}\right].$$

Finite additivity of μ obtains:

$$
\begin{aligned}
\mu\left[X^{-1}\left(\prod_{i=1}^n (a_i, b_i]\right)\right] &= \mu\left[X^{-1}A_{(b_1,...,b_n)}\right] - \mu\left[X^{-1}\left(\bigcup_{j=1}^n A_{x^{(j)}}\right)\right] \qquad (1)\\
&= F(b_1,...,b_n) - \mu\left[\bigcup_{j=1}^n X^{-1}\left(A_{x^{(j)}}\right)\right].
\end{aligned}
$$

Applying 6.13 to the measure of the union:

$$
\begin{aligned}
&-\mu\left[\bigcup_{j=1}^n X^{-1}\left(A_{x^{(j)}}\right)\right]\\
={}&-\sum_{j=1}^n \mu\left[X^{-1}\left(A_{x^{(j)}}\right)\right] + \sum_{i<j}\mu\left[X^{-1}\left(A_{x^{(i)}}\right)\bigcap X^{-1}\left(A_{x^{(j)}}\right)\right]\\
&-\sum_{i<j<k}\mu\left[X^{-1}\left(A_{x^{(i)}}\right)\bigcap X^{-1}\left(A_{x^{(j)}}\right)\bigcap X^{-1}\left(A_{x^{(k)}}\right)\right] +\\
&\cdots + (-1)^n\mu\left[\bigcap_{j=1}^n X^{-1}\left(A_{x^{(j)}}\right)\right].
\end{aligned}
$$

Now:

$$-\sum_{j=1}^n \mu\left[X^{-1}\left(A_{x^{(j)}}\right)\right] = -\sum_{j=1}^n F(x^{(j)}),$$

noting that the sign for this summation is -1, consistent with each $x^{(j)}$ having one a_j component.

Next:

$$X^{-1}\left(A_{x^{(i)}}\right)\bigcap X^{-1}\left(A_{x^{(j)}}\right) = X^{-1}\left(A_{x^{(i,j)}}\right),$$

where $x_l^{(i,j)} = b_l$ for $l \neq i,j$ and $x_l^{(i,j)} = a_l$ for $l = i,j$. Thus:

$$\sum_{i<j}\mu\left[X^{-1}\left(A_{x^{(i)}}\right)\bigcap X^{-1}\left(A_{x^{(j)}}\right)\right] = \sum_{i<j}\mu\left[X^{-1}\left(A_{x^{(i,j)}}\right)\right] = \sum_{i<j} F(x^{(i,j)}),$$

noting that the sign for this summation is $+1$, consistent with each $x^{(i,j)}$ having two a_j components.

With the same notational convention:

$$
\begin{aligned}
&-\sum_{i<j<k}\mu\left[X^{-1}\left(A_{x^{(i)}}\right)\bigcap X^{-1}\left(A_{x^{(j)}}\right)\bigcap X^{-1}\left(A_{x^{(k)}}\right)\right]\\
={}&-\sum_{i<j<k}\mu\left[X^{-1}\left(A_{x^{(i,j,k)}}\right)\right]\\
\equiv{}&-\sum_{i<j<k} F(x^{(i,j,k)}),
\end{aligned}
$$

where again the sign of this sum is consistent with each $x^{(i,j,k)}$ having three a_j components.

Continuing and making substitutions into (1):

$$\mu\left[X^{-1}\left(\prod_{i=1}^n (a_i, b_i]\right)\right] = \sum_x sgn(x)F(x), \qquad (2)$$

where each $x = (x_1,...,x_n)$ is one of the 2^n vertices of $\prod_{i=1}^n (a_i, b_i]$, so $x_i = a_i$ or $x_i = b_i$, and $sgn(x)$ is -1 if the number of components of x that equal a_i is odd, and $+1$ otherwise. Thus (6.11) is proved because by definition of measure:

$$\mu\left[X^{-1}\left(\prod_{i=1}^n (a_i, b_i]\right)\right] \geq 0.$$

That F induces a unique probability measure μ_F that satisfies (6.12) for all bounded right semi-closed rectangles in $\mathbb{R}^n$, is a restatement of Propositions I.8.15 and I.8.16.

Finally, (6.14) follows by uniqueness. By (2) and (6.12):

$$\mu\left[X^{-1}\left(\prod_{i=1}^{n}(a_i, b_i]\right)\right] = \mu_F\left(\prod_{i=1}^{n}(a_i, b_i]\right),$$

for all $\prod_{i=1}^{n}(a_i, b_i] \in \mathcal{A}'_B$. Since both $\mu\left[X^{-1}(\cdot)\right]$ and μ_F are well-defined measures on $\mathcal{B}(\mathbb{R}^n)$, uniqueness assures that this identity extends to this sigma algebra. ∎

6.2.2 Random Vectors from Distribution Functions

As noted in the introduction to this chapter, Book I addressed the connections between Borel measures and their associated distribution functions. In the context of probability theory, every probability measure μ on $(\mathbb{R}^n, \mathcal{B}(\mathbb{R}^n))$ gives rise to a distribution function F_μ that is n-increasing and continuous from above, and this function is unique if we specify that $F_\mu(-\infty) = 0$ and $F_\mu(\infty) = 1$ defined as limits (Proposition I.8.10). Conversely, every such function F gives rise to a unique probability measure μ_F on $(\mathbb{R}, \mathcal{B}(\mathbb{R}))$ (Proposition I.8.15).

The question investigated in this section is as follows. Given an n-increasing and continuous from above function $F : \mathbb{R}^n \to \mathbb{R}$ with $F(-\infty) = 0$ and $F(\infty) = 1$ defined as limits, does there exist a probability space $(\mathcal{S}, \mathcal{E}, \mu)$ and a random vector X so that F is the joint distribution function of X? Equivalently, given a probability measure μ' on $(\mathbb{R}^n, \mathcal{B}(\mathbb{R}^n))$, does there exist a probability space $(\mathcal{S}, \mathcal{E}, \mu)$ and a random variable X so that μ' is the Borel measure induced by the joint distribution function of X? In other words, with $\mu' = \mu_F$?

As above, the reader is encouraged to verify the details of the equivalence of these questions, that if either is answered affirmatively, so too is the other.

Proposition 6.10 ($(\mathcal{S}, \mathcal{E}, \mu)$ and X induced by d.f. $F : \mathbb{R}^n \to \mathbb{R}$) *Given a function $F : \mathbb{R}^n \to \mathbb{R}$ that is n-increasing and continuous from above, with $F(-\infty) = 0$ and $F(\infty) = 1$ defined as limits, there exists a probability space $(\mathcal{S}, \mathcal{E}, \mu)$ and a random vector X so that F is the joint distribution function of X.*

Proof. *Let $(\mathcal{S}, \mathcal{E}, \mu) \equiv (\mathbb{R}^n, \mathcal{B}(\mathbb{R}^n), \mu_F)$, the probability space of Proposition I.8.15. With m denoting Lebesgue measure on $\mathbb{R}^n$, define:*

$$X : (\mathbb{R}^n, \mathcal{B}(\mathbb{R}^n), \mu_F) \to (\mathbb{R}^n, \mathcal{B}(\mathbb{R}^n), m),$$

by $X(y) = y$. Then X is measurable, since $X^{-1}(A) = A$ for all $A \in \mathcal{B}(\mathbb{R}^n)$.

The distribution function of X is defined at $y \equiv (y_1, ..., y_n)$ by:

$$
\begin{aligned}
F_X(y) &\equiv \mu_F\left[\bigcap_{j=1}^{n} X_j^{-1}(-\infty, y_j]\right] \\
&= \mu_F\left[\prod_{j=1}^{n}(-\infty, y_j]\right] \\
&= \overline{F}_{\mu_F}(y).
\end{aligned}
$$

Here, $\overline{F}_{\mu_F}(y)$ denotes the distribution function associated with the Borel measure μ_F. This function is defined in I.(8.2).

By Corollary I.8.17, there exists $c \in \mathbb{R}$ so that:

$$\overline{F}_{\mu_F}(y) = F(y) + c.$$

Since μ_F is a probability measure, it follows that $\overline{F}_{\mu_F}(-\infty) = 0$ and $\overline{F}_{\mu_F}(\infty) = 1$, defined as limits. As F has these same limits by assumption, we conclude that $c = 0$ and thus $F_X(y) = F(y)$. ∎

6.2.3   Marginal and Conditional Distribution Functions

With most of the work done in the previous sections, we simply make the necessary observations here to justify the application of earlier results to marginal and conditional distribution functions. We focus on Borel measures induced by such distribution functions, as addressed in Propositions 6.3 and 6.9.

Proposition 6.11 (μ_{F_I} induced by marginal d.f. $F_I : \mathbb{R}^m \to \mathbb{R}$) *Let $X_j : \mathcal{S} \longrightarrow \mathbb{R}$ be random variables on $(\mathcal{S}, \mathcal{E}, \mu)$, $j = 1, 2, ..., n$, and $F \colon \mathbb{R}^n \to \mathbb{R}$ the joint distribution function defined in (3.15). For any index set $I = \{i_1, ..., i_m\} \subset \{1, 2, ..., n\}$, the marginal distribution function $F_I(x_1, x_2, ..., x_m)$ defined in (3.19) induces a unique probability measure μ_{F_I} on the Borel sigma algebra $\mathcal{B}(\mathbb{R}^m)$, which satisfies (6.12) for all bounded right semi-closed rectangles in $\mathbb{R}^m$.*
Proof. *By Proposition 3.36, $F_I(x_1, x_2, ..., x_m)$ is the joint distribution function of $X_I \equiv (X_{i_1}, X_{i_2}, ..., X_{i_m})$, and thus this is a corollary of Proposition 6.9.* ∎

Proposition 6.12 ($\mu_{F_{J|B}}$ induced by conditional d.f. $F_{J|B} : \mathbb{R}^n \to \mathbb{R}$) *Let $X_j : \mathcal{S} \longrightarrow \mathbb{R}$ be random variables on $(\mathcal{S}, \mathcal{E}, \mu)$, $j = 1, 2, ..., n$, and $F \colon \mathbb{R}^n \to \mathbb{R}$ the joint distribution function defined in (3.15). For any index set $J = \{j_1, ..., j_m\} \subset \{1, 2, ..., n\}$ and $B \in \mathcal{B}(\mathbb{R}^m)$ with $\mu\left[X_J^{-1}(B)\right] \neq 0$ for $X_J = (X_{j_1}, X_{j_2}, ..., X_{j_m})$, the conditional distribution function $F_{J|B}(x_1, x_2, ..., x_n)$ defined in (3.21) induces a unique probability measure $\mu_{F_{J|B}}$ on the Borel sigma algebra $\mathcal{B}(\mathbb{R}^n)$, which satisfies (6.12) for all bounded right semi-closed rectangles in $\mathbb{R}^n$.*
Proof. *By Proposition 3.41, $F_{J|B}(x_1, x_2, ..., x_n)$ is the joint distribution function of $(X_1, X_2, ..., X_n)$ defined on the probability space $(\mathcal{S}, \mathcal{E}, \mu(\cdot | X_J^{-1}(B)))$, and thus this is a corollary of Proposition 6.9.* ∎

7

Copulas and Sklar's Theorem

In this chapter we introduce an interesting and important theory which studies the linkages between a joint distribution function $F(x_1, x_2, ..., x_n)$, and the collection of one variable marginal distribution functions $\{F_j(x)\}_{j=1}^n$.

7.1 Fréchet Classes

Let $\{F_j(x)\}_{j=1}^n$ be a collection of n univariate distribution functions. For example, these could be associated with random variables $\{X_j\}_{j=1}^n$ defined on a probability space $(\mathcal{S}, \mathcal{E}, \mu)$. This probability space and random variables are assured to exist by Proposition 3.6, but these are not essential for this section other than as a notational convenience in some proofs.

Definition 7.1 (Fréchet class $\mathfrak{F}(F_1, F_2, ..., F_n)$) *Given* $\{F_j(x)\}_{j=1}^n$, *the **Fréchet class** $\mathfrak{F}(F_1, F_2, ..., F_n)$ is the collection of joint distribution functions $F(x_1, x_2, ..., x_n)$, so that $F_{\{j\}}(x_j) = F_j(x_j)$, where $F_{\{j\}}$ is the marginal distribution function of F associated with $I = \{j\}$ as given in Definition 3.34.*

This class is named for **Maurice Fréchet** (1878–1973) who was interested in what came to be known as **Fréchet's problem,** of characterizing the properties of the joint distribution functions in a given Fréchet class. These classes can also be defined more generally by a collection of not necessarily univariate marginal distribution functions, but then the first question to be resolved is one of consistency of these marginal distributions.

For the current collection with univariate marginals, consistency is never a problem in the sense that $\mathfrak{F}(F_1, F_2, ..., F_n)$ always contains the distribution function:

$$F(x_1, x_2, ..., x_n) = \prod_{j=1}^n F_j(x_j).$$

By Proposition 3.53, this is the distribution function of the random vector $(X_1, X_2, ..., X_n)$ under the assumption that these random variables are independent.

The first thing that can be said about any $F \in \mathfrak{F}(F_1, F_2, ..., F_n)$ is that F satisfies the **Fréchet bounds** of the next proposition. These bounds are also known as the **Fréchet-Hoeffding bounds**, named for **Wassily Hoeffding** (1914–1991).

Proposition 7.2 (Fréchet-Hoeffding bounds) *If* $F \in \mathfrak{F}(F_1, F_2, ..., F_n)$, *then:*

$$\max\left\{0, \sum_{j=1}^n F_j(x_j) - (n-1)\right\} \le F(x_1, x_2, ..., x_n) \le \min_j F_j(x_j). \qquad (7.1)$$

Proof. *First:*

$$\bigcap_{j=1}^n X_j^{-1}(-\infty, x_j] \subset X_j^{-1}(-\infty, x_j], \text{ all } j,$$

and thus by definition in (3.15) and monotonicity of measures:

$$F(x_1, x_2, ..., x_n) \equiv \mu\left[\bigcap_{j=1}^{n} X_j^{-1}(-\infty, x_j]\right]$$
$$\leq \min_j\{\mu\left[X_j^{-1}(-\infty, x_j]\right]\} = \min_j F_j(x_j).$$

For the lower bound, by De Morgan's laws:

$$\mu\left[\bigcap_{j=1}^{n} X_j^{-1}(-\infty, x_j]\right] = 1 - \mu\left[\bigcup_{j=1}^{n} X_j^{-1}(x_j, \infty)\right].$$

Then subadditivity of measures on general unions obtains:

$$\mu\left[\bigcup_{j=1}^{n} X_j^{-1}(x_j, \infty)\right] \leq \sum_{j=1}^{n} \mu\left[X_j^{-1}(x_j, \infty)\right]$$
$$= \sum_{j=1}^{n} [1 - F_j(x_j)],$$

and the result follows. ∎

Another important result that has implications for the continuity of $F(x_1, x_2, ..., x_n)$ is the following.

Proposition 7.3 (Bound for $|F(x) - F(y)|$) *If $F \in \mathfrak{F}(F_1, F_2, ..., F_n)$, then with $x = (x_1, x_2, ..., x_n)$ and $y = (y_1, y_2, ..., y_n)$:*

$$|F(x) - F(y)| \leq \sum_{j=1}^{n} |F_j(x_j) - F_j(y_j)|. \tag{7.2}$$

Proof. *Define $x^{(0)} = x$, and for $k = 1, ..., n$, define $x^{(k)}$ so that $x_j^{(k)} = y_j$ for $j \leq k$ and $x_j^{(k)} = x_j$ for $j > k$. Note that $x^{(k)}$ and $x^{(k-1)}$ differ only in the kth component, and that $x^{(n)} = y$. By the triangle inequality,*

$$|F(x) - F(y)| \leq \sum_{k=1}^{n} \left|F(x^{(k)}) - F(x^{(k-1)})\right|.$$

The proof is completed by showing that for each k:

$$\left|F(x^{(k)}) - F(x^{(k-1)})\right| \leq |F_k(x_k) - F_k(y_k)|.$$

Consider the first term:

$$\left|F(x^{(1)}) - F(x^{(0)})\right| \equiv |F(y_1, x_2, ..., x_n) - F(x_1, x_2, ..., x_n)|.$$

If F is the joint distribution function of random variables $\{X_j\}_{j=1}^{n}$ defined on a probability space $(\mathcal{S}, \mathcal{E}, \mu)$, then if $x_1 \leq y_1$:

$$F(y_1, x_2, ..., x_n) - F(x_1, x_2, ..., x_n) = \mu\left[X_1^{-1}[(x_1, y_1]] \bigcap \bigcap_{j=2}^{n} X_j^{-1}[(-\infty, x_j]]\right]$$
$$\leq \mu\left[X_1^{-1}[(x_1, y_1]]\right]$$
$$= F_1(y_1) - F_1(x_1).$$

If $y_1 \leq x_1$ we reverse the subtraction, and so in either case the inequality of absolute values holds.

The same derivation works for the other terms, proving (7.2). ∎

Corollary 7.4 (Continuous $\{F_j(x_j)\}_{j=1}^n \Rightarrow$ continuous $F(x)$) *If $F \in \mathfrak{F}(F_1, F_2, ..., F_n)$ and $\{F_j(x_j)\}_{j=1}^n$ are continuous functions, then so too is $F(x_1, x_2, ..., x_n)$.*
Proof. *This follows from (7.2), since $y \to x$ implies $y_j \to x_j$ for all j.* ∎

For the next result, recall that by Proposition 6.1, each $F_j(x_j)$ has at most countably many discontinuities, and thus the associated continuity sets are **dense** in $\mathbb{R}$. The reader should prove this last statement as an exercise.

Corollary 7.5 (Continuity set for $F(x)$) *Let $\{A_j\}_{j=1}^n \subset \mathbb{R}$ denote the continuity sets for $\{F_j(x_j)\}_{j=1}^n$, and thus $\mathbb{R} - A_j$ is at most countably infinite for each j. If $A \subset \mathbb{R}^n$ is the continuity set for $F(x_1, x_2, ..., x_n)$, then:*

$$\prod_{j=1}^n A_j \subset A. \tag{7.3}$$

Thus A is dense in $\mathbb{R}^n$, though $\mathbb{R}^n - A$ need not be countably infinite.
Proof. *First, (7.3) follows from (7.2), since if each F_j is continuous at x_j, then F must be continuous at $x \equiv (x_1, x_2, ..., x_n)$.*

To see that A is dense, let $x \in \mathbb{R}^n$. We must prove that for any $r > 0$:

$$B_r(x) \bigcap A \neq \emptyset,$$

where $B_r(x)$ denotes the open ball about x of radius r:

$$B_r(x) = \{y | |x - y| < r\}.$$

First, for any $\epsilon > 0$,

$$(x_j - \epsilon, x_j + \epsilon) \bigcap A_j \neq \emptyset, \text{ all } j,$$

by density of $\{A_j\}_{j=1}^n$. Thus:

$$\prod_{j=1}^n (x_j - \epsilon, x_j + \epsilon) \bigcap \prod_{j=1}^n A_j \neq \emptyset,$$

and then by (7.3):

$$\prod_{j=1}^n (x_j - \epsilon, x_j + \epsilon) \bigcap A \neq \emptyset.$$

Finally, $\prod_{j=1}^n (x_j - \epsilon, x_j + \epsilon) \subset B_{\sqrt{n}\epsilon}(x)$ and so:

$$B_{\sqrt{n}\epsilon}(x) \bigcap A \neq \emptyset.$$

As a simple example that $\mathbb{R}^n - A$ need not be countably infinite, let $F(x_1, x_2) = F_1(x_1)F_2(x_2)$ with $A_1 = \mathbb{R}$ and $A_2 = \mathbb{R} - \{0\}$. Then $F(x_1, x_2)$ has continuity set:

$$A = \mathbb{R} \times (-\infty, 0) \bigcup \mathbb{R} \times (0, \infty) \bigcup \{x_1 | F_1(x_1) = 0\} \times \{0\},$$

and so:

$$\mathbb{R}^2 - A = \{(x_1, 0) | F_1(x_1) \neq 0\}.$$

This example also illustrates that the inclusion in (7.3) can in general be strict. ∎

7.2   Copulas and Sklar's Theorem

For the study of Fréchet's problem, there is a special Fréchet class which will prove to contain the key to the general case.

Definition 7.6 (Fréchet class $\mathfrak{F}(U_1, U_2, ..., U_n)$) *Given distribution functions $\{U_j(x)\}_{j=1}^n$ associated with continuous, uniformly distributed random variables with range $[0,1]$, so $U_j(x) = x$ for $x \in [0,1]$, the class $\mathfrak{F}(U_1, U_2, ..., U_n)$ is the **Fréchet class of joint distributions with uniform marginal distributions**. These joint distribution functions are called **copulas**.*

Sklar's theorem identifies a fundamental relationship between $\mathfrak{F}(U_1, U_2, ..., U_n)$ and $\mathfrak{F}(F_1, F_2, ..., F_n)$ for arbitrary marginals $\{F_j(x)\}_{j=1}^n$. This result is named for **Abe Sklar** (1925–2020), who published these results in 1959.

Proposition 7.7 (Sklar's Theorem) *For any $F \in \mathfrak{F}(F_1, F_2, ..., F_n)$, there exists a copula $C \in \mathfrak{F}(U_1, U_2, ..., U_n)$ so that:*

$$F(x_1, x_2, ..., x_n) = C(F_1(x_1), F_2(x_2), ..., F_n(x_n)). \tag{7.4}$$

*The distribution function C that satisfies (7.4) is called **a copula associated with** F, and also **a copula of the random vector** $X = (X_1, X_2, ..., X_n)$ when $F(x_1, x_2, ..., x_n) = F_X(x_1, x_2, ..., x_n)$, the joint distribution function of X, with associated marginals $\{F_j(x)\}_{j=1}^n$.*
   In addition, for any $C \in \mathfrak{F}(U_1, U_2, ..., U_n)$ and marginals $\{F_j(x)\}_{j=1}^n$, if $F(x_1, x_2, ..., x_n)$ is defined as in (7.4), then $F \in \mathfrak{F}(F_1, F_2, ..., F_n)$.
Proof. *This will be proved in stages below.* ∎

While the proof of this result will take some work, we can enjoy a corollary quickly.

Corollary 7.8 (Sklar's Theorem - Marginal DFs) *Let $F \in \mathfrak{F}(F_1, F_2, ..., F_n)$ and $C \in \mathfrak{F}(U_1, U_2, ..., U_n)$ be given as in Proposition 7.7. For $I = \{i_1, ..., i_m\} \subset \{1, 2, ..., n\}$, the marginal distribution function $F_I(x_I) \equiv F_I(x_{i_1}, x_{i_2}, ..., x_{i_m})$ is given on $\mathbb{R}^m$ by:*

$$F_I(x_I) = C_I(F_{i_1}(x_{i_1}), F_{i_2}(x_{i_2}), ..., F_{i_m}(x_{i_m})), \tag{7.5}$$

where with $u_I \equiv (u_{i_1}, u_{i_2}, ..., u_{i_m})$ and $u_J \equiv (u_{j_1}, u_{j_2}, ..., u_{j_{n-m}})$ for $j_k \in J \equiv \widetilde{I}$:

$$C_I(u_I) \equiv \lim_{u_J \to 1} C(u_1, u_2, ..., u_n). \tag{7.6}$$

In other words, $C_I \in \mathfrak{F}(U_{i_1}, U_{i_2}, ..., U_{i_m})$, and thus all marginal distributions of a copula are copulas.
   Further, if

$$I' = \{i'_1, ..., i'_l\} \subset I = \{i_1, ..., i_m\},$$

then:

$$(C_I)_{I'}(u_{I'}) = C_{I'}(u_{I'}). \tag{7.7}$$

In other words, the marginal distributions of C_I agree with the comparably defined marginal distributions of C.

Proof. *By (3.19):*

$$
\begin{aligned}
F_I\left(x_I\right) & \equiv \lim_{x_J \to \infty} F(x_1, x_2, ..., x_n) \\
& = \lim_{x_J \to \infty} C(F_1\left(x_1\right), F_2\left(x_2\right), ..., F_n\left(x_n\right)) \\
& = C_I(F_{i_1}\left(x_{i_1}\right), F_{i_2}\left(x_{i_2}\right), ..., F_{i_m}\left(x_{i_m}\right)),
\end{aligned}
$$

since for any $j_k \in J$, $F_{j_k}\left(x_{j_k}\right) \to 1$ as $x_{j_k} \to \infty$.

To prove that $C_I \in \mathfrak{F}(U_{i_1}, U_{i_2}, ..., U_{i_m})$, recall that $C(u_1, u_2, ..., u_n)$ is a joint distribution function, and thus $C_I\left(u_I\right)$ is a marginal distribution function of C by Definition 3.34. To prove that $C_I\left(u_I\right)$ has uniform marginal distributions and is thus a copula, consider $(C_I)_{\{i_1\}}\left(u_{i_1}\right)$, the marginal distribution function of $C_I\left(u_I\right)$ with respect to the first variate. Other than notation, the other marginals are proved similarly.

Letting $u_{I'} \equiv (u_{i_2}, ..., u_{i_m})$, we have by definition:

$$
\begin{aligned}
(C_I)_{\{i_1\}}\left(u_{i_1}\right) & = \lim_{u_{i'} \to 1} C_I(u_{i_1}, u_{i_2}, ..., u_{i_m}) \\
& = \lim_{u_{I'} \to 1} \lim_{u_J \to 1} C(u_1, u_2, ..., u_n). \tag{1}
\end{aligned}
$$

Now $C \in \mathfrak{F}(U_1, U_2, ..., U_n)$ and thus has uniform marginals. With $K \equiv I' \bigcup J$, it follows that the marginal distribution $C_K\left(u_{i_1}\right)$ is well defined by:

$$
C_K\left(u_{i_1}\right) = \lim_{u_K \to 1} C(u_1, u_2, ..., u_n). \tag{2}
$$

Comparing (1) with (2), the existence of the latter limit proves the existence of the former for any such partition of K. Recalling that $C_K\left(u_{i_1}\right) = u_{i_1}$ completes the proof.

The proof of (7.7) is similar to the above proof and left as an exercise. ∎

The language above, referring to "a" copula associated with F rather "the" copula, is intended to connote that copula functions need not be uniquely defined as distributions on $[0, 1]^n$. However, copulas are uniquely defined in the more limited sense of the next result. See also Example 7.11 below.

Proposition 7.9 (On uniqueness of C) *If a copula function C associated with F exists, it is uniquely defined on the range of the marginal distribution vector $Rng(F_1, F_2, ..., F_n)$, defined by:*

$$
Rng(F_1, F_2, ..., F_n) \equiv \prod_{j=1}^{n} Rng[F_j] \subset [0, 1]^n.
$$

Thus if $\{F_j(x)\}_{j=1}^{n}$ are continuous, then any copula function C associated with F is uniquely defined on a set that contains $(0, 1)^n$.

Proof. *If C_1 and C_2 are copula functions associated with a given F, and $(u_1, u_2, ..., u_n) \in Rng(F_1, F_2, ..., F_n)$, meaning that there exists $(x_1, x_2, ..., x_n)$ so that:*

$$
(u_1, u_2, ..., u_n) = (F_1\left(x_1\right), F_2\left(x_2\right), ..., F_n\left(x_n\right)),
$$

then by the respective representations for $F(x_1, x_2, ..., x_n)$ in (7.4) it follows that:

$$
C_1(u_1, u_2, ..., u_n) = C_2(u_1, u_2, ..., u_n).
$$

If $\{F_j(x)\}_{j=1}^{n}$ are continuous, then by (6.3) and (6.4) of Proposition 6.1 and the intermediate value theorem, $(0, 1) \subset Rng[F_j]$ for all j, and the result follows. ∎

Remark 7.10 (On copulas) *Sklar named the special joint distributions $C \in \mathfrak{F}(U_1, U_2, ..., U_n)$ **copulas,** and thus the near-universal notation of C for these distribution functions. Some new to this subject might wonder why those small architectural structures, often seen on the top of a dome or roof, provide the intuitive model for a copula. But these structures are in fact **cupolas,** a word from Latin which means "a small cask."*

*The word **copula** also comes from Latin and means "something that connects or links other things." It is commonly used in linguistics as a synonym for a "linking verb," which links the subject and the subject complement of a sentence. For example, "Sklar **was** professor emeritus of applied mathematics."*

This word is consequently fittingly applied to C, as it serves to connect a joint distribution function with its marginal distribution functions.

It is not difficult to illustrate the non-uniqueness of copulas outside the scope of Proposition 7.9.

Example 7.11 (Non-uniqueness of C) *Let X and Y be binomial variates defined on a probability space $(\mathcal{S}, \mathcal{E}, \mu)$ with range $\{-1, 1\}$, and a joint distribution function specified by:*

$$F(x,y) = \begin{cases} 1, & (x,y) = (1,1), \\ 0.6, & (x,y) = (-1,1), \\ 0.4, & (x,y) = (1,-1), \\ 0.2, & (x,y) = (-1,-1). \end{cases}$$

It is worth a moment for the reader to confirm that this sparse specification is enough to uniquely define $F(x,y) \equiv \mu\left((X,Y)^{-1}\left[(-\infty, x] \times (-\infty, y]\right]\right)$ on $\mathbb{R}^2$. Moreover, this well-defined function is indeed a distribution function and satisfies the 2-increasing condition in (6.11), and the continuity from above requirement of (6.10).

The associated marginal distribution functions are specified by:

$$F_X(x) = \begin{cases} 0.6, & x = -1, \\ 1, & x = 1, \end{cases} \qquad F_Y(y) = \begin{cases} 0.4, & y = -1, \\ 1, & y = 1. \end{cases}$$

Thus $Rng(F_X, F_Y)$, as defined in Proposition 7.9, contains nine points in $\mathbb{R}^2$:

$$Rng(F_X, F_Y) = \left\{ \begin{array}{ccc} (0,0), & (0,0.4), & (0,1), \\ (0.6,0), & (0.6,0.4), & (0.6,1), \\ (1,0), & (1,0.4), & (1,1), \end{array} \right\}$$

and any copula function C for F is uniquely defined on these points.

*Now every copula on $\mathbb{R}^2$ satisfies $C(u,1) = u$ and $C(1,v) = v$, since these are the marginal distributions of C, as well as $C(0,0) = 0$ by definition. Thus a copula function C for **this** distribution function F is additionally constrained only by its values on the three points: $(0,0.4)$, $(0.6,0)$, and $(0.6,0.4)$. Then to have:*

$$F(x,y) = C(F_X(x), F_Y(y))$$

on the associated (x,y)-points requires that:

$$C(0,0.4) = C(0.6,0) = C(0.6,0.4) = 0.2.$$

This example illustrates that discontinuities in the marginal distribution functions contribute to the nonuniqueness of C. For example, the discontinuities at $x = -1, y = -1$ allow C to be almost arbitrarily defined when $0 < x < 0.6$ and $0 < y < 0.4$ subject to the constraint that C be a distribution function with uniform marginal distributions.

Though these are immediate consequences of the results of the prior section, we highlight two important properties of copulas.

Proposition 7.12 (Fréchet-Hoeffding bounds; bound for $|C(u) - C(v)|$)
If $C \in \mathfrak{F}(U_1, U_2, ..., U_n)$, then:

    1. *Fréchet-Hoeffding bounds:*

$$\max\left\{0, \sum\nolimits_{j=1}^{n} u_j - (n-1)\right\} \leq C(u_1, u_2, ..., u_n) \leq \min_j u_j. \tag{7.8}$$

    2. *C is **Lipschitz continuous** on $[0,1]^n$, and for $u, v \in [0,1]^n$:*

$$|C(u) - C(v)| \leq \sum\nolimits_{j=1}^{n} |u_j - v_j|. \tag{7.9}$$

Proof. *Immediate from Propositions 7.2 and 7.3.* ∎

Remark 7.13 (Lipschitz continuity) *The notion of Lipschitz continuity is named for **Rudolf Lipschitz** (1832–1903).*
A function $f : \mathbb{R} \to \mathbb{R}$ is said to be Lipschitz continuous if there exists a constant $L > 0$ so that for all $x, x' \in \mathbb{R}$:
$$|f(x) - f(x')| \leq L|x - x'|.$$

More generally, a function f between metric spaces, $f : (X, d_X) \to (Y, d_Y)$ is said to be Lipschitz continuous if there exists a constant $L > 0$ so that for all $x, x' \in X$:
$$d_Y[f(x), f(x')] \leq L d_X[x, x'].$$

*In the above application, $(X, d_X) = (\mathbb{R}^n, d_1)$ and $(Y, d_Y) = (\mathbb{R}, |\cdot|)$, where the d_1-distance function is induced by the l_1-norm on $\mathbb{R}^n$ and defined on the right of (7.9), while $|\cdot|$ denotes the standard distance function on $\mathbb{R}$. See Chapters 3 and 6 of **Reitano** (2010) for more on l_p-norms on $\mathbb{R}^n$ and $\mathbb{R}^{\mathbb{N}}$, and Book V for and the integral-based generalizations of these norms to L_p-norms.*

Not only can a given joint distribution be associated with more than one copula, but a given copula can be associated with more than one joint distribution function. The following result states that a copula C for the random vector $(X_1, X_2, ..., X_n)$ is also a copula for $(G_1(X_1), G_2(X_2), ..., G_n(X_n))$ if $\{G_j\}_{j=1}^{n}$ are strictly increasing functions. Note that this statement is well defined since increasing functions are Borel measurable, and thus $(G_1(X_1), G_2(X_2), ..., G_n(X_n))$ is a random vector.

The transformation:

$$(X_1, X_2, ..., X_n) \to (G_1(X_1), G_2(X_2), ..., G_n(X_n)),$$

preserves ranking properties of these variates since all G_j are increasing functions. Thus this result is sometimes referred to as the **rank-invariance of copulas.**

Proposition 7.14 (Rank-Invariance of Copulas) *If C is a copula for the random vector $(X_1, X_2, ..., X_n)$ and $\{G_j\}_{j=1}^{n}$ are strictly increasing functions, then C is also a copula for the random vector $(G_1(X_1), G_2(X_2), ..., G_n(X_n))$.*
Proof. *If C is a copula for the random vector $(X_1, X_2, ..., X_n)$, then by (7.4):*

$$F(x_1, x_2, ..., x_n) = C(F_1(x_1), F_2(x_2), ..., F_n(x_n)).$$

Let $F_G(x_1, x_2, ..., x_n)$ denote the distribution function of the random vector $(G_1(X_1), G_2(X_2), ..., G_n(X_n))$, which is well defined as noted above. Then since each G_j is strictly increasing, each G_j^{-1} is well defined.

Assume that $(X_1, X_2, ..., X_n)$ is defined on a probability space $(\mathcal{S}, \mathcal{E}, \mu)$ for notational convenience. Then:

$$
\begin{aligned}
F_G(x_1, x_2, ..., x_n) &= \mu\left[\{G_j(X_j) \le x_j, \; all\; j\}\right] \\
&= \mu\left[\{X_j \le G_j^{-1}(x_j), \; all\; j\}\right] \\
&= F(G_1^{-1}(x_1), G_2^{-1}(x_2), ..., G_n^{-1}(x_n)) \\
&= C(F_1(G_1^{-1}(x_1)), F_2(G_2^{-1}(x_2)), ..., F_n(G_n^{-1}(x_n))).
\end{aligned}
$$

Now $F_j(G_j^{-1}(x_j))$ is the distribution function of $G_j(X_j)$ since:

$$\{G_j(X_j) \le x_j\} = \{X_j \le G_j^{-1}(x_j)\},$$

and hence C is also a copula of $(G_1(X_1), G_2(X_2), ..., G_n(X_n))$. ∎

7.2.1   Identifying Copulas

In cases where the joint distribution function $F(x_1, x_2, ..., x_n)$ is known, the associated copula can sometimes be derived by algebraic or numerical methods. For example, from the 2-dimensional normal distribution and its normal marginal distributions one can estimate the normal copula function. A corollary benefit of such an exercise is to obtain a collection of copulas that can then be tested in a given application for which the marginal distributions have been estimated, but the theoretical joint distribution function is unknown. This copula testing problem can arise in the following way.

First, the data set on which the marginals are based provides some information on this joint distribution function. In particular, if this sample data set is represented by $\left\{\left(x_1^{(j)}, x_2^{(j)}, ..., x_n^{(j)}\right)\right\}_{j=1}^N$, then the distribution function can be estimated:

$$F\left(x_1^{(k)}, x_2^{(k)}, ..., x_n^{(k)}\right) \approx N_k/N.$$

Here N_k is the number of sample points with:

$$\left(x_1^{(j)}, x_2^{(j)}, ..., x_n^{(j)}\right) \le \left(x_1^{(k)}, x_2^{(k)}, ..., x_n^{(j)}\right),$$

where this vector inequality denotes component-wise inequalities. These estimates then yield what in effect is a generalized version of Example 7.11 above, and thus can only uniquely define the associated copula on the range of the marginal distribution vector as defined in Proposition 7.9.

Ignoring estimation errors in the above data estimates, C is defined for $1 \le j \le N$ by:

$$C\left(F_1\left(x_1^{(j)}\right), F_2\left(x_2^{(j)}\right), ..., F_n\left(x_n^{(j)}\right)\right) = F\left(x_1^{(k)}, x_2^{(k)}, ..., x_n^{(k)}\right),$$

where the marginal distributions $\{F_i\}_{i=1}^n$ are defined empirically by, or estimated from, $\{x_i^{(j)}\}$ for $i = 1, .., n$.

This is referred to as an **empirical copula**. Hence, in such applications, the objective is to test a variety of known copula functions in search of a model that fits well enough to justify probability estimates outside the original data set. As there is no practical way to identify the "best copula" among **all** copulas, the usual approach is to identify the best copula only within a given **parametric class of copulas** using maximum likelihood or other methods.

Specifically, if $\{F_k\,(x_k)\}_{k=1}^n$ are marginal distribution functions estimated from the individual variates of the data set, and C is a "trial" copula, then one models a theoretical joint distribution function by (7.4):

$$F(x_1, x_2, ..., x_n) = C(F_1\,(x_1)\,, F_2\,(x_2)\,, ..., F_n\,(x_n)).$$

The second result in Sklar's theorem and proved in Proposition 7.16 assures that given any copula function C, $F(x_1, x_2, ..., x_n)$ so defined is indeed a distribution function with the specified marginal distributions.

Thus we are left to evaluate for various C how well such a specification matches the empirical estimates above for vectors in the original data set.

Example 7.15

1. *In two variables $x_1, x_2 \geq 0$ let:*

$$F(x_1, x_2) = (1 + e^{-x_1} + e^{-x_2})^{-1}.$$

This is a distribution function because it is continuous and hence continuous from above as defined in (6.10), and satisfies the 2-increasing condition in (6.11) because this function has a positive density function as can be derived by differentiation and using Riemann integration theory.

The marginal distribution functions are given by $F(x_1) = (1 + e^{-x_1})^{-1}$ and $F(x_2) = (1 + e^{-x_2})^{-1}$, and thus by algebraic manipulations the copula function associated with this joint distribution is:

$$C(u_1, u_2) = u_1 u_2/(u_1 + u_2 - u_1 u_2).$$

2. *As an example of a **parametric class of copulas**, called the **Gumbel class** in Section 7.4, let:*

$$F(x_1, x_2; \theta) = \exp\left[-(e^{-\theta x_1} + e^{-\theta x_2})^{1/\theta}\right], \quad \theta \geq 1.$$

Then $F(x_1) = \exp\left[-e^{-x_1}\right]$ and $F(x_2) = \exp\left[-e^{-x_2}\right]$ are independent of θ, and:

$$C(u_1, u_2 : \theta) = \exp\left[-\left[(-\ln u_1)^\theta + (-\ln u_2)^\theta\right]^{1/\theta}\right].$$

Thus we have a family of copulas parametrized by θ, which we can exploit in a data fitting exercise.

7.3 Partial Results on Sklar's Theorem

In this section we prove the general version of part 2 of Sklar's theorem. Specifically, given $C \in \mathfrak{F}(U_1, U_2, ..., U_n)$ and arbitrary marginal distributions, $\{F_j\}_{j=1}^n$, define:

$$F(x_1, x_2, ..., x_n) \equiv C(F_1\,(x_1)\,, F_2\,(x_2)\,, ..., F_n\,(x_n))$$

as in (7.4). We prove that F is a joint distribution function with the given marginal distributions, and hence $F \in \mathfrak{F}(F_1, F_2, ..., F_n)$.

In addition, we prove a special case of part 1 of Sklar's theorem, related to the existence of a copula C given $F(x_1, x_2, ..., x_n)$. This is the case of continuous marginal distributions, for which the copula C can be explicitly and uniquely defined.

Proposition 7.16 (Sklar's Theorem, part 2) *Given* $C \in \mathfrak{F}(U_1, U_2, ..., U_n)$ *and distribution functions* $\{F_j(x)\}_{j=1}^n$:

$$C(F_1(x_1), F_2(x_2), ..., F_n(x_n)) \in \mathfrak{F}(F_1, F_2, ..., F_n). \tag{7.10}$$

In other words:

$$F(x_1, x_2, ..., x_n) \equiv C(F_1(x_1), F_2(x_2), ..., F_n(x_n)),$$

is a distribution function with the given marginal distributions.

Proof. *First,* $F(x_1, x_2, ..., x_n)$ *is a well-defined function because* $F_j(x_j) \in [0, 1]$ *for all* x_j. *Also,* $\{F_j(x_j)\}_{j=1}^n$ *are the marginal distribution functions of* F. *For example, if* $x_2, ..., x_n \to \infty$, *then* $F_2(x_2) \to 1, ..., F_n(x_n) \to 1$ *and by (3.19):*

$$F_{\{1\}}(x_1) \equiv C(F_1(x_1), 1, ..., 1).$$

But then $F_{\{1\}}(x_1) = F_1(x_1)$ *since a copula has uniform marginal distribution functions by definition.*

To prove that $F(x)$ *is a joint distribution function, it is sufficient to prove that* $F(x)$ *is continuous from above as in (6.10), and satisfies the n-increasing condition in (6.11). Then by Proposition 6.10, F is a joint distribution function. For this result, it should be verified as an exercise that* $F(-\infty) = 0$ *and* $F(\infty) = 1$, *defined as limits.*

That F is continuous from above follows because each $F_j(x_j)$ *is monotonic and right continuous, and the distribution function C is continuous from above. Specifically, given* $x = (x_1, ..., x_n)$ *and a sequence* $x^{(m)} = (x_1^{(m)}, ..., x_n^{(m)})$ *with* $x_j^{(m)} \geq x_j$ *for all j and m, and* $x^{(m)} \to x$ *as* $m \to \infty$, *it follows by monotonicity and right continuity of F_j that* $F_j\left(x_j^{(m)}\right) \geq F_j(x_j)$ *for all j and m, and* $F_j\left(x_j^{(m)}\right) \to F_j(x_j)$ *as* $m \to \infty$ *for all j. Thus since C is a distribution function and continuous from above:*

$$
\begin{aligned}
F\left(x_1^{(m)}, x_2^{(m)}, ..., x_n^{(m)}\right) &\equiv C\left(F_1\left(x_1^{(m)}\right), ..., F_n\left(x_n^{(m)}\right)\right) \\
&\to C\left(F_1(x_1), ..., F_n(x_n)\right) \\
&= F(x_1, x_2, ..., x_n).
\end{aligned}
$$

To prove the n-increasing condition in (6.11), let $\prod_{i=1}^n (a_i, b_i] \subset \mathbb{R}^n$ *be a bounded right semi-closed rectangle. It must be proved that:*

$$\sum_x sgn(x) F(x) \geq 0,$$

where each $x = (x_1, ..., x_n)$ *is one of the 2^n vertices of* $\prod_{i=1}^n (a_i, b_i]$, *and* $sgn(x)$ *is defined as -1 if the number of components of x that equal a_i is odd, and as $+1$ otherwise.*

Given such a vertex x, let $y = (F_1(x_1), F_2(x_2), ..., F_n(x_n))$, *where now each component equals* $F_i(a_i)$ *or* $F_i(b_i)$. *Then:*

$$\sum_x sgn(x) F(x) = \sum_y sgn(y) C(y),$$

where $sgn(y)$ *is defined as -1 if the number of components of y that equal $F_i(a_i)$ is odd,* *and as +1 otherwise, and so* $sgn(y) = sgn(x)$. *Since* C *is a distribution function, we obtain from Proposition 6.9 that if* $F_i(a_i) < F_i(b_i)$ *for all* i:

$$\sum_y sgn(y)C(y) = \mu_C\left[\prod_{i=1}^n (F_i(a_i), F_i(b_i))\right] \geq 0,$$

where μ_C *is the Borel measure induced by* C.

Nonnegativity of this signed summation is preserved in the case where some or all $F_i(a_i) = F_i(b_i)$, *by continuity from above of* μ_C. *Specifically, if* $F_i(a_i) \leq F_i(b_i) < c_i$ *for all* i, *then as just proved:*

$$\mu_C\left[\prod_{i=1}^n (F_i(a_i), c_i]\right] \geq 0.$$

If $\left\{c_i^{(k)}\right\}_{k=1}^\infty$ *is a decreasing sequence for each* i *with limit* $F_i(b_i)$, *then* $A^{(k)} \equiv \prod_{i=1}^n (F_i(a_i), c_i^{(k)}]$ *is a nested sequence of sets with:*

$$\bigcap_{k=1}^\infty A^{(k)} = \prod_{i=1}^n (F_i(a_i), F_i(b_i)].$$

Thus by continuity from above:

$$\mu_C\left[\prod_{i=1}^n (F_i(a_i), F_i(b_i))\right] = \lim_{k\to\infty} \mu_C\left[\prod_{i=1}^n (F_i(a_i), c_i^{(k)}]\right] \geq 0.$$

∎

Remark 7.17 (Identifying copulas) *This result justifies the discussion in Section 7.2.1 related to testing and using copulas obtained from one application in another application, where* $\{F_j(x)\}_{j=1}^n$ *have been estimated.*

In the special case of **continuous marginal distributions**, part 1 of Sklar's theorem is easier to prove because a copula C associated with a joint distribution function F can be explicitly identified. And in this case we show that C is essentially unique and thus is *the* copula associated with F. For this statement, recall the **left continuous inverse** of a distribution as given in Definition 3.12.

Proposition 7.18 (Sklar's Theorem, part 1 - Continuous marginals) *If:*

$$F(x_1, x_2, ..., x_n) \in \mathfrak{F}(F_1, F_2, ..., F_n),$$

where $\{F_j(x)\}_{j=1}^n$ *are continuous distribution functions, then a copula associated with* F *is given by:*

$$C(u_1, u_2, ..., u_n) = F(F_1^*(u_1), F_2^*(u_2), ..., F_n^*(u_n)), \tag{7.11}$$

where $F_j^*(u)$ *denotes the left continuous inverse of the marginal distribution function* $F_j(x)$. *Further, the copula defined in (7.11) is unique on a set which contains* $(0,1)^n$.

If each $F_j(x)$ *is also strictly increasing, then* $F_j^*(u) = F_j^{-1}(u)$ *and thus:*

$$C(u_1, u_2, ..., u_n) = F(F_1^{-1}(u_1), F_2^{-1}(u_2), ..., F_n^{-1}(u_n)). \tag{7.12}$$

Proof. *For notational convenience, assume that* F *is the joint distribution function of* $X = (X_1, X_2, ..., X_n)$, *defined on some probability space* $(\mathcal{S}, \mathcal{E}, \mu)$. *Since* $\{F_j(x)\}_{j=1}^n$ *are continuous, they are Borel measurable by Proposition I.3.11, and thus*

$(F_1(X_1), F_2(X_2), ..., F_n(X_n))$ *is a random vector on* $(\mathcal{S}, \mathcal{E}, \mu)$. *Define* $C(u_1, u_2, ..., u_n)$ *as the distribution function of this random vector:*

$$C(u_1, u_2, ..., u_n) = \mu\left[\bigcap_{j=1}^n \{F_j(X_j) \leq u_j\}\right], \tag{1}$$

where $\{F_j(X_j) \leq u_j\} \equiv \{s | F_j(X_j(s)) \leq u_j\}$.

Letting $u_2, ..., u_n \to 1$, *then* $\{F_j(X_j) \leq u_j\} \to \mathcal{S}$ *for* $j \geq 2$ *and thus by Definition 3.34, the marginal distribution function* $C_{\{1\}}(u_1, 1, ..., 1)$ *is given by:*

$$C_{\{1\}}(u_1, 1, ..., 1) = F_{F_1(X_1)}(u_1).$$

In other words, the marginal distribution function $C_{\{1\}}(u_1, 1, ..., 1)$ *is the distribution function of the random variable* $F_1(X_1)$. *The same derivation and conclusion hold similarly for all such marginal distribution functions of* C.

By part 1 of Proposition 4.8, $F_{F_1(X_1)}(u_1) = u_1$ *if and only if* F_1 *is continuous, and thus by assumption* $C_{\{1\}}(u_1, 1, ..., 1) = u_1$. *The same is true for all the marginal distribution functions of* C. *As these marginal distributions are those of a continuous, uniformly distributed random variable with range* $[0, 1]$, *and* C *is a distribution function by definition in* (1), *it follows that* $C \in \mathfrak{F}(U_1, U_2, ..., U_n)$ *and thus* C *is a copula.*

To prove that this copula C *is that given in (7.11), note that since each* F_j *is increasing:*

$$\{X_j \leq F_j^*(u_j)\} = \{F_j(X_j) \leq F_j[F_j^*(u_j)]\} = \{F_j(X_j) \leq u_j\}.$$

For the last step, $F_j[F_j^*(u_j)] = u_j$ *for* $u_j \in (0, 1)$ *by part 1 of Proposition 3.22 and continuity of* F_j. *Combining this with the representation above for* C:

$$C(u_1, u_2, ..., u_n) = \mu\left[\bigcap_{j=1}^n \{X_j \leq F_j^*(u_j)\}\right] = F(F_1^*(u_1), F_2^*(u_2), ..., F_n^*(u_n)),$$

which is (7.11).

To prove that C *so defined is a copula associated with* F, *we must verify (7.4). By the definition of* C *in (7.11), it must be verified that:*

$$F(x_1, x_2, ..., x_n) \equiv F(F_1^*(F_1(x_1)), F_2^*(F_2(x_2)), ..., F_n^*(F_n(x_n))). \tag{2}$$

The expression on the right can be restated in terms of μ *as:*

$$F(F_1^*(F_1(x_1)), F_2^*(F_2(x_2)), ..., F_n^*(F_n(x_n))) = \mu\left[\bigcap_{j=1}^n \{X_j \leq F_j^*(F_j(x_j))\}\right].$$

To prove (2) it is enough to show that outside a set of μ-*measure 0:*

$$\{X_j \leq F_j^*(F_j(x_j))\} = \{X_j \leq x_j\},$$

and this would be proved by definition if:

$$F_j[F_j^*(F_j(x_j))] = F_j(x_j).$$

But by continuity of F_j, $F_j[F_j^*(u_j)] = u_j$ *for all* $u_j \in (0, 1)$ *by part 1 of Proposition 3.22, and we let* $u_j \equiv F_j(x_j)$. *Thus* C *is a copula associated with* F.

Uniqueness of C *on a set which contains* $(0, 1)^n$ *is Proposition 7.9 by continuity of* $\{F_j(x)\}_{j=1}^n$.

Finally, if each $F_j(x)$ *is continuous and also strictly increasing, then* $F_j^*(u) = F_j^{-1}(u)$ *by Proposition 3.22, and (7.12) follows.* ∎

7.4 Examples of Copulas

Before turning to the proof of part 2 of Sklar's theorem in the case of general marginal distribution functions, we discuss several common examples. Some of these examples reflect intuitive ideas to be formalized in later chapters and/or books. These examples are often categorized into the **explicit copulas,** for which the formula for $C(u_1, u_2, ..., u_n)$ is explicitly given, and the **implicit copulas,** for which there is no simple closed form representation, but these copulas are implied by well-known multivariate distribution functions.

Example 7.19

1. **Independence Copula:** *Perhaps the simplest copula is:*

$$C^I(u_1, u_2, ..., u_n) = u_1 u_2 ... u_n,$$

   which by (3.30) equals the joint distribution function of independent, uniformly distributed random variates, $\{U_j\}_{j=1}^n$, with range $[0,1]$.

2. **Comonotonicity Copula:** *By (7.1), this copula is the upper bound of all possible copulas:*

$$C^{Co}(u_1, u_2, ..., u_n) = \min_j u_j.$$

   It also equals the joint distribution function of uniformly distributed variates where $U_1 = U_2 = \cdots = U_n$ almost surely. While this distribution function can also in theory be defined by $\max_j u_j$, due to almost sure equality, this is not then a copula because the marginal distribution functions are not uniform distributions.

3. **Countermonotonicity Copula:** *It can be proved that the lower bound in (7.1) is not a copula for $n \geq 3$. However, for each fixed $(u_1, u_2, ..., u_n)$:*

$$\inf_C C(u_1, u_2, ..., u_n) = \max\left\{0, \sum_{j=1}^n u_j - (n-1)\right\},$$

   where the infimum is defined over all copulas. In the case $n = 2$, this lower bound is the countermonotonicity copula:

$$C^{Cn}(u_1, u_2) = \max\{0, u_1 + u_2 - 1\},$$

   and equals the distribution function of uniform variates with $U_1 = 1 - U_2$ almost surely.

4. **Gaussian Copula:** *As was the case for the normal probability measure, the Gaussian copula is named for **Carl Friedrich Gauss** (1777–1855) because it is related to the multivariate normal distribution function to be discussed in Book VI. It is defined by:*

$$C^G(u_1, ..., u_n) = \int_{-\infty}^{\Phi^{-1}(u_n)} \cdots \int_{-\infty}^{\Phi^{-1}(u_1)} c^G(t_1, ..., t_n) dt_1 ... dt_n.$$

   The copula density function is defined by:

$$c^G(t_1, ..., t_n) = \frac{1}{(2\pi)^{n/2} [\det R]^{1/2}} \exp\left[-\frac{1}{2} t^T R^{-1} t\right], \tag{7.13}$$

   where $t = (t_1, ..., t_n)$ is identified with a column matrix, and t^T the row matrix "transpose" of t. The matrix R is an $n \times n$ symmetric matrix with element R_{ij} equal to the *correlation between U_i and U_j, which we formally define in Book IV, and $\det R$ denotes the determinant of this matrix. This density will be studied in detail in Book VI.*

The upper limits of integration, $\Phi^{-1}(u_j)$, denote the inverse of the normal distribution function. Recalling (1.24), this inverse function $\Phi^{-1}(u)$ is defined for $u \in (0,1)$ by

$$\Phi\left(\Phi^{-1}(u)\right) = u,$$

or:

$$\int_{-\infty}^{\Phi^{-1}(u)} \phi(x)dx = u,$$

where $\phi(x) = \frac{1}{\sqrt{2\pi}}\exp\left(-x^2/2\right)$.

5. **Student T Copula:** *The univariate Student T distribution is discussed in Book IV in the context of ratios of independent random variables. There is also a multivariate version often used as a copula function which is introduced here for completeness. For this copula there is a Student T parameter $\nu > 0$, called the **degrees of freedom**, and then as for the Gaussian copula:*

$$C^T(u_1, ..., u_n) = \int_{-\infty}^{T_\nu^{-1}(u_n)} \cdots \int_{-\infty}^{T_\nu^{-1}(u_1)} c^T(t_1, ..., t_n)dt_1...dt_n.$$

The copula density function is defined by:

$$c^T(t_1, ..., t_n) = \frac{\Gamma((\nu+n)/2)}{(\pi\nu)^{n/2}\Gamma(\nu/2)[\det R]^{1/2}}\left(1 + \left(t^T R^{-1} t\right)/\nu\right)^{-(\nu+n)/2}, \tag{7.14}$$

*and $\Gamma(y)$ is the **gamma function** defined:*

$$\Gamma(y) = \int_0^\infty x^{y-1}e^{-x}dx. \tag{7.15}$$

As above, $T_\nu^{-1}(u_j)$ is the inverse of the univariate Student T distribution function with $\nu > 0$ degrees of freedom, and defined as in the normal case but with the Student T density function given by:

$$f_T(t) = \frac{\Gamma((\nu+1)/2)}{\sqrt{\pi\nu}\Gamma(\nu/2)}\left(1 + \frac{t^2}{\nu}\right)^{-(\nu+1)/2}. \tag{7.16}$$

*It is named for **William Sealy Gosset** (1876–1937) who published under the pen name of Student.*

7.4.1   Archimedean Copulas

A copula is said to be an Archimedean copula if there exists a function, $\psi_\theta(t)$, called the **Archimedean generator function** or simply the **generator function,** so that with the generalized inverse or **pseudo-inverse** $\psi_\theta^{[-1]}$ defined in (7.18) below:

$$C^A(u_1, u_2, ..., u_n) = \psi_\theta\left(\psi_\theta^{[-1]}(u_1) + \psi_\theta^{[-1]}(u_2) + \cdots \psi_\theta^{[-1]}(u_n)\right). \tag{7.17}$$

The generator function $\psi_\theta(t)$, where θ is understood as a parameter, satisfies:

- $\psi_\theta : [0, \infty) \to [0, 1]$ is continuous;

- $\psi_\theta(0) = 1$ and $\lim_{t \to \infty} \psi_\theta(t) = 0$;

- ψ_θ is strictly decreasing on $[0, \psi_\theta^{[-1]}(0)]$, and the generalized inverse is defined by:

$$\psi_\theta^{[-1]}(u) = \begin{cases} \psi_\theta^{-1}(u), & 0 < u \leq 1, \\ \inf\{t | \psi_\theta(t) = 0\}, & u = 0. \end{cases} \tag{7.18}$$

Remark 7.20 (On $\psi_\theta^{[-1]}(u)$) *Note that $\psi_\theta^{-1}(u)$ is well defined on $(0,1]$ since ψ_θ is continuous and strictly decreasing. Then $\psi_\theta^{[-1]}(0) = \infty$ if $\psi_\theta(t) > 0$ for all t, and otherwise $\psi_\theta^{[-1]}(0) = \lim_{u \to 0} \psi_\theta^{-1}(u)$. This definition allows for generator functions like $\psi_\theta(t) = e^{-\theta t}$, for which $\psi_\theta^{-1}(0) = \infty$ and which decrease on $[0, \infty)$ if $\theta > 0$, as well as generator functions such as $\psi_\theta(t) = (1 - t/\theta)^\theta$, for which $\psi_\theta^{-1}(0) = \theta$ and which decrease on $[0, \theta]$ if $\theta > 0$.*

Notation: Because $\psi_\theta^{[-1]}(u)$ is equal to $\psi_\theta^{-1}(u)$ in all cases with one exception, we will in the interest of notational simplicity, and hopefully with little chance for confusion, remove the brackets and use $\psi_\theta^{-1}(u)$ below.

Notation: It should be noted that the representation in (7.17) is not universal, and that in some references the roles of ψ_θ and $\psi_\theta^{[-1]}$ are reversed. The above convention is consistent with the papers noted next.

Exercise 7.21

1. *Verify that the above assumptions on $\psi_\theta(t)$ assure that $C^A(u_1, u_2, ..., u_n)$ as defined in (7.17) is continuous from above, and has the necessary uniform marginal distributions to be a copula.*

2. *Verify that $\psi_\theta(t) = e^{-\theta t}$ produces the **independence copula** for any $\theta > 0$.*

While Exercise 7.21 confirms that the basic assumptions on $\psi_\theta(t)$ assure that $C^A(u_1, u_2, ..., u_n)$ is continuous from above and has the necessary uniform marginal distributions, more is needed to ensure the n-increasing condition in (6.11). We state without proof two important results. The first is from **C. H. Kimberling** (1974), and the second from **Alexander J. McNeil** and **Johanna Nešlehová** (2009).

Proposition 7.22 (Kimberling) *The function ψ_θ above generates an Archimedean copula in any dimension n if and only if ψ_θ is **completely monotone**, which means that:*

- $\psi_\theta \in C^\infty(0, \infty)$, *i.e., is infinitely differentiable;*

- $(-1)^k \psi_\theta^{(k)}(t) \geq 0$ *for all $k = 1, 2, ...$, where $\psi_\theta^{(k)}$ denotes the kth derivative.*

Remark 7.23 (Completely monotone) *This definition of completely monotone can often be readily applied when a given potential candidate function $\psi_\theta(t)$ is proposed, but it is natural to wonder how, other than by trial and error, we might identify a list of potential candidate functions.*

*Another famous result by **Sergei Natanovich Bernstein** (1880–1968), noted for **Bernstein's inequality** in the proof of Proposition 5.3, addresses this question. Called **Bernstein's theorem on monotone functions**, it states that $\psi(t)$ is completely monotone if and only if there exists a finite Borel measure μ on $[0, \infty)$ so that:*

$$\psi(t) = \int_0^\infty e^{-st} d\mu(s).$$

*This expression is called the **Laplace transform of** μ, named for **Pierre-Simon Laplace** (1749–1827), and is an expression that will be seen again in Book IV in the context of **moment generating functions**. While the definition and properties of such integrals will not be addressed until Book V, it is not difficult to gain some insights into some important special cases of Bernstein's result.*

If μ is a finite Borel measure on $[0, \infty)$, it follows from Proposition I.5.7 that there exists an associated right continuous increasing function F_μ so that for any right semi-closed interval $(a, b] \subset [0, \infty)$:

$$\mu[(a, b]] = F_\mu(b) - F_\mu(a),$$

where $\mu[\{0\}] = F_\mu(0)$ can be defined arbitrarily. Such F_μ is continuous except for an at most countable set of points, and thus is continuous almost everywhere.

*It will be seen in Book III that such F_μ is in fact differentiable almost everywhere. Then by a version of the fundamental theorem of calculus for the Lebesgue integral developed there, if F_μ is also **absolutely continuous** then:*

$$F_\mu(x) = F_\mu(0) + (\mathcal{L}) \int_0^x F_\mu'(y) dy.$$

As an example, in the special case where F_μ has a continuous derivative, it is automatically absolutely continuous and then this identity is valid using a Riemann integral.

When μ is a probability measure with $\mu[[0, \infty)] = 1$, then F_μ is a distribution function and $F_\mu' = f_\mu$ the associated density function. Of necessity, such f_μ is nonnegative and integrates to 1 over $[0, \infty)$. In this special case of absolutely continuous F_μ, a change of variable justified in Book V allows a more familiar expression for $\psi(t)$:

$$\psi(t) = (\mathcal{L}) \int_0^\infty e^{-st} f_\mu(s) ds. \tag{1}$$

When F_μ is continuously differentiable, then f is continuous and this result is valid as a Riemann integral.

Bernstein's result now assures that for any such nonnegative f_μ with unit Lebesgue integral, $\psi(t)$ defined in (1) is completely monotone. More familiarly, for such nonnegative continuous f_μ with unit Riemann integral, $\psi(t)$ defined as a Riemann integral in (1) is completely monotone.

Proposition 7.24 (McNeil and Nešlehová) *The function ψ_θ above generates an Archimedean copula in dimension n if and only if ψ_θ is n-**monotone**, which means that:*

- $\psi_\theta \in C^{n-2}(0, \infty)$, *i.e., is continuously differentiable to order $n - 2$;*

- $(-1)^k \psi_\theta^{(k)}(t) \geq 0$ *for $k = 1, 2, ..., n - 2$;*

- $\psi_\theta^{(n-2)}(t)$ *is decreasing and convex.*

Remark 7.25 *Note that if ψ_θ has additional derivatives for $k = n - 1, n$, the last condition is equivalent to assuming the second condition is satisfied for such k since for differentiable $\psi_\theta^{(n-2)}(t)$ to be decreasing requires a negative first derivative, and if twice differentiable, convexity requires a positive second derivative.*

Example 7.26 (Archimedean Copulas) *Popular examples of Archimedean copulas include the following. See the references.*

1. **Gumbel Copula:** *Named for* **Emil Julius Gumbel** *(1891–1966), who is also noted in the section below on extreme value theory, this copula uses the completely monotone generator:*

$$\psi_\theta(t) = \exp\left(-t^{1/\theta}\right), \ \theta \geq 1.$$

This can be verified by observing the Taylor series expansion. This generator produces the parametric Gumbel copula family:

$$C^G(u_1, u_2, ..., u_n; \theta) = \exp\left[-\left(\sum\nolimits_{j=1}^{n} (-\ln u_j)^\theta\right)^{1/\theta}\right].$$

Note that when $\theta = 1$, the Gumbel copula is the independence copula, while the limit as $\theta \to \infty$ produces the comonotonicity copula as can be verified by factoring $(-\ln u^{\min})$ out of the summation where $u^{\min} \equiv \min(u_1, u_2, ..., u_n)$.

This family of copulas is often referred to as the **Gumbel-Hougaard Copula,** *also named for* **Philip Hougaard,** *for generalizing the earlier bivariate Gumbel model to this multivariate context.*

2. **Clayton Copula:** *Named for* **David George Clayton,** *this copula uses the generator:*

$$\psi_\theta(t) = (1 + \theta t)^{-1/\theta},$$

which is confirmed by differentiation to be completely monotone if $\theta \geq 0$, and n-monotone for $\theta \geq -1/(n-1)$. Note that $\psi_\theta(t)$ is defined for $\theta = 0$ by the limit:

$$\lim_{\theta \to 0} (1 + \theta t)^{-1/\theta} = e^{-t},$$

and so the Clayton generator converges as $\theta \to 0$ to the Gumbel generator with Gumbel parameter $\theta^G = 1$. This generator produces the parametric Clayton copula family:

$$C^C(u_1, u_2, ..., u_n; \theta) = \left(\sum\nolimits_{j=1}^{n} u_j^{-\theta} - (n-1)\right)^{-1/\theta}.$$

Thus as was the case for the Gumbel copula, the limiting case, here as $\theta \to 0^+$, produces the independence copula. This follows because:

$$\lim_{\theta \to 0^+} \ln \left(\sum\nolimits_{j=1}^{n} u_j^{-\theta} - (n-1)\right)^{-1/\theta} = -f'(0),$$

defined in terms of the right derivative at 0 of:

$$f(x) = \ln \left(1 + \sum\nolimits_{j=1}^{n} \left(u_j^{-x} - 1\right)\right).$$

In addition, the limit as $\theta \to \infty$ is again the comonotonicity copula as can be verified by taking a logarithm of the expression $\left(\sum_{j=1}^{n} u_j^{-\theta} - (n-1)\right)^{-1/\theta}$ and applying **L'Hôpital's rule** *twice, and then doing some algebra. This rule, named for* **Guillaume de l'Hôpital** *(1661–1704), has several formulations but for this derivation states:*

Exercise 7.27 (L'Hôpital's rule) *If differentiable $f(x), g(x) \to \infty$ as $x \to \infty$ and $f'(x)/g'(x) \to L$, then $f(x)/g(x) \to L$. Another version states that if differentiable $f(x), g(x) \to 0$ as $x \to c$ and $f'(x)/g'(x) \to L$, then $f(x)/g(x) \to L$.*

A deeper result proved in the above referenced **McNeil** *and* **Nešlehová** *paper is that when* $\theta = -1/(n-1)$, *the associated Clayton copula satisfies:*

$$C^C(u_1, u_2, ..., u_n) \leq C^A(u_1, u_2, ..., u_n),$$

for all Archimedean copulas C^A *and all* $(u_1, u_2, ..., u_n)$.

3. **Frank Copula:** *Named for* **Maurice J. Frank,** *this copula uses the generator:*

$$\psi_\theta(t) = -\frac{1}{\theta} \ln\left(1 - \left(1 - e^{-\theta}\right) e^{-t}\right),$$

which is confirmed to be strictly monotone for $\theta > 0$ *by differentiation. This function is well defined for* $\theta = 0$ *as a limit, and is then given by* $-g'(0) = e^{-t}$, *defined in terms of the right derivative at 0 of:*

$$g(x) = \ln\left(1 - \left(1 - e^{-x}\right) e^{-t}\right).$$

This generator produces the parametric Frank copula family:

$$C^F(u_1, u_2, ..., u_n; \theta) = -\frac{1}{\theta} \ln\left(1 - \left[\prod_{j=1}^n \left(1 - e^{-\theta u_j}\right)\right] \Big/ \left(1 - e^{-\theta}\right)^{n-1}\right).$$

The independence copula is again produced as the limiting case as $\theta \to 0$, *using estimates based on Taylor series, or with more work by observing that this limit equals* $-f'(0)$, *defined in terms of the right derivative at 0 of:*

$$f(x) = \ln\left(1 - \left[\prod_{j=1}^n \left(1 - e^{-x u_j}\right)\right] \Big/ \left(1 - e^{-x}\right)^{n-1}\right).$$

In addition, the limit as $\theta \to \infty$ *is again the comonotonicity copula and derived by proving that after subtracting* $u' \equiv u^{\min}$ *from* C^F, *and using the identity:*

$$u' = -\frac{1}{\theta} \ln\left(1 - \left(1 - e^{-\theta u'}\right)\right),$$

that the limit is 0.

7.4.2   Extreme Value Copulas

Extreme value theory is studied in Chapter 9 of this book and continued in Book IV. The primary results there are developed for one-variable distribution functions, which in the context of this section will represent the marginal distribution functions of the associated joint distribution. Here we introduce the basic definitions in a multivariate context. We will return to this topic in Section 9.6 after these marginal distributions are identified, when more formal conclusions will be possible.

Let $X \equiv (X_1, X_2, ..., X_n)$ be a random vector defined on some probability space $(\mathcal{S}, \mathcal{E}, \mu)$, with joint distribution function $F(x_1, x_2, ..., x_n)$ and marginal distributions $\{F_j(x_j)\}_{j=1}^n$. Let $\{(X_{1k}, X_{2k}, ..., X_{nk})\}_{k=1}^m$ be a **random sample** from this random vector, which means as in Example 4.6 that this collection is "approximately" independent and distributed as X.

By approximately **distributed as** X is meant that for $A \in \mathcal{B}(\mathbb{R}^n)$:

$$\Pr\left[(X_{1k}, X_{2k}, ..., X_{nk}) \in A\right] \approx \mu\left[X^{-1}(A)\right], \tag{7.19}$$

where this probability is defined as m_A/m, with m_A the count of sample points in A.

By approximately **independent** is meant that given $\{A_j\}_{j=1}^{m'} \subset \mathcal{B}(\mathbb{R}^n)$ for $m' \leq m$, and subcollection $\{ (X_{1k_j}, X_{2k_j}, ..., X_{nk_j}) \}_{j=1}^{m'}$:

$$\Pr\left[(X_{1k_j}, X_{2k_j}, ..., X_{nk_j}) \in A_j, \text{ all } j\right] \approx \prod_{j=1}^{m'} \Pr\left[(X_{1k_j}, X_{2k_j}, ..., X_{nk_j}) \in A_j\right],$$

and so by the distributional assumption:

$$\Pr\left[(X_{1k_j}, X_{2k_j}, ..., X_{nk_j}) \in A_j, \text{ all } j\right] \approx \prod_{j=1}^{m'} \mu\left[X^{-1}(A_j)\right]. \tag{7.20}$$

By probability we again mean in terms of relative counts, meaning in this case $\prod_{j=1}^{m'} m_{A_j}/m$, where m_{A_j} is the count of sample points in A_j.

Given this random sample, define the random vector of **component-wise maximum variates** by:

$$\left(X_1^{(m)}, X_2^{(m)}, ..., X_n^{(m)}\right) \equiv (\max_{k \leq m} X_{1k}, \max_{k \leq m} X_{2k}, ..., \max_{k \leq m} X_{nk}).$$

Each variate $X_j^{(m)}$ is defined consistently with the 1-dimensional maximum variates M_m defined in (9.5).

Over-simply stated, extreme value theory investigates distributional properties of such maximum variates, once properly scaled and centered, as $m \to \infty$.

If $F^{(m)}(x_1, x_2, ..., x_n)$ denotes the distribution function of this component-wise maximum random vector, then since:

$$\max_{k \leq m} X_{ik} \leq x_i \text{ if and only if } X_{ik} \leq x_i \text{ for all } k,$$

it follows that:

$$F^{(m)}(x_1, x_2, ..., x_n) = F^m(x_1, x_2, ..., x_n). \tag{7.21}$$

Here F^m denotes the *mth* power of the distribution function of X. By the same logic, if $F_j^{(m)}(x_j)$ is the distribution function of $X_j^{(m)} \equiv \max_{k \leq m} X_{jk}$, then:

$$F_j^{(m)}(x_j) = F_j^m(x_j). \tag{7.22}$$

Now let $C(u_1, u_2, ..., u_n)$ be a copula associated with $F(x_1, x_2, ..., x_n)$, meaning as in Proposition 7.7:

$$F(x_1, x_2, ..., x_n) = C\left(F_1(x_1), F_2(x_2), ..., F_n(x_n)\right).$$

Let $C^{(m)}(u_1, u_2, ..., u_n)$ denote a copula associated with $F^{(m)}(x_1, x_2, ..., x_n)$.

By Sklar's theorem applied to $F^{(m)}$:

$$F^{(m)}(x_1, x_2, ..., x_n) = C^{(m)}\left(F_1^{(m)}(x_1), F_2^{(m)}(x_2), ..., F_n^{(m)}(x_n)\right),$$

while taking the *mth* power of Sklar's theorem applied to F:

$$F^m(x_1, x_2, ..., x_n) = C^m\left(F_1(x_1), F_2(x_2), ..., F_n(x_n)\right).$$

These copula expressions agree by (7.21), and then by (7.22) it follows that the copula associated with $\left(X_1^{(m)}, X_2^{(m)}, ..., X_n^{(m)}\right)$ is given by:

$$C^{(m)}(u_1, u_2, ..., u_n) = C^m\left(u_1^{1/m}, u_2^{1/m}, ..., u_n^{1/m}\right). \tag{7.23}$$

An extreme value copula is now defined as a copula that equals $\lim_{m \to \infty} C^{(m)}$.

Definition 7.28 (Extreme value copula) *A copula* $C^{EV}(u_1, u_2, ..., u_n)$ *is an **extreme value copula** if for some copula* C:

$$C^{EV}(u_1, u_2, ..., u_n) = \lim_{m \to \infty} C^m \left(u_1^{1/m}, u_2^{1/m}, ..., u_n^{1/m} \right). \tag{7.24}$$

In this case, we say that the copula C *is in the **domain of attraction** of copula* C^{EV}.

Remark 7.29 (On $\lim_{m \to \infty}$) *Note that although (7.24) is stated with m implied to be an integer, when this limit exists, the same limit is produced with real $r \to \infty$.*
  To see this, note that if $m \leq r \leq m + 1$:

$$
\begin{aligned}
C^{m+1} \left(u_1^{1/m}, u_2^{1/m}, ..., u_n^{1/m} \right) &\leq C^r \left(u_1^{1/r}, u_2^{1/r}, ..., u_n^{1/r} \right) \\
&\leq C^m \left(u_1^{1/m+1}, u_2^{1/m+1}, ..., u_n^{1/m+1} \right).
\end{aligned}
$$

Each inequality is justified by $0 \leq C \leq 1$, or, $0 \leq u_j \leq 1$ and that C is increasing in each variable separately.
  The outer terms now have the same limit, since for example:

$$C^{m+1} \left(u_1^{1/m}, u_2^{1/m}, ..., u_n^{1/m} \right) = \left[C^m \left(u_1^{1/m}, u_2^{1/m}, ..., u_n^{1/m} \right) \right]^{(m+1)/m},$$

and thus the limit in (7.24) can also be defined over the real numbers.

Example 7.30 (On existence 1) *An obvious question is, how can we prove that extreme value copulas exist? Another is, how would we recognize an extreme value copula, or verify that a given copula was of this type?*
  A partial answer is that any copula C, which for all real $r > 0$:

$$C^r \left(u_1^{1/r}, u_2^{1/r}, ..., u_n^{1/r} \right) = C(u_1, u_2, ..., u_n), \tag{7.25}$$

is an extreme value copula. Since then, $C^{EV}(u_1, u_2, ..., u_n) = C(u_1, u_2, ..., u_n)$.

Definition 7.31 (Max-stable copula) *A copula* $C^{MS}(u_1, u_2, ..., u_n)$ *is said to be **max-stable** if (7.25) is satisfied for all real $r > 0$. Thus, every max-stable copula is an extreme value copula, and is also in its own domain of attraction.*

It is not initially obvious that max-stable copulas even exist, even though max-stable copulas are extreme value copulas. A simple but perhaps surprising result is the following.

Proposition 7.32 (Extreme value $\Leftrightarrow$ Max-stable) *A copula is extreme value if and only if it is max-stable.*
Proof. *Only one direction needs to be addressed, so assume C is an extreme value copula and that C_0 is the copula in the domain of attraction of C as in (7.24). Then, given real $r > 0$, it follows from (7.24) and Remark 7.29 that:*

$$
\begin{aligned}
C^r(u_1^{1/r}, u_2^{1/r}, ..., u_n^{1/r}) &= \lim_{m \to \infty} C_0^{mr} \left(u_1^{1/mr}, u_2^{1/mr}, ..., u_n^{1/mr} \right) \\
&= \lim_{s \to \infty} C_0^s \left(u_1^{1/s}, u_2^{1/s}, ..., u_n^{1/s} \right) \\
&= C(u_1, u_2, ..., u_n).
\end{aligned}
$$

Thus, C is max-stable. ∎

Remark 7.33 (On existence 2) *The utility of this result, if it is not apparent, is that it is often straightforward to test if a given copula C is max-stable, and this then assures that it is an extreme value copula. To verify that C satisfies (7.24) directly is a challenge, and one made the more difficult by the fact that we must first identify a copula in the domain of attraction of C.*

Example 7.34 (Extreme Value Copulas) *On the question of existence, here are a few examples and a general characterization:*

1. **Independence Copula:** *It is apparent that $C^I(u_1, u_2, ..., u_n) = u_1 u_2 ... u_n$ satisfies (7.25) and is thus an extreme value copula.*

2. **Comonotonicity Copula:** *As noted above, this copula is the upper bound of all possible copulas by (7.1). Since $C^{Co}(u_1, u_2, ..., u_n) = \min_j u_j$ also satisfies (7.25), it is thus also an extreme value copula.*

3. **Gumbel-Hougaard Copula:** *Recall that as defined above, for $\theta \geq 1$:*

$$C^G(u_1, u_2, ..., u_n) = \exp\left[-\left(\sum\nolimits_{j=1}^{n}(-\ln u_j)^\theta\right)^{1/\theta}\right].$$

   It is left as an exercise to verify that (7.25) is satisfied.

   *This is a unique example by a 1989 result of **Christian Genest** and **Louis-Paul Rivest**. They proved that every Archimedean copula that is in the extreme value class is in the Gumbel-Hougaard family.*

4. **The Pickands characterization:** *Based on an earlier 1977 characterization by **Laurens de Haan** and **Sidney Resnick**, in 1981 **James Pickands** derived the characterization that $C^{EV}(u_1, u_2, ..., u_n)$ is an extreme value copula if and only if for $(u_1, u_2, ..., u_n) \neq (1, 1, ..., 1)$:*

$$\begin{aligned} &C^{EV}(u_1, u_2, ..., u_n) \hspace{5cm} (7.26) \\ &= \exp\left[\left(\sum\nolimits_{j=1}^{n} \ln u_j\right) A\left(\frac{\ln u_1}{\sum_{j=1}^{n} \ln u_j}, \frac{\ln u_2}{\sum_{j=1}^{n} \ln u_j}, \cdots, \frac{\ln u_n}{\sum_{j=1}^{n} \ln u_j}\right)\right]. \end{aligned}$$

   See Section 9.6.3 for more details.

   *Here, the **Pickands dependence function** $A(w_1, w_2, ..., w_n)$ is defined on the **simplex**:*

$$\Delta_{n-1} \equiv \{(w_1, w_2, ..., w_n) | w_j \geq 0, \sum\nolimits_{j=1}^{n} w_j = 1\},$$

   is convex, and satisfies:

$$\max w_j \leq A(w_1, w_2, ..., w_n) \leq 1.$$

   In general for $n > 2$, convexity of A and satisfaction of these bounds does not assure that the function C^{EV} in (7.26) is a copula, which is to say, these two conditions alone do not characterize all possible Pickands dependence functions.

   However, the bounding functions produce familiar copulas. Specifically, if $A(w_1, w_2, ..., w_n) \equiv 1$, then $C^{EV}(u_1, u_2, ..., u_n) = u_1 u_2 ... u_n$, the independence copula, while if $A(w_1, w_2, ..., w_n) \equiv \max w_j$, then $C^{EV}(u_1, u_2, ..., u_n) = \min w_j$, the comonotonicity copula. It is left as an exercise to determine $A(w_1, w_2, ..., w_n)$ for the Gumbel-Hougaard copula.

Example 7.35 (Estimating $A(w_1, w_2)$) *If $n = 2$, then $\Delta_1 \equiv \{(1 - t, t)|t \in [0, 1]\}$. if $A_0(t) \equiv A(1 - t, t)$ is convex on $[0, 1]$ and satisfies the above bounds:*

$$\max(t, 1 - t) \leq A_0(t) \leq 1,$$

then $C^{EV}(u, v)$ defined in (7.26) is always an extreme value copula. The copula can be equivalently expressed for $(u, v) \in (0, 1]^2 - \{(1, 1)\}$ by:

$$C^{EV}(u, v) = \exp\left[(\ln uv) A_0\left(\frac{\ln v}{\ln(uv)}\right)\right] = (uv)^{A_0[\ln v/\ln(uv)]}.$$

For this bivariate case, a number of approaches have been introduced to estimate the function $A_0(t)$ needed for this copula, and we develop Pickands' approach next as this introduces the key insight.

Let $\left\{\left(X_j^{(m)}, Y_j^{(m)}\right)\right\}_{j=1}^N$ be a random sample of maximum variates as defined in the introduction to this section:

$$\left(X_j^{(m)}, Y_j^{(m)}\right) \equiv (\max_{k \leq m} X_k^{(j)}, \max_{k \leq m} Y_k^{(j)}).$$

As an example, each pair could represent the maximum daily loss on two equity indexes in a given year, tracked over N years.

Next, let $\{(U_j, V_j)\}_{j=1}^N$ be defined as the uniformly distributed variates associated with this sample, so $U_j = F(X_j^{(m)})$ and $V_j = G(Y_j^{(m)})$, where the marginal distributions F and G are assumed to be known. Define for $0 < t < 1$:

$$\xi_j(t) = \min\left(-\frac{\ln U_j}{1 - t}, -\frac{\ln V_j}{t}\right),$$

and extend mathematically to $\xi_j(0) = -\ln U_j$ and $\xi_j(1) = -\ln V_j$. Then a calculation produces:

$$
\begin{aligned}
\Pr\left[\xi_j(t) > x\right] &= \Pr\left[U_j < e^{-(1-t)x}, \ V_j < e^{-tx}\right] \\
&= C^{EV}(e^{-(1-t)x}, e^{-tx}) \\
&= e^{-xA_0(t)}.
\end{aligned}
$$

Recalling (1.21), it follows that for each t, $\{\xi_j(t)\}_{j=1}^N$ has an exponential distribution with exponential parameter $\lambda = A_0(t)$.

It will be seen in the Book IV discussion on moments that the mean of this exponential distribution equals $1/\lambda$, and thus the Pickands' estimator for $A_0(t)$ is given by:

$$1/\widehat{A}_0(t) = \sum_{j=1}^N \xi_j(t)/N.$$

7.5 General Result on Sklar's Theorem

Extending Proposition 7.18, we now turn to the general proof of part 1 of Sklar's theorem. This result asserts the existence of a copula $C(u_1, u_2, ..., u_n)$ in the case of a joint distribution function $F(x_1, x_2, ..., x_n)$ with "general," meaning not necessarily continuous, marginal distribution functions $\{F_j(x)\}_{j=1}^n$. Since all distribution functions are right continuous by Proposition 6.1, any such discontinuities for marginal distributions must be left discontinuities. Further, since distribution functions are increasing, the number of such left discontinuities is at most countable.

Part 2 of Sklar's theorem was already proved in the general case in Proposition 7.16. Specifically, for any copula $C(u_1, u_2, ..., u_n)$ and distribution functions $\{F_j(x)\}_{j=1}^n$:

$$F(x_1, x_2, ..., x_n) \equiv C(F_1(x_1), F_2(x_2), ..., F_n(x_n)),$$

is a joint distribution function, with marginal distributions $\{F_j(x)\}_{j=1}^n$.

Before introducing the necessary machinery to prove the general existence result of Sklar's theorem, it is worth a moment to contemplate why neither the original definition of $C(u_1, u_2, ..., u_n)$ in the proof of Proposition 7.18, nor the identity in (7.11), yields the correct answer in the general case.

1. In the proof of Proposition 7.18, $C(u_1, u_2, ..., u_n)$ was defined as the distribution function of the random vector $(F_1(X_1), F_2(X_2), ..., F_n(X_n))$, and it was then shown that C had uniform marginal distributions and thus was a copula. However, **the distribution function of the random vector $(F_1(X_1), F_2(X_2), ..., F_n(X_n))$ is not a copula in the general case.**

   For $C(u_1, u_2, ..., u_n)$ so defined to be a copula only requires that this distribution function have uniform marginal distributions. In this general case as in the above proof, it can be concluded that the marginal distribution of $U_1 \equiv F_1(X_1)$ say, satisfies:

   $$C_{\{1\}}(u_1, 1, ..., 1) = F_{F_1(X_1)}(u_1).$$

   By Proposition 4.8, $F_{F_1(X_1)}(u_1) \leq u_1$ in general, and $F_{F_1(X_1)}(u_1) = u_1$ if and only if F_1 is continuous. And this is true for all marginals.

   Thus defining C as the distribution function of the random vector $(F_1(X_1), F_2(X_2), ..., F_n(X_n))$ does not produce a copula in the general case, because this function will not in general have uniform marginal distributions as is required. By Proposition 4.8, C so defined will be a copula if and only if the original marginal distributions, $\{F_j(x)\}_{j=1}^n$, are continuous.

2. Once C was defined as the distribution function of the random vector $(F_1(X_1), F_2(X_2), ..., F_n(X_n))$ in the proof of Proposition 7.18, it was then proved as stated in (7.11) that:

   $$C(u_1, u_2, ..., u_n) = F(F_1^*(u_1), F_2^*(u_2), ..., F_n^*(u_n)). \tag{1}$$

   Thus as another possible route to a general solution, perhaps it is possible to simply **define** C directly in this way. This definition certainly makes sense in the general case since $F_j^*(u_j)$ is well defined without any continuity assumptions on F_j. However, **once again C as defined in (1) does not produce a copula in the general case.**

   Two problems arise with this approach. The first is a special case of the second, but worth noting separately.

   (a) **The marginal distributions of C as defined in (1) are not uniform in the general case, and thus C is not a copula.**

   Letting $u_2, ..., u_n \to 1$ say, it follows from (1) that since $F_j^*(u_j) \to \infty$ as $u_j \to 1$:

   $$C_{\{1\}}(u_1, 1, ..., 1) = F_1(F_1^*(u_1)).$$

However, by part 1 of Proposition 3.22, $F(F^*(u_1)) = u_1$ for all $u_1 \in (0,1)$ if and only if F is continuous. In the case of general marginal distributions, the function $C(u_1, u_2, ..., u_n)$ defined in (1) will thus not be a copula.

(b) **The function C in (1) does not satisfy the defining property in (7.4) in the general case. In other words, C does not reproduce $F(x_1, x_2, ..., x_n)$.**

For C in (1) to reproduce $F(x_1, x_2, ..., x_n)$ requires that:

$$F(x_1, x_2, ..., x_n) = F(F_1^*(F_1(x_1)), F_2^*(F_2(x_2)), ..., F_n^*(F_n(x_n))). \qquad (2)$$

This identity would be satisfied if $F_j^*(F_j(x_j)) = x_j$ for all j, but by Proposition 3.22, this would require F_j to be strictly increasing, an assumption not required even in the continuous case. More generally, the identity in (2) would be satisfied as shown in the proof of Proposition 7.18 if for all j:

$$F_j\left[F_j^*(F_j(x_j))\right] = F_j(x_j).$$

By Proposition 3.22, this result would be satisfied if either $F_j^*(F_j(x_j)) = x_j$, which as noted requires that F_j be strictly increasing, or if $F_j(F_j^*(y_j)) = y_j$, which requires that F_j be continuous. In the general case, $y_j \leq F_j(F_j^*(y_j))$ by (3.10), and $F_j^*(F_j(x_j)) \leq x_j$ by (3.7).

Thus, although the formula for C in (1) is well defined in the general case, it does not replicate the original distribution $F(x_1, x_2, ..., x_n)$.

There are a number of proofs of part 1 of Sklar's theorem in the case of general marginal distributions. A relatively recent proof was published in 2009 by **Ludger Rüschendorf** using the notion of a **distributional transform**. This tool appears to have been introduced in the late 1960s and widely used since the early 1980s in connection with the study of stochastic orderings. The essence of Rüschendorf's proof is to use this transform to define a continuous version of a general marginal distribution function $F_j(x)$, which works well with the original left continuous inverse function in the following sense.

Let $F(x)$ be a general marginal distribution function of a random variable X defined on a probability space $(\mathcal{S}, \mathcal{E}, \mu)$. An associated continuous distribution function F^c will be defined and proved to be uniformly distributed on $(0,1)$ in Proposition 7.40. More importantly, when the left continuous inverse of the original distribution function F is applied to this uniform variate defined on $(0,1)$, it produces the original random variable X **in distribution.**

In other words, defining:

$$X' \equiv F^*(F^c),$$

then

$$F_{X'}(x) = F(x).$$

Thus the proof of the general Sklar existence result will follow from the case of continuous marginal distributions, once a few details are checked.

7.5.1   The Distributional Transform

Let $X : (\mathcal{S}, \mathcal{E}, \mu) \to (\mathbb{R}, \mathcal{B}(\mathbb{R}), m)$ be a random variable with distribution function F defined as usual by

$$F(x) = \mu\left[X^{-1}(-\infty, x]\right].$$

As always, m denotes Lebesgue measure. With μ_F the Borel (probability) measure induced by F (Proposition 6.3), $(\mathbb{R}, \mathcal{B}(\mathbb{R}), \mu_F)$ is a probability space and:

$$F : (\mathbb{R}, \mathcal{B}(\mathbb{R}), \mu_F) \to ((0,1), \mathcal{B}((0,1)), m),$$

is an increasing and right continuous function by Proposition 6.1. Hence F is Borel measurable and thus a random variable on this probability space.

Next consider the **left limit function** F^-:

$$F^-(x) \equiv \lim_{y \to x^-} \mu \left[X^{-1}(-\infty, y] \right],$$

where $y \to x^-$ denotes that $y < x$ and $y \to x$. Given an increasing sequence $\{y_j\}_{j=1}^\infty$ that converges to x, the collection of sets $\{X^{-1}(-\infty, y_j]\}_{j=1}^\infty$ is nested with union $X^{-1}(-\infty, x)$, and so by continuity from below:

$$F^-(x) = \mu \left[X^{-1}(-\infty, x) \right].$$

Hence $F^-(x)$ is an increasing function and also a random variable on $(\mathbb{R}, \mathcal{B}(\mathbb{R}), \mu_F)$.

Finally, define:

$$V : ([0, 1], \mathcal{B}([0, 1]), m) \to ([0, 1], \mathcal{B}([0, 1]), m),$$

as the identity random variable:

$$V(y) = y.$$

Notation 7.36 *The left limit of a function f at x is sometimes denoted $f^-(x)$ as above, and sometimes $f(x^-)$. Similarly, the right limit is denoted $f^+(x)$, and sometimes $f(x^+)$.*

With this setup, the **distributional transform of** X is defined as follows. Recall the development of product measure spaces in Chapter I.7, and particularly Proposition I.7.20.

Definition 7.37 (Distributional transform) *Given the notation above, the distributional transform of X, denoted $F^c(x, y)$, is defined as a function on the product probability space:*

$$(\mathbb{R} \times [0, 1], \sigma \left(\mathcal{B}(\mathbb{R}) \times \mathcal{B}([0, 1]) \right), \mu_F \times m) \equiv (\mathbb{R}, \mathcal{B}(\mathbb{R}), \mu_F) \times ([0, 1], \mathcal{B}([0, 1]), m),$$

by:

$$F^c(x, y) = F^-(x) + V(y) \left[F(x) - F^-(x) \right]. \tag{7.27}$$

Remark 7.38 (On the distributional transform) *A few comments on this transform are in order:*

1. *The above product space is constructed in Chapter I.7, and $\sigma \left(\mathcal{B}(\mathbb{R}) \times \mathcal{B}([0, 1]) \right)$ denotes the sigma algebra generated by the associated measurable rectangles, denoted $\sigma(\mathcal{A})$ in the proof of Proposition I.7.20.*

2. *Using (6.2) of Proposition 6.1, and with $\Pr$ denoting probabilities defined in $(\mathcal{S}, \mathcal{E}, \mu)$:*

$$F^c(x, y) = \Pr[X < x] + y \Pr[X = x]. \tag{7.28}$$

   In other words:

$$F^c(x, y) = \mu[X^{-1}(-\infty, x)] + y\mu[X^{-1}(x)].$$

3. *Each of F, F^- and V are random variables on this product space and defined in the natural way by:*

$$F(x, y) = F(x), \quad F^-(x, y) = F^-(x), \quad V(x, y) = V(y).$$

   All have range space $([0, 1], \mathcal{B}([0, 1]), m)$ by definition, while for measurability, let $A \in$

$\mathcal{B}([0,1])$. *Letting $F^{-1}[A]$ denote the pre-image of F in the product space and $F_0^{-1}[A]$ the pre-image of F in $(\mathbb{R}, \mathcal{B}(\mathbb{R}), \mu_F)$:*

$$F^{-1}[A] = F_0^{-1}[A] \times [0,1],$$

a measurable rectangle in this product space. The other variates are handled similarly.

Thus F^c is a random variable on this product space (see Proposition I.3.30, and note that the proofs generalize to arbitrary measurable functions).

4. *Both F and F^- are independent of V on this product space. To see this, recall the definition of independent random variables in (3.26). For example with F and V, this requires that for all $A_1, A_2 \in \mathcal{B}([0,1])$:*

$$(\mu_F \times m)\left(F^{-1}(A_1) \bigcap V^{-1}(A_2)\right)$$
$$= (\mu_F \times m)\left(F^{-1}(A_1)\right) \cdot (\mu_F \times m)\left(V^{-1}(A_2)\right).$$

But in the notation of item 3, $F^{-1}(A_1) = F_0^{-1}(A_1) \times [0,1]$, and using analogous notation, $V^{-1}(A_2) = \mathbb{R} \times V_0^{-1}(A_2)$. So the intersection set is $F_0^{-1}(A_1) \times V_0^{-1}(A_2)$, and this identity then follows from the definition of product measure on rectangles:

$$(\mu_F \times m)\left(F^{-1}(A_1) \bigcap V^{-1}(A_2)\right)$$
$$= (\mu_F \times m)\left(F_0^{-1}(A_1) \times V_0^{-1}(A_2)\right)$$
$$= \mu_F\left(F_0^{-1}(A_1)\right) \cdot m\left(V_0^{-1}(A_2)\right)$$
$$= (\mu_F \times m)\left(F^{-1}(A_1)\right) \cdot (\mu_F \times m)\left(V^{-1}(A_2)\right).$$

The same derivation holds for F^- and V.

5. *Now $F^c(x,y)$ can also be rewritten:*

$$F^c(x,y) = F(x) - (1 - V(y))\left[F(x) - F^-(x)\right].$$

Since $0 \leq V \leq 1$ and $F^-(x) \leq F(x)$ it follows that:

$$F^c(x,y) \leq F(x),$$

and:

$$F^c(x,y) = F(x) \text{ if and only if } x \text{ is a continuity point, or } y = 1.$$

Example 7.39 (A distributional transform) *Let $X : (\mathcal{S}, \mathcal{E}, \mu) \to (\mathbb{R}, \mathcal{B}(\mathbb{R}), m)$ be a random variable with $X(\mathcal{S})$ contained in $[-1,1]$ and distribution function:*

$$F(x) = \begin{cases} 0, & x < -1, \\ 1/2, & -1 \leq x \leq 0, \\ 1/2 + x/2, & 0 \leq x \leq 1, \\ 1, & x > 1. \end{cases}$$

This distribution displays the 3 primary behaviors of such functions containing a discontinuity, a continuous but nonincreasing component, and a continuous and strictly increasing component.

It follows from (7.28) that:

$$F^c(x,y) = \begin{cases} 0, & x < -1, \\ y/2, & x = -1, \\ 1/2, & -1 < x \leq 0, \\ 1/2 + x/2, & 0 \leq x \leq 1, \\ 1, & x > 1. \end{cases}$$

As noted in 5 of Remark 7.38, $F^c(x,y) = F(x)$ unless $x = -1$, the only discontinuity point of F, and $F^c(-1,y) < F(-1)$ unless $y = 1$.

It is apparent that for this example, the range of F^c is $[0,1]$. In a more general case, for example if X is a random variable with distribution function in (1.17) with the standard normal probability density in (1.23), the range of F is $(0,1)$ and so too is the range of F^c.

We now show as an example of the general case in Proposition 7.40, that F^c is uniformly distributed on $(0,1)$. It then follows that $F^*(F^c(x,y))$ is a random variable with distribution function $F(x)$ by part 2 of Proposition 4.8. Said another way:

$$F^*(F^c(x,y)) = X,$$

in distribution.

To prove that F^c is uniformly distributed on $(0,1)$, choose $\alpha \in (0,1)$ and consider $A_\alpha \equiv (F^c)^{-1}(0,\alpha]$. Since $A_\alpha \subset (\mathbb{R}, \mathcal{B}(\mathbb{R}), \mu_F) \times ([0,1], \mathcal{B}([0,1]), m)$, it must be shown that $(\mu_F \times m)[A_\alpha] = \alpha$. Now:

$$A_\alpha = \begin{cases} (-\infty, -1) \times [0,1] \bigcup \{-1\} \times [0, 2\alpha], & 0 < \alpha \leq 1/2, \\ (-\infty, 2\alpha - 1] \times [0,1], & 1/2 < \alpha < 1. \end{cases}$$

Thus for $0 < \alpha \leq 1/2$:

$$\begin{aligned} (\mu_F \times m)[A_\alpha] &= (\mu_F \times m)((-\infty, -1) \times [0,1]) \\ &\quad + (\mu_F \times m)(\{-1\} \times [0, 2\alpha]) \\ &= 0 + 1/2\,(2\alpha) = \alpha. \end{aligned}$$

A similar calculation obtains the same measure for $1/2 < \alpha < 1$.

Thus $(\mu_F \times m)[A_\alpha] = \alpha$ for all α, and so F^c is uniformly distributed on $(0,1)$. Then as noted above, $F^*(F^c(x,y)) = X$ in distribution.

The next proposition follows the 2009 proof by **Ludger Rüschendorf** and generalizes this example. It is a worthwhile exercise to apply the general approach of this proof to the above example to better understand how this construction works.

Proposition 7.40 ($F^*(F^c(x,y)) = X$, in distribution) *Let $X : (\mathcal{S}, \mathcal{E}, \mu) \to (\mathbb{R}, \mathcal{B}(\mathbb{R}), m)$ be a random variable with distribution function $F(x)$, and define $F^c(x,y)$ on $(\mathbb{R}, \mathcal{B}(\mathbb{R}), \mu_F) \times ([0,1], \mathcal{B}([0,1]), m)$ as in (7.27).*

Then F^c is uniformly distributed on $(0,1)$, and thus $F^*(F^c(x,y))$ is a random variable with distribution function $F(x)$. In other words, $F^*(F^c(x,y)) = X$ in distribution.

Proof. *As noted above, the distributional result on $F^*(F^c(x,y))$ follows from part 2 of Proposition 4.8 once it is proved that F^c is uniformly distributed on $(0,1)$. To prove this, we must generalize the construction for the above example so as to indirectly identify discontinuities.*

*To this end, choose $\alpha \in (0,1)$ and consider $A_\alpha \equiv (F^c)^{-1}(0,\alpha]$. If $F(x) < \alpha$, then since $F^c(x,y) \leq F(x)$ by 5 of Remark 7.38, it follows that $(-\infty, x] \times [0,1] \subset A_\alpha$. Define $q_\alpha^-(X)$, the **lower α-quantile**, by:*

$$q_\alpha^-(X) = \sup\{x | F(x) < \alpha\}.$$

For every x in the set $\{x|F(x) < \alpha\}$ that defines $q_\alpha^-(X)$, it thus follows that $(-\infty, x] \times [0,1] \subset A_\alpha$, and so:

$$(-\infty, q_\alpha^-(X)) \times [0,1] \subset A_\alpha. \tag{1}$$

But it need not follow that $(-\infty, q_\alpha^-(X)] \times [0,1] \subset A_\alpha$, since F need not be left continuous at $q_\alpha^-(X)$. Equivalently, it need not be the case that for $(x,y) \in (-\infty, q_\alpha^-(X)] \times [0,1]$ that:

$$F^c(x,y) = F^-(x) + V(y)\left[F(x) - F^-(x)\right] \le \alpha.$$

With F^- defined as in (7.27), and $\Pr$ as in 2 of Remark 7.38, let:

$$q = F^-(q_\alpha^-(X)) = \Pr\{X < q_\alpha^-(X)\},$$

and:

$$\beta = F(q_\alpha^-(X)) - F^-(q_\alpha^-(X)) = \Pr\{X = q_\alpha^-(X)\}.$$

Consider the cases $\beta > 0$ and $\beta = 0$.

    1. *If $\beta > 0$, then:*

$$A_\alpha = (-\infty, q_\alpha^-(X)) \times [0,1] \bigcup \{q_\alpha^-(X)\} \times [0, (\alpha - q)/\beta],$$

    since $F^c(x,y) < \alpha$ on the first rectangle by (1), while $F^c(x,y) = \alpha$ on the second rectangle by construction.

    In addition, $(\mu_F \times m)[A_\alpha] = \alpha$ since these rectangles are disjoint with respective measures q and $\alpha - q$.

    2. *If $\beta = 0$, then:*

$$A_\alpha = (-\infty, q_\alpha^-(X)] \times [0,1]$$

    by (1), and since $q_\alpha^-(X)$ is a continuity point:

$$F^c(q_\alpha^-(X), y) = F(q_\alpha^-(X)) \le \alpha.$$

    Also by continuity:

$$(\mu_F \times m)[A_\alpha] = \Pr[X \le q_\alpha^-(X)] = \alpha.$$

    In both cases:

$$(\mu_F \times m)[F^c \le \alpha] = (\mu_F \times m)[A_\alpha] = \alpha,$$

and thus F^c is uniformly distributed on $(0,1)$. ■

7.5.2   Sklar's Theorem - The General Case

With the aid of the result in Proposition 7.40, the general proof of part 1 of Sklar's theorem is now relatively simple.

Proposition 7.41 (Sklar's Theorem, part 1 - General Case) *If $F(x_1, x_2, ..., x_n) \in \mathfrak{F}(F_1, F_2, ..., F_n)$ where $\{F_j(x)\}_{j=1}^n$ are given distribution functions, then there exists a copula C associated with F, and so:*

$$F(x_1, x_2, ..., x_n) = C(F_1(x_1), F_2(x_2), ..., F_n(x_n)).$$

In addition, $C(u_1, u_2, ..., u_n)$ can be defined as the joint distribution function of

$(F_1^c(x_1, y), F_2^c(x_2, y), ..., F_n^c(x_n, y))$, *where* $F_j^c(x_j, y)$ *denotes the distributional transform of* X_j.

Proof. *By Proposition 7.40,* $F_j^* \left[F_j^c(x_j, y) \right] = X_j$ *in distribution for all* j*, and then by (3.8):*

$$
\begin{aligned}
F(x_1, x_2, ..., x_n) &= \mu \left[\bigcap_j \{ X_j \leq x_j \} \right] \\
&= \mu \left[\bigcap_j \{ F_j^* \left[F_j^c(x_j, y) \right] \leq x_j \} \right] \\
&= \mu \left[\bigcap_j \{ F_j^c(x_j, y) \leq F_j(x_j) \} \right].
\end{aligned}
$$

Now let $C(u_1, u_2, ..., u_n)$ *be defined as the joint distribution function of* $(F_1^c(x_1, y), F_2^c(x_2, y), ..., F_n^c(x_n, y))$*, and so by definition:*

$$
\mu \left[\bigcap_j \{ F_j^c(x_j, y) \leq F_j(x_j) \} \right] = C(F_1(x_1), F_2(x_2), ..., F_n(x_n)).
$$

Combining with the prior result, (7.4) is satisfied.

That C *is a copula associated with* F *now follows because the marginal distributions of* C *are the distribution functions of the* $F_j^c(x_j, y)$*-variates, which are uniformly distributed by Proposition 7.40.* ∎

7.6 Tail Dependence and Copulas

In this section we study various measures of tail dependency in random vectors and the implications for the associated copulas. We begin with bivariate random vectors, to introduce ideas and examples. We then turn to the general multivariate case, which indirectly introduces another copula. This is the survival copula, the topic of the last section.

7.6.1 Bivariate Tail Dependence

Given random variables X_1 and X_2 defined on a probability space $(\mathcal{S}, \mathcal{E}, \mu)$, the notion of **tail dependence** is introduced as a measure of the extent to which these variates exhibit extreme outcomes simultaneously. Such outcomes are traditionally defined in terms of the probabilities of the simultaneous events $X_j > F_j^*(t)$ as $t \to 1$, or of $X_j \leq F_j^*(t)$ as $t \to 0$. In other words, simultaneous extreme events are defined in terms of the **quantiles** of the distribution functions, which are then defined in terms of the left continuous inverse functions. Since the joint probability of either of these simultaneous events converges to zero in all cases, one is usually interested in conditional probabilities. See below.

While the following definitions appear asymmetrical in terms of the roles of X_1 and X_2, see Remark 7.45 below for a symmetrical version in the case of continuous marginals.

Definition 7.42 (Bivariate tail dependence measures) *Let* X_1 *and* X_2 *be random variables defined on a probability space* $(\mathcal{S}, \mathcal{E}, \mu)$ *with respective distribution functions* F_1 *and* F_2.

*The **lower tail dependence measure** or **coefficient** or **index** λ_L, is defined when the limit exists by:*

$$
\lambda_L = \lim_{t \to 0^+} \mu \left[X_1^{-1}(-\infty, F_1^*(t)] \bigcap X_2^{-1}(-\infty, F_2^*(t)] \right] \Big/ \mu \left[X_2^{-1}(-\infty, F_2^*(t)] \right]. \tag{7.29}
$$

*The **upper tail dependence measure** or **coefficient** or **index** λ_U, is defined when the limit exists by:*

$$\lambda_U = \lim_{t \to 1^-} \mu\left[X_1^{-1}(F_1^*(t), \infty) \bigcap X_2^{-1}(F_2^*(t), \infty)\right] \Big/ \mu\left[X_2^{-1}(F_2^*(t), \infty)\right]. \qquad (7.30)$$

*The variates X_1 and X_2 are said to be **lower tail dependent** if $\lambda_L \in (0,1]$ and **lower tail independent** if $\lambda_L = 0$. Similarly, X_1 and X_2 are said to be **upper tail dependent** if $\lambda_U \in (0,1]$ and **upper tail independent** if $\lambda_U = 0$.*

The reader will note that by Definition 1.31, the tail dependency measures can be stated in terms of the conditional probabilities:

$$\begin{aligned} \lambda_L &= \lim_{t \to 0^+} \mu\left[X_1^{-1}(-\infty, F_1^*(t)) \,\middle|\, X_2^{-1}(-\infty, F_2^*(t))\right], \\ \lambda_U &= \lim_{t \to 1^-} \mu\left[X_1^{-1}(F_1^*(t), \infty) \,\middle|\, X_2^{-1}(F_2^*(t), \infty)\right]. \end{aligned} \qquad (7.31)$$

Thus it is important to verify that these conditional probabilities are well defined, which is to say that $\mu\left[X_2^{-1}(-\infty, F_2^*(t))\right] \neq 0$, and similarly, $\mu\left[X_2^{-1}(F_2^*(t), \infty)\right] \neq 0$.

First by Proposition 3.19:

$$\begin{aligned} \mu\left[X_2^{-1}(-\infty, F_2^*(t))\right] &= \mu\left[\{X_2 \leq F_2^*(t)\}\right] \\ &= F_2(F_2^*(t)) \\ &\geq t. \end{aligned}$$

Thus the conditional probability underlying the definition of λ_L is always well defined, though there remains the question of the existence of this limit.

On the other hand:

$$\begin{aligned} \mu\left[X_2^{-1}(F_2^*(t), \infty)\right] &= \mu\left[\{X_2 > F_2^*(t)\}\right] \\ &= 1 - F_2(F_2^*(t)). \end{aligned}$$

Again by Proposition 3.19, $F_2(F_2^*(t)) = t$, except for at most countably many points. As $F_2(F_2^*(t))$ is nondecreasing in t, it follows that the conditional probability underlying the definition of λ_U is well defined, though there remains the question of the existence of this limit.

In the special case where F_1 and F_2 are continuous, these tail dependence measures have relatively simple expressions.

Proposition 7.43 (Bivariate tail dependence measures and copulas) *Given random variables X_1 and X_2 defined on a probability space $(\mathcal{S}, \mathcal{E}, \mu)$ with respective continuous distribution functions F_1 and F_2, and a copula C associated with the joint distribution function F, then when the above limits exist:*

$$\lambda_L = \lim_{t \to 0^+} \frac{C(t,t)}{t}, \qquad \lambda_U = \lim_{t \to 1^-} \frac{1 - 2t + C(t,t)}{1 - t}. \qquad (7.32)$$

If $C(t,t)$ is right differentiable at $t = 0$, respectively left differentiable at $t = 1$, then:

$$\lambda_L = \left.\frac{dC(t,t)}{dt}\right|_{t=0}, \qquad \text{respectively,} \quad \lambda_U = 2 - \left.\frac{dC(t,t)}{dt}\right|_{t=1}. \qquad (7.33)$$

Finally, if $C(t,t)$ is continuously differentiable in a neighborhood of $t = 0$, respectively $t = 1$, then:

$$\lambda_L = \lim_{t \to 0^+} \frac{dC(t,t)}{dt}, \qquad \text{respectively,} \quad \lambda_U = 2 - \lim_{t \to 1^-} \frac{dC(t,t)}{dt}. \qquad (7.34)$$

Proof. *By continuity, both F_1 and F_2 have range $(0, 1)$ and $F_j(F_j^*(y)) = y$ for $y \in (0, 1)$ by Proposition 3.22. Thus as above:*

$$\mu\left[X_2^{-1}(-\infty, F_2^*(t))\right] = F_2\left(F_2^*(t)\right) = t.$$

Analogously, $\mu\left[X_2^{-1}(F_2^*(t), \infty)\right] = 1 - t$. Thus the measures of these sets are positive for all $t \in (0, 1)$, and the conditional probabilities in Definition 7.42 are well defined.

With F denoting the joint distribution function of (X_1, X_2) and C a copula associated with F:

$$\begin{aligned} \lambda_L &= \lim_{t \to 0^+} \mu\left[X_1^{-1}(-\infty, F_1^*(t)) \bigcap X_2^{-1}(-\infty, F_2^*(t))\right] / \mu\left[X_2^{-1}(-\infty, F_2^*(t))\right] \\ &= \lim_{t \to 0^+} F\left(F_1^*(t), F_2^*(t)\right) / F_2\left(F_2^*(t)\right) \\ &= \lim_{t \to 0^+} C\left(F_1\left(F_1^*(t)\right), F_2\left(F_2^*(t)\right)\right) / F_2\left(F_2^*(t)\right). \end{aligned}$$

The first expression in (7.32) now follows since $F_j\left(F_j^*(t)\right) = t$ for all $t \in (0, 1)$ by continuity.

For notational simplicity, let $A \equiv X_1^{-1}(-\infty, F_1^*(t)]$ and $B \equiv X_2^{-1}(-\infty, F_2^*(t)]$. Then:

$$\mathcal{S} = \left(A \bigcap B\right) \bigcup \left(\tilde{A} \bigcap B\right) \bigcup \left(A \bigcap \tilde{B}\right) \bigcup \left(\tilde{A} \bigcap \tilde{B}\right),$$

as a disjoint union, and so:

$$\mu\left(A \bigcap B\right) = 1 - \mu\left(\tilde{A} \bigcap B\right) - \mu\left(A \bigcap \tilde{B}\right) - \mu\left(\tilde{A} \bigcap \tilde{B}\right).$$

Substituting $\mu\left(\tilde{A} \bigcap B\right) = \mu\left(\tilde{A}\right) - \mu\left(\tilde{A} \bigcap \tilde{B}\right)$ and $\mu\left(A \bigcap \tilde{B}\right) = \mu\left(\tilde{B}\right) - \mu\left(\tilde{A} \bigcap \tilde{B}\right)$ obtains:

$$\mu\left(A \bigcap B\right) = 1 - \mu\left(\tilde{A}\right) - \mu\left(\tilde{B}\right) + \mu\left(\tilde{A} \bigcap \tilde{B}\right).$$

Returning to the definitions of A and B:

$$\begin{aligned} \lambda_U &\\ &= \lim_{t \to 1^-} \mu\left[X_1^{-1}(F_1^*(t), \infty) \bigcap X_2^{-1}(F_2^*(t), \infty)\right] / \mu\left[X_2^{-1}(F_2^*(t), \infty)\right] \\ &= \lim_{t \to 1^-} \left[1 - F_1\left(F_1^*(t)\right) - F_2\left(F_2^*(t)\right) + F\left(F_1^*(t), F_2^*(t)\right)\right] / \left[1 - F_2\left(F_2^*(t)\right)\right] \\ &= \lim_{t \to 1^-} \left[1 - F_1\left(F_1^*(t)\right) - F_2\left(F_2^*(t)\right) + C\left(F_1\left(F_1^*(t)\right), F_2\left(F_2^*(t)\right)\right)\right] / \\ &\qquad \left[1 - F_2\left(F_2^*(t)\right)\right]. \end{aligned}$$

The second expression in (7.32) now follows as above.

If C is differentiable at $t = 0$ or $t = 1$ as assumed, the expressions in (7.33) reflect the definition of derivative applied to (7.32), since $C(0, 0) = 0$ and $C(1, 1) = 1$. Similarly, by definition of continuously differentiable, (7.34) follows. ∎

Exercise 7.44 (Alternative λ_U) *Show that if $C(t, t)$ is left continuously differentiable at $t = 1$, then λ_U can be expressed:*

$$\lambda_U = 2 - \lim_{t \to 1^-} \frac{\ln C(t, t)}{\ln t}.$$

*Hint: **l'Hôpital's rule** of Exercise 7.27.*

Remark 7.45 (On tail dependence) *Some observations on the definitions of tail dependence.*

1. **Independent random variables:** *If X_1 and X_2 are independent random variables, then they are both upper and lower tail independent by Definition 3.47.*

2. **Asymmetry in the definition:** *While there appears to be an asymmetry in the definition of the **tail dependence measures** in terms of the roles of X_1 and X_2, it follows from Proposition 7.43 that in the case of continuous marginal distribution functions, that the values of these measures are unchanged if the roles are reversed.*

3. **Generalization:** *Both of the above comments apply in the general case of Definition 7.47 of the next section, again by Definition 3.47 for 1, and by Proposition 7.48 for 2.*

4. **Independence vs. Tail Independence:** *Perhaps surprisingly, the converse if 1 is false. See parts 3 and 4 of Example 7.46, which show that dependent random variables may be tail independent.*

Example 7.46 *Recalling some of the examples introduced in Section 7.4:*

1. **Independence Copula:** *Since $C^I(t,t) = t^2$, it follows from (7.33) that $\lambda_L = \lambda_U = 0$, which implies upper and lower tail independence. This result is what we would have anticipated since here $C^I(u_1, u_2)$ is the copula for independent random variables with continuous distributions.*

2. **Comonotonicity Copula:** *Here, $C^{Co}(t,t) = t$, and so $\lambda_L = \lambda_U = 1$, which implies (strong) upper and lower tail dependence. This is consistent with $C^{Co}(u_1, u_2)$ being defined as the joint distribution function of uniformly distributed variates with $U_1 = U_2$, almost surely.*

3. **Countermonotonicity Copula:** *With $C^{Cn}(t,t) = \max\{0, 2t-1\}$, it follows that $C^{Cn}(t,t) = 0$ in a neighborhood of $t = 0$ and $C^{Cn}(t,t) = 2t-1$ in a neighborhood of $t = 1$. Thus $\lambda_L = \lambda_U = 0$. This tail independence is consistent with $C^{Cn}(t,t)$ being the distribution function of uniform variates with $U_1 = 1 - U_2$ almost surely.*

4. **Gaussian Copula:** *When $n = 2$, the matrix R of correlation coefficients as defined in Book IV is given by:*

$$R = \begin{pmatrix} 1 & \rho \\ \rho & 1 \end{pmatrix}, \qquad R^{-1} = \begin{pmatrix} 1/(1-\rho^2) & -\rho/(1-\rho^2) \\ -\rho/(1-\rho^2) & 1/(1-\rho^2) \end{pmatrix}.$$

Hence by (7.13):

$$c^G(t_1, t_2) = \frac{1}{2\pi[1-\rho^2]^{1/2}} \exp\left[-\left(t_1^2 - 2\rho t_1 t_2 + t_2^2\right)/\left(2(1-\rho^2)\right)\right],$$

and thus

$$C^G(t,t) = \int_{-\infty}^{\Phi^{-1}(t)} \int_{-\infty}^{\Phi^{-1}(t)} c^G(t_1, t_2) dt_1 dt_2.$$

*Quite surprisingly, a 1960 result of **Masaaki Sibuya** yields that in the limit, $\lambda_L = \lambda_U = 0$ for $|\rho| < 1$. See Example 9.68.*

5. **Student T Copula:** *Let the matrix R be defined as for the Gaussian copula. While parametrized similarly, this is a correlation matrix for the variates only when* $\nu > 2$. *Then as in (7.14):*

$$c^T(t_1, t_2) = \frac{1}{2\pi[1-\rho^2]^{1/2}}\left[1 + \left(t_1^2 - 2\rho t_1 t_2 + t_2^2\right)/\left(\nu(1-\rho^2)\right)\right]^{-(\nu+2)/2}.$$

An identity was used to simplify the numerical coefficient, in that from $\Gamma(1+\alpha) = \alpha\Gamma(\alpha)$ *it follows that*

$$\Gamma(\nu/2 + 1)/[\nu\Gamma(\nu/2)] = 1/2.$$

Then:

$$C^T(t,t) = \int_{-\infty}^{T_\nu^{-1}(t)} \int_{-\infty}^{T_\nu^{-1}(t)} c^T(t_1, t_2) dt_1 dt_2,$$

and a long exercise in differential calculus yields that in the limit, if $|\rho| < 1$:

$$\lambda_L = \lambda_U = 2T_{\nu+1}\left(-\sqrt{\nu+1}\sqrt{\frac{1-\rho}{1+\rho}}\right).$$

Here $T_{\nu+1}$ *denotes the Student T distribution function defined with density in (7.16) but with* $\nu + 1$ *degrees of freedom.*

Hence for $|\rho| < 1$, $\lambda_L, \lambda_U \to 0$ *as* $\nu \to \infty$ *consistent with the observation that* $c^T(t_1, t_2) \to c^G(t_1, t_2)$ *in the limit. That this convergence of integrands implies the convergence of the associated integrals requires the integration tools of Book III, and specifically the **bounded convergence theorem**.*

Because of the symmetry of the Student T distribution, this expression is sometimes written as:

$$\lambda_L = \lambda_U = 2\widetilde{T}_{\nu+1}\left(\sqrt{\nu+1}\sqrt{\frac{1-\rho}{1+\rho}}\right),$$

where $\widetilde{T}_{\nu+1} \equiv 1 - T_{\nu+1}$ *is the Student T **survival function**.*

6. **Archimedean Copulas:** *By (7.17),* $C^A(t,t) = \psi_\theta\left(2\psi_\theta^{-1}(t)\right)$ *and so formulaically:*

$$\frac{dC^A(t,t)}{dt} = \frac{2\psi_\theta'\left(2\psi_\theta^{-1}(t)\right)}{\psi_\theta'(\psi_\theta^{-1}(t))},$$

since $\psi_\theta\left(\psi_\theta^{-1}(t)\right) = t$ *implies that:*

$$\psi_\theta'\left(\psi_\theta^{-1}(t)\right)\frac{d\psi_\theta^{-1}(t)}{dt} = 1.$$

Letting $s = \psi_\theta^{-1}(t)$, *note that since* $\psi_\theta : [0, \infty) \to [0, 1]$, $\psi_\theta(0) = 1$, *and* $\lim_{t\to\infty} \psi_\theta(t) = 0$, *it follows that:*

$$s \to \begin{cases} \infty, & t \to 0, \\ 0, & t \to 1. \end{cases}$$

Thus:

$$\lambda_L = 2\lim_{s\to\infty}\frac{\psi_\theta'(2s)}{\psi_\theta'(s)}, \qquad \lambda_U = 2\left[1 - \lim_{s\to0}\frac{\psi_\theta'(2s)}{\psi_\theta'(s)}\right]. \tag{7.35}$$

The tail dependence measures depend on the functional form of the generator function $\psi_\theta(t)$, *and those introduced above are evaluated directly below, with details left as exercises.*

(a) **Gumbel Copula:** *Defined by* $\psi_\theta(t) = \exp\left(-t^{1/\theta}\right)$, $\theta \geq 1$, *it follows that* $C^G(t,t) = t^{2^{1/\theta}}$, $dC^G(t,t)/dt = 2^{1/\theta}t^{2^{1/\theta}-1}$, *and thus:*

$$\lambda_L = 0, \qquad \lambda_U = 2 - 2^{1/\theta},$$

and so $0 \leq \lambda_U < 1$.

(b) **Clayton Copula:** *Defined by* $\psi_\theta(t) = (1 + \theta t)^{-1/\theta}$, $\theta \geq 0$, *it follows that* $C^C(t,t) = \left(2t^{-\theta} - 1\right)^{-1/\theta}$ *for* $\theta > 0$, *while* $C^C(t,t) = t^2$ *for* $\theta = 0$, *defined as a limit. This latter value is the same as that for the Gumbel copula when* $\theta^G = 1$. *Hence* $dC^C(t,t)/dt = 2\left(2 - t^\theta\right)^{-\frac{1}{\theta}(\theta+1)}$ *and* $dC^C(t,t)/dt = 2t$ *in the respective cases of* $\theta > 0$ *and* $\theta = 0$. *In all cases it follows that:*

$$\lambda_U = 0, \qquad \lambda_L = 2^{-1/\theta},$$

and so $0 \leq \lambda_L < 1$.

(c) **Frank Copula:** *Defined with* $\psi_\theta(t) = -\frac{1}{\theta}\ln\left(1 - \left(1 - e^{-\theta}\right)e^{-t}\right)$, $\theta \geq 0$, *it follows that* $C^F(t,t) = -\frac{1}{\theta}\ln\left(1 - \left(1 - e^{-\theta t}\right)^2 / \left(1 - e^{-\theta}\right)\right)$ *for* $\theta > 0$, *while* $C^F(t,t) = t^2$ *for* $\theta = 0$, *defined as a limit. This is again the same as the Gumbel copula when* $\theta^G = 1$. *Hence for* $\theta > 0$:

$$\frac{dC^F(t,t)}{dt} = 2\frac{e^{-t\theta}\left(e^{-t\theta} - 1\right)}{\left(e^{-t\theta} - 1\right)^2 + \left(e^{-\theta} - 1\right)},$$

while $dC^F(t,t)/dt = 2t$ *for* $\theta = 0$. *In all cases it follows that:*

$$\lambda_U = \lambda_L = 0.$$

7.6.2   Multivariate Tail Dependence and Copulas

When $F(x_1, x_2, ..., x_n)$ is a distribution function of $n > 2$ variates, the above definitions of the tail index can be generalized, again using conditional probability statements. As above, the conditional probabilities underlying these tail dependence measures are always well defined. Also as above, there is an apparent asymmetry in the definition to be discussed below.

Definition 7.47 (Multivariate tail dependence measures) *Let* $\{X_j\}_{j=1}^n$ *be random variables defined on a probability space* $(\mathcal{S}, \mathcal{E}, \mu)$ *with respective distribution functions* $\{F_j(x_j)\}_{j=1}^n$.

*The **lower tail dependence measure** or **coefficient** or **index** λ_L, is defined when the limit exists by:*

$$\lambda_L = \lim_{t \to 0^+} \mu\left[\bigcap_{j=1}^n X_j^{-1}(-\infty, F_j^*(t))\right] \Big/ \mu\left[X_n^{-1}(-\infty, F_n^*(t))\right]. \tag{7.36}$$

*The **upper tail dependence measure** or **coefficient** or **index** λ_U, is defined when the limit exists by:*

$$\lambda_U = \lim_{t \to 1^-} \mu\left[\bigcap_{j=1}^n X_j^{-1}(F_j^*(t), \infty)\right] \Big/ \mu\left[X_n^{-1}(F_n^*(t), \infty)\right]. \tag{7.37}$$

The variates $\{X_j\}_{j=1}^n$ *are said to be **lower tail dependent** if* $\lambda_L \in (0, 1]$ *and **lower tail independent** if* $\lambda_L = 0$. *Similarly,* $\{X_j\}_{j=1}^n$ *are said to be **upper tail dependent** if* $\lambda_U \in (0, 1]$ *and **upper tail independent** if* $\lambda_U = 0$.

We now generalize Proposition 7.43, deriving expressions for λ_L and λ_U when the marginal distributions are continuous, or locally differentiable. It will be noted that the copula-based representations for λ_L have generalized naturally, while those for λ_U have entered into a realm of complexity only hinted at above. These results will be discussed again and simplified in Remark 7.56, once survival copulas have been introduced and investigated.

Proposition 7.48 (Multivariate tail dependence measures and copulas) *Given random variables $\{X_j\}_{j=1}^n$ defined on a probability space $(\mathcal{S}, \mathcal{E}, \mu)$ with continuous distribution functions $\{F_j(x_j)\}_{j=1}^n$, and a copula C associated with the joint distribution function F, then when the limit exists:*

$$\lambda_L = \lim_{t \to 0^+} \frac{C(t, t, ..., t)}{t}. \tag{7.38}$$

For each k with $1 \le k \le n$, let J_k denote the set of $\binom{n}{k}$ distinct, increasing k-tuples of indexes from $(1, ..., n)$:

$$J_k \equiv \{I_k \equiv (i_{j_1}, ..., i_{j_k}) \subset (1, ..., n)\}.$$

Then when the limit exists:

$$\lambda_U = \lim_{t \to 1^-} \frac{\overline{C}(t, t, ..., t)}{1 - t}, \tag{7.39}$$

where with C_{I_k} denoting the various marginal copulas of Corollary 7.8:

$$\overline{C}(t_1, t_2, ..., t_n)$$
$$\equiv 1 - \sum_{I_1 \in J_1} C_{I_1}(t_{j_1}) + \sum_{I_2 \in J_2} C_{I_2}(t_{j_1}, t_{j_2}) -$$
$$.... + (-1)^{n-1} \sum_{I_{n-1} \in J_{n-1}} C_{I_{n-1}}(t_{j_1}, t_{j_2}, ..., t_{j_{n-1}}) \tag{7.40}$$
$$+ (-1)^n C(t_{j_1}, t_{j_2}, ..., t_{j_n}).$$

If $C(t, t, ..., t)$ is right differentiable at $t = 0$, respectively, $\overline{C}(t, t, ..., t)$ is left differentiable at $t = 1$, then:

$$\lambda_L = \left. \frac{dC(t, t, ..., t)}{dt} \right|_{t=0}, \qquad \lambda_U = - \left. \frac{d\overline{C}(t, t, ..., t)}{dt} \right|_{t=1}. \tag{7.41}$$

Further, if $C(t, t, ..., t)$ is continuously differentiable in a neighborhood of $t = 0$, respectively, $\overline{C}(t, t, ..., t)$ is continuously differentiable in a neighborhood of $t = 1$, then:

$$\lambda_L = \lim_{t \to 0^+} \frac{dC(t, t, ..., t)}{dt}, \qquad \lambda_U = - \lim_{t \to 1^-} \frac{d\overline{C}(t, t, ..., t)}{dt}. \tag{7.42}$$

Proof. *That the conditional probabilities underlying the definitions of λ_L and λ_U are well defined is proved as in Proposition 7.43.*

With F the joint distribution function of $(X_1, X_2, ..., X_n)$ and C a copula associated with F:

$$\begin{aligned}
\lambda_L &= \lim_{t \to 0^+} \mu \left[\bigcap_{j=1}^n X_j^{-1}(-\infty, F_j^*(t)) \right] \bigg/ \mu \left[X_n^{-1}(-\infty, F_n^*(t)) \right] \\
&= \lim_{t \to 0^+} F(F_1^*(t), F_2^*(t), ..., F_n^*(t)) / F_n(F_n^*(t)) \\
&= \lim_{t \to 0^+} C(F_1(F_1^*(t)), F_2(F_2^*(t)), ..., F_n(F_n^*(t))) / F_n(F_n^*(t)).
\end{aligned}$$

The expression in (7.38) now follows since $F_j(F_j^*(t)) = t$ for all $t \in (0, 1)$ by continuity and Proposition 3.22.

For (7.39) we require the inclusion-exclusion formula in (6.13) of Proposition 6.8. For notational convenience, we subscript t variates to better reveal the formula for $\overline{C}(t_1, t_2, ..., t_n)$, but then we set $t_j = t$ for all j.

Define:

$$\overline{C}(t_1, t_2, ..., t_n) \equiv \mu\left[\bigcap_{j=1}^{n} X_j^{-1}(F_j^*(t_j), \infty)\right], \tag{1}$$

and for notational simplicity let $\tilde{A}_j \equiv X_j^{-1}(F_j^*(t_j), \infty)$, and thus $A_j = X_j^{-1}(-\infty, F_j^*(t_j)]$.

By De Morgan's laws:

$$\mu\left[\bigcap_{j=1}^{n} \tilde{A}_j\right] = 1 - \mu\left[\bigcup_{j=1}^{n} A_j\right],$$

and then by (6.13):

$$\begin{aligned}
\mu\left[\bigcap_{j=1}^{n} X_j^{-1}(F_j^*(t_j), \infty)\right] &= 1 - \mu\left[\bigcup_{j=1}^{n} A_j\right] \\
&= 1 - \sum_{j=1}^{n} \mu[A_j] \\
&\quad + \sum_{i<j} \mu\left[A_i \bigcap A_j\right] ... \\
&\quad ... + (-1)^n \mu\left[\bigcap_{j=1}^{n} A_j\right].
\end{aligned}$$

Note that the various summations in this formula are equivalent to summations over the various J_k index collections defined above. For example:

$$\sum_{i<j<k} \mu\left[A_i \bigcap A_j \bigcap A_k\right] = \sum_{I_3 \in J_3} \mu\left[A_{i_1} \bigcap A_{i_2} \bigcap A_{i_3}\right].$$

Also note that for any term in any summation, if $I_k \equiv (i_{j_1}, ..., i_{j_k})$ and $F_{I_k}(x_{i_1}, ..., x_{i_k})$ is the associated marginal distribution function:

$$\begin{aligned}
\mu\left[A_{j_1} \bigcap A_{j_2} .... \bigcap A_{j_k}\right] &= \mu\left[\bigcap_{i=1}^{k} X_{j_i}^{-1}(-\infty, F_{j_i}^*(t_{j_i})]\right] \\
&= F_{I_k}(F_{j_1}^*(t_{i_1}), ..., F_{j_k}^*(t_{j_k})) \\
&= C_{I_k}\left(F_{j_1}\left(F_{j_1}^*(t_{j_1})\right), ..., F_{j_k}\left(F_{j_k}^*(t_{j_k})\right)\right) \\
&= C_{I_k}(t_{j_1}, ..., t_{j_k}).
\end{aligned}$$

Here $F_{j_i}\left(F_{j_i}^*(t_{j_i})\right) = t_{j_i}$ by continuity, and C_{I_k} is the associated marginal copula function of Corollary 7.8. Since $\mu\left[X_n^{-1}(F_n^*(t), \infty)\right] = 1 - t$ as in Proposition 7.43, this completes the proof of (7.39) after setting $t_{j_i} = t$ for all j_i.

If C is differentiable at $t = 0$ or $\overline{C}$ is differentiable at $t = 1$ as assumed, the expressions in (7.41) reflect the definition of derivative applied to (7.38) and (7.39), since $C(0, 0, ..., 0) = 0$ and $\overline{C}(1, 1, ..., 1) = 0$. For this last identity, note that $C_{I_k}(1, ..., 1) = 1$ for all I_k by definition of copula. Thus by the binomial formula:

$$\overline{C}(1, 1, ..., 1) = \sum_{k=0}^{n} (-1)^k \binom{n}{k} = (1-1)^n.$$

Similarly, by definition of continuously differentiable, (7.42) follows. ∎

Exercise 7.49 *Determine $\overline{C}(t, t)$ in the bivariate case and confirm that the results of Propositions 7.43 and 7.48 agree.*

7.6.3   Survival Functions and Copulas

While it might have initially been postulated by the notation that $\overline{C}(t, t, ..., t)$ in (7.41) is a copula, the conclusion in the prior proof that $\overline{C}(1, 1, ..., 1) = 0$ proved otherwise. On the other hand, $\overline{C}(0, 0, ..., 0) = 1$ since $C_{I_k}(0, ..., 0) = 0$ for all I_k by definition of this marginal copula. Thus, perhaps it is possible that $\overline{C}(t, t, ..., t)$ is a copula, but is parametrized "backwards." It turns out that this is the case.

With $\overline{C}(t_1, t_2, ..., t_n)$ given in (7.40), define:

$$\widehat{C}(t_1, t_2, ..., t_n) \equiv \overline{C}(1 - t_1, 1 - t_2, ..., 1 - t_n), \tag{7.43}$$

and note that $\widehat{C}(0, 0, ..., 0) = 0$ and $\widehat{C}(1, 1, ..., 1) = 1$. We will prove over several steps that $\widehat{C}$ is a copula.

First, $\widehat{C}(t_1, t_2, ..., t_n)$ has uniform marginal distributions. For notational simplicity, we check only the first marginal. Let $t' = (t_2, ..., t_n)$:

$$
\begin{aligned}
\widehat{C}_{\{1\}}(t_1) &= \lim_{t' \to 1} \widehat{C}(t_1, t_2, ..., t_n) \\
&= \lim_{t' \to 1} \overline{C}(1 - t_1, 1 - t_2, ..., 1 - t_n).
\end{aligned}
$$

Looking at the formula for $\overline{C}(t_1, t_2, ..., t_n)$, it follows that for $k \geq 2$:

$$\lim_{t' \to 1} C_{I_k}(1 - t_{j_1}, ..., 1 - t_{j_k}) = \begin{cases} C_{I_k}(0, ..., 0), & j_i \neq 1 \text{ for all } i, \\ C_{I_k}(1 - t_1, 0, ..., 0), & j_i = 1 \text{ for } i = 1. \end{cases}$$

Recalling that $C_{I_k}(t_{j_1}, ..., t_{j_k})$ is the joint distribution of uniformly distributed $(U_{j_1}, ..., U_{j_k})$, we conclude that $\lim_{t' \to 1} C_{I_k}(1 - t_{j_1}, ..., 1 - t_{j_k}) = 0$ in all cases.

When $k = 1$, this limit is $C_{\{1\}}(1 - t_1) = 1 - t_1$, and $C_{\{j\}}(0) = 0$ for $j \geq 2$.

Combining obtains:

$$\widehat{C}_{\{1\}}(t_1) = 1 - C_{\{1\}}(1 - t_1) = t_1,$$

and thus $\widehat{C}(t_1, t_2, ..., t_n)$ has uniform marginal distributions.

We next show that $\widehat{C}$ is continuous from above.

Proposition 7.50 ($\widehat{C}(t_1, t_2, ..., t_n)$ is continuous from above) *Given random variables* $\{X_j\}_{j=1}^n$ *defined on a probability space* $(\mathcal{S}, \mathcal{E}, \mu)$ *with respective continuous distribution functions* $\{F_j(x_j)\}_{j=1}^n$ *and a copula C associated with the joint distribution function F, let* $\widehat{C}(t_1, t_2, ..., t_n)$ *be defined as in (7.43) and (7.40). Then $\widehat{C}$ is continuous from above.*

Proof. *Recalling the n-increasing condition of Definition 6.7, let $t = (t_1, ..., t_n)$ and a sequence $t^{(m)} = (t_1^{(m)}, ..., t_n^{(m)})$ be given, with $t_i^{(m)} \geq t_i$ for all i and m and $t^{(m)} \to t$ as $m \to \infty$. To prove that $\widehat{C}(t) = \lim_{m \to \infty} \widehat{C}(t^{(m)})$, use the definition of $\widehat{C}$ in terms of $\overline{C}$ in (7.43), and then the definition of $\overline{C}$ in (1) of the proof of Proposition 7.48. We then obtain:*

$$
\begin{aligned}
\widehat{C}(t^{(m)}) &= \overline{C}\left(1 - t_1^{(m)}, 1 - t_2^{(m)}, ..., 1 - t_n^{(m)}\right) \\
&= \mu\left[\bigcap_{j=1}^n X_j^{-1}\left(F_j^*\left(1 - t_j^{(m)}\right), \infty\right)\right]. \tag{1}
\end{aligned}
$$

Now $t_j^{(m)} \to t_j$ is decreasing, and this implies that $1 - t_j^{(m)} \to 1 - t_j$ is increasing, and since F_j^* is left continuous this obtains that $F_j^*\left(1 - t_j^{(m)}\right) \to F_j^*(1 - t_j)$ for all j. For notational simplicity let $A_k = \bigcap_{j=1}^n X_j^{-1}(F_j^*\left(1 - t_j^{(k)}\right), \infty)$ and define a sequence of sets $\{B_m\}_{m=1}^\infty$ by:

$$B_m \equiv \bigcap_{k=1}^m A_k.$$

Then $B_m \subset A_m$ for all m and $\bigcap_{m=1}^{\infty} B_m = \bigcap_{m=1}^{\infty} A_m$.

Also, $\{B_m\}_{m=1}^{\infty}$ is a nested sequence, $B_{m+1} \subset B_m$ for all m, and so by continuity from above of μ:

$$\mu\left(\bigcap_{m=1}^{\infty} B_m\right) = \lim_{m\to\infty} \mu(B_m). \tag{2}$$

By convergence of $F_j^*\left(1 - t_j^{(m)}\right) \to F_j^*(1 - t_j)$:

$$\bigcap_{m=1}^{\infty} B_m = \bigcap_{j=1}^{n} X_j^{-1}(F_j^*(1 - t_j), \infty).$$

This convergence also implies that for any m, there exists $N \geq m$ so that $A_k \subset B_m \subset A_m$ for all $k \geq N$. By monotonicity of μ, $\mu(A_k) \leq \mu(B_m) \leq \mu(A_m)$ for all $k \geq N$, and thus $\lim_{m\to\infty} \mu(B_m) = \lim_{m\to\infty} \mu(A_m)$.

Combining into (2):

$$\mu\left(\bigcap_{j=1}^{n} X_j^{-1}(F_j^*(1 - t_j), \infty)\right) = \lim_{m\to\infty} \mu\left(\bigcap_{j=1}^{n} X_j^{-1}(F_j^*\left(1 - t_j^{(m)}\right), \infty)\right),$$

and by (1) this obtains $\widehat{C}(t) = \lim_{m\to\infty} \widehat{C}(t^{(m)})$. Thus, $\widehat{C}$ is continuous from above. ∎

The final property that needs to be verified to prove that $\widehat{C}(t_1, t_2, ..., t_n)$ is a copula is that this function is n-increasing, and for this a technical result is needed. First a definition.

Definition 7.51 (Survival functions and copulas) *If X is a random variable on a probability space $(\mathcal{S}, \mathcal{E}, \mu)$, the **survival function (s.f.)** associated with X, denoted S or S_X, is defined on $\mathbb{R}$ by:*

$$S(x) = \mu[X^{-1}(x, \infty)]. \tag{7.44}$$

*If $X_j : \mathcal{S} \longrightarrow \mathbb{R}$ are random variables on this probability space, $j = 1, 2, ..., n$, with **random vector** $X = (X_1, X_2, ..., X_n)$:*

$$X : \mathcal{S} \longrightarrow \mathbb{R}^n,$$

defined on $s \in \mathcal{S}$ by:

$$X(s) = (X_1(s), X_2(s), ..., X_n(s)),$$

*the **joint survival function (s.f.)** associated with X, denoted S or S_X, is defined on $(x_1, x_2, ..., x_n) \in \mathbb{R}^n$ by:*

$$S(x_1, x_2, ..., x_n) = \mu\left[\bigcap_{j=1}^{n} X_j^{-1}(x_j, \infty)\right]. \tag{7.45}$$

*Given $S(x_1, x_2, ..., x_n)$ and $I = \{i_1, ..., i_m\} \subset \{1, 2, ..., n\}$, let $x_J \equiv (x_{j_1}, x_{j_2}, ..., x_{j_{n-m}})$ for $j_k \in J \equiv \widetilde{I}$. The **marginal survival function** $S_I(x_I) \equiv S_I(x_{i_1}, x_{i_2}, ..., x_{i_m})$ is defined on $\mathbb{R}^m$ by (Note: $x_J \to -\infty$):*

$$S_I(x_I) \equiv \lim_{x_J \to -\infty} S(x_1, x_2, ..., x_n). \tag{7.46}$$

*Taking I equal to the n singleton sets, the collection $\{S_j(x_j)\}_{j=1}^{n}$ comprises the **marginal survival functions** of $S(x_1, x_2, ..., x_n)$.*

*A copula $C \in \mathfrak{F}(U_1, U_2, ..., U_n)$ is called **a survival copula associated with** $S(x_1, x_2, ..., x_n)$, and also **a survival copula of the random vector** $X = (X_1, X_2, ..., X_n)$ when S is the joint survival function associated with X, if:*

$$S(x_1, x_2, ..., x_n) = C(S_1(x_1), ..., S_n(x_n)). \tag{7.47}$$

Remark 7.52 (Survival functions) *A few clarifying comments are in order.*

1. *In the same way that the distribution function of X can be expressed $F(x) = \mu\{X \le x\}$, so too $S(x) = \mu\{X > x\}$, and thus:*

$$S(x) = 1 - F(x).$$

A consequence of this is that a survival function is continuous if and only if the associated distribution function is continuous. This also applies to the one-variable marginal distributions in (7.47).

2. *Similarly, just as $F(x_1, x_2, ..., x_n) = \mu\{X_j \le x_j, \text{ all } j\}$, we have $S(x_1, x_2, ..., x_n) = \mu\{X_j > x_j, \text{ all } j\}$, but as is readily appreciated for $n \ge 2$:*

$$S(x_1, x_2, ..., x_n) \ne 1 - F(x_1, x_2, ..., x_n).$$

3. *As is the case for $F(x_1, x_2, ..., x_n)$ in (3.17), it is sometimes convenient to express:*

$$S(x_1, x_2, ..., x_n) = \mu\left[X^{-1}\left(\prod\nolimits_{j=1}^{n}(x_j, \infty)\right)\right], \tag{7.48}$$

where $X \equiv (X_1, ..., X_n)$.

4. *While $S_I(x_I)$ is called a marginal survival function, it is an exercise to prove that this limit is indeed a survival function, and in particular it is the joint survival function of $X_I \equiv (X_{i_1}, X_{i_2}, ..., X_{i_m})$.*

5. *Although we do not develop all the details, there is also a version of **Sklar's theorem** for joint survival functions, named for **Abe Sklar** (1925–2020), who published this in 1959. The results are the same as above, that a survival copula always exists, and when the marginal survival functions are continuous, this copula is unique on $(0, 1)^n$.*

To set the stage for the next result, recall the development in Proposition 6.9.

Given $\{X_j\}_{j=1}^{n}$ defined a probability space $(\mathcal{S}, \mathcal{E}, \mu)$, the associated joint distribution function $F(x_1, x_2, ..., x_n)$ induces a unique probability measure μ_F on the Borel sigma algebra $\mathcal{B}(\mathbb{R}^n)$ which satisfies (6.12):

$$\mu_F\left(\prod\nolimits_{i=1}^{n}(a_i, b_i]\right) = \sum\nolimits_{x} sgn(x)F(x),$$

for all bounded right semi-closed rectangles in $\mathbb{R}^n$. In this expression, each $x = (x_1, ..., x_n)$ in the summation is one of the 2^n vertices of this rectangle, so $x_i = a_i$ or $x_i = b_i$, and $sgn(x)$ is defined as -1 if the number of a_i-components of x is odd, and $+1$ otherwise.

The next result is that $\mu_F\left(\prod_{i=1}^{n}(a_i, b_i]\right)$ can also be calculated with the joint survival function, but with $sgn(x)$ redefined. Again we deploy the inclusion-exclusion formula of Proposition 6.8.

Proposition 7.53 ($\mu_F\left(\prod_{i=1}^{n}(a_i, b_i]\right)$ with $S(x)$) *Let $\{X_j\}_{j=1}^{n}$ be defined on a probability space $(\mathcal{S}, \mathcal{E}, \mu)$ with joint survival function $S(x)$ and joint distribution function $F(x)$. Then with μ_F the induced probability measure on $\mathcal{B}(\mathbb{R}^n)$:*

$$\mu_F\left(\prod\nolimits_{i=1}^{n}(a_i, b_i]\right) = \sum\nolimits_{x} sgn'(x)S(x), \tag{7.49}$$

for all bounded right semi-closed rectangles in $\mathbb{R}^n$. In this expression, each $x = (x_1, ..., x_n)$

in the summation is one of the 2^n vertices of this rectangle, so $x_i = a_i$ or $x_i = b_i$, and $sgn'(x)$ is defined as -1 if the number of b_i-components of x is odd, and $+1$ otherwise.
Proof. *Let $\prod_{i=1}^n (a_i, b_i] \subset \mathbb{R}^n$ be a bounded right semi-closed rectangle. To simplify notation, for any point $x = (x_1, ..., x_n)$ let B_x or $B_{(x_1, ..., x_n)}$ denote the rectangle:*

$$B_x \equiv \prod_{j=1}^n (x_j, \infty).$$

We claim that:

$$B_{(a_1, ..., a_n)} = \prod_{i=1}^n (a_i, b_i] \bigcup \bigcup_{j=1}^n B_{x^{(j)}},$$

where for each j in the union, $x_i^{(j)} = a_i$ for $i \neq j$ and $x_j^{(j)} = b_j$.

To prove this note that if $x \in A_{(a_1, ..., a_n)}$, then $x_j > a_j$ for all j. Thus if $x \notin \prod_{i=1}^n (a_i, b_i]$, then $x_j > b_j$ for at least one j, and so $x \in \bigcup_{j=1}^n B_{x^{(j)}}$.

As $B_{(a_1, ..., a_n)}$ is a disjoint union of these two sets, $X^{-1}[B_{(a_1, ..., a_n)}]$ is again a disjoint union, where $X = (X_1, ..., X_n)$:

$$X^{-1}[B_{(a_1, ..., a_n)}] = X^{-1}\left[\prod_{i=1}^n (a_i, b_i]\right] \bigcup X^{-1}\left[\bigcup_{j=1}^n B_{x^{(j)}}\right].$$

Finite additivity of μ obtains:

$$\mu\left[X^{-1}\left(\prod_{i=1}^n (a_i, b_i]\right)\right] = \mu\left[X^{-1}(B_{(a_1, ..., a_n)})\right] - \mu\left[X^{-1}\left(\bigcup_{j=1}^n B_{x^{(j)}}\right)\right] \qquad (1)$$

$$= S(a_1, ..., a_n) - \mu\left[\bigcup_{j=1}^n X^{-1}(B_{x^{(j)}})\right].$$

Applying (6.13) to the measure of the union:

$$-\mu\left[\bigcup_{j=1}^n X^{-1}(B_{x^{(j)}})\right]$$

$$= -\sum_{j=1}^n \mu\left[X^{-1}(B_{x^{(j)}})\right] + \sum_{i<j} \mu\left[X^{-1}(B_{x^{(i)}})\bigcap X^{-1}(B_{x^{(j)}})\right]$$

$$- \sum_{i<j<k} \mu\left[X^{-1}(B_{x^{(i)}})\bigcap X^{-1}(B_{x^{(j)}})\bigcap X^{-1}(B_{x^{(k)}})\right] +$$

$$\cdots + (-1)^n \mu\left[\bigcap_{j=1}^n X^{-1}(B_{x^{(j)}})\right].$$

Now:

$$-\sum_{j=1}^n \mu\left[X^{-1}(B_{x^{(j)}})\right] = -\sum_{j=1}^n S(x^{(j)}),$$

noting that the sign for this summation is -1, consistent with each $x^{(j)}$ having one b_j component.

Next:

$$X^{-1}(B_{x^{(i)}})\bigcap X^{-1}(B_{x^{(j)}}) = X^{-1}(B_{x^{(i,j)}}),$$

where $x_l^{(i,j)} = a_l$ for $l \neq i, j$ and $x_l^{(i,j)} = b_l$ for $l = i, j$. Thus:

$$\sum_{i<j} \mu\left[X^{-1}(B_{x^{(i)}})\bigcap X^{-1}(B_{x^{(j)}})\right] = \sum_{i<j} \mu\left[X^{-1}(B_{x^{(i,j)}})\right]$$

$$= \sum_{i<j} S(x^{(i,j)}).$$

The sign for this summation is $+1$, consistent with each $x^{(i,j)}$ having two b_j components.

With the same notational convention:

$$-\sum_{i<j<k} \mu \left[X^{-1}\left(B_{x^{(i)}}\right) \bigcap X^{-1}\left(B_{x^{(j)}}\right) \bigcap X^{-1}\left(B_{x^{(k)}}\right) \right]$$

$$= -\sum_{i<j<k} \mu \left[X^{-1}\left(B_{x^{(i,j,k)}}\right) \right]$$

$$\equiv -\sum_{i<j<k} S(x^{(i,j,k)}),$$

where again the sign of this sum is consistent with each $x^{(i,j,k)}$ having three b_j components. Continuing and making substitutions into (1):

$$\mu \left[X^{-1}\left(\prod_{i=1}^{n}(a_i, b_i] \right) \right] = \sum_x sgn'(x)S(x), \qquad (2)$$

where each $x = (x_1, ..., x_n)$ is one of the 2^n vertices of $\prod_{i=1}^{n}(a_i, b_i]$, and $sgn'(x)$ is -1 if the number of components of x that equal b_j is odd, and $+1$ otherwise.

The identity in (7.49) now follows from (2) and (6.14). ∎

With the result of (7.49) in hand, we are now ready to prove that $\widehat{C}(t_1, t_2, ..., t_n)$ in (7.43) satisfies the n-increasing condition and is thus a copula.

Proposition 7.54 ($\widehat{C} \in \mathfrak{F}(U_1, U_2, ..., U_n)$) *Given random variables $\{X_j\}_{j=1}^{n}$ defined on a probability space $(\mathcal{S}, \mathcal{E}, \mu)$ with respective continuous distribution functions $\{F_j(x_j)\}_{j=1}^{n}$ and a copula C associated with the joint distribution function F, let $\widehat{C}(t_1, t_2, ..., t_n)$ be defined as in (7.43) and (7.40).*

Then $\widehat{C} \in \mathfrak{F}(U_1, U_2, ..., U_n)$. In other words, $\widehat{C}$ is a copula.

Proof. *At the beginning of the section it was seen that $\widehat{C}(0, 0, ..., 0) = 0$ and $\widehat{C}(1, 1, ..., 1) = 1$, and that $\widehat{C}(t_1, t_2, ..., t_n)$ has uniform marginal distributions. This function is also continuous from above by Proposition 7.50. To prove that $\widehat{C}(t_1, t_2, ..., t_n)$ is a distribution function by Proposition 6.10, we only need to verify the n-increasing condition of Definition 6.7.*

To this end, let the bounded measurable rectangle $A \equiv \prod_{j=1}^{n}(s_j, t_j] \subset (0,1)^n$ be given, so $s_j < t_j$ for all j. We must prove that:

$$\sum_x sgn(x)\widehat{C}(x) \geq 0,$$

where each $x = (x_1, ..., x_n)$ in the summation is one of the 2^n vertices of this rectangle, so $x_i = s_i$ or $x_i = t_i$, and $sgn(x)$ is defined as -1 if the number of s_i-components of x is odd, and $+1$ otherwise.

As in (1) in the proof of Proposition 7.50, but using the notation of (7.48):

$$\begin{aligned} \widehat{C}(x) &= \overline{C}(1 - x_1, 1 - x_2, ..., 1 - x_n) \\ &= \mu \left[X^{-1}\left(\prod_{j=1}^{n}\left(F_j^*(1 - x_j), \infty\right) \right) \right] \\ &= S\left(F_1^*(1 - x_1), ..., F_n^*(1 - x_n)\right), \end{aligned} \qquad (1)$$

where S is the joint survival function of Definition 7.51. Thus:

$$\sum_x sgn(x)\widehat{C}(x) = \sum_x sgn(x)S\left(F_1^*(1 - x_1), ..., F_n^*(1 - x_n)\right).$$

Now let $A' \equiv \prod_{i=1}^{n}(F_j^*(1 - t_j), F_j^*(1 - s_j)]$. Then by substitution:

$$\sum_x sgn(x)\widehat{C}(x) = \sum_y sgn'(y)S(y_1, ..., y_n),$$

where each $y = (y_1, ..., y_n)$ in the summation is one of the 2^n vertices of the rectangle A', so $y_i = F_i^* (1 - s_i)$ or $y_i = F_i^* (1 - t_i)$, and $\text{sgn}'(y)$ is defined as -1 if the number of $F_i^* (1 - s_i)$-components of y is odd, and $+1$ otherwise. Note that $\text{sgn}(x) = \text{sgn}'(y)$ as both reflect the parity of the s_i-components.

Finally, we have by (7.49) that:

$$\sum_x \text{sgn}(x)\widehat{C}(x) = \mu_F \left(\prod_{i=1}^n \left(F_j^* (1 - t_j), F_j^* (1 - s_j) \right] \right) \geq 0.$$

Thus $\widehat{C}$ satisfies the n-increasing condition and is therefore a copula. ∎

The final result of this section is to prove that not only is $\widehat{C}$ is a copula, but it is the survival copula associated with X. Recall item 1 of Remark 7.52, that one-variable marginal survival functions are continuous if and only if the associated one-variable marginal distribution functions are continuous.

Proposition 7.55 ($\widehat{C}$ is a survival copula) *Given random variables $\{X_j\}_{j=1}^n$ defined on a probability space $(\mathcal{S}, \mathcal{E}, \mu)$ with respective continuous survival functions $\{S_j(x_j)\}_{j=1}^n$ and a joint survival function $S(x_1, x_2, ..., x_n)$. Then $\widehat{C}(t_1, t_2, ..., t_n)$ as defined as in (7.43) and (7.40) is a survival copula associated with $S(x_1, x_2, ..., x_n)$ as in (7.47):*

$$S(x_1, x_2, ..., x_n) = C\left(S_1(x_1), ..., S_n(x_n) \right).$$

Proof. *By (1) of the prior proof, and recalling that $1 - S_j(x_j) = F_j(x_j)$:*

$$
\begin{aligned}
C\left(S_1(x_1), ..., S_n(x_n) \right) &= \mu \left[X^{-1} \left(\prod_{j=1}^n \left(F_j^* (1 - S_j(x_j)), \infty \right) \right) \right] \\
&= \mu \left[X^{-1} \left(\prod_{j=1}^n \left(F_j^* (F_j(x_j)), \infty \right) \right) \right] \\
&= \mu \left[X^{-1} \left(\prod_{j=1}^n (x_j, \infty) \right) \right] \\
&= S(x_1, x_2, ..., x_n).
\end{aligned}
$$

In the third line, recall that $F_j^* (F_j(x_j)) = x_j$ for continuous marginal distributions by Proposition 3.22. ∎

Remark 7.56 (On λ_U) *Recall (7.39), that for continuous marginal distribution functions:*

$$\lambda_U = \lim_{t \to 1^-} \frac{\overline{C}(t, t, ..., t)}{1 - t}.$$

By (7.43), $\overline{C}(t, t, ..., t) = \widehat{C}(1 - t, 1 - t, ..., 1 - t)$, and thus by a substitution and change of variable:

$$\lambda_U = \lim_{t \to 0+} \frac{\widehat{C}(t, t, ..., t)}{t}.$$

This looks identical to the formula for λ_L in (7.38), but now using a copula for $S(x_1, x_2, ..., x_n)$ rather that a copula for $F(x_1, x_2, ..., x_n)$.

Similarly, if $\widehat{C}(t, t, ..., t)$ is right differentiable at $t = 0$, then:

$$\lambda_U = \left. \frac{d\widehat{C}(t, t, ..., t)}{dt} \right|_{t=0},$$

while if continuously differentiable in a neighborhood of $t = 0$, then:

$$\lambda_U = \lim_{t \to 0+} \frac{d\widehat{C}(t, t, ..., t)}{dt}.$$

In all cases, these results are formulaically identical to the respective results for λ_L, but now using the survival copula.

8

Weak Convergence

In this chapter we introduce the definition of **weak convergence** of a sequence of distribution functions $\{F_n(x)\}_{n=1}^{\infty}$ to $F(x)$, and also **weak convergence** of a sequence of measures $\{\mu_n\}_{n=1}^{\infty}$ to μ. For function sequences this will add to the notions of **uniform** and **pointwise** convergence of Definition I.3.40, and **almost everywhere** convergence in the sense of Definition I.3.17. We begin with a few examples, discussing both the distribution functions and their associated measures. Recall the discussion on distribution functions in the introduction to Chapter 6.

Let $F(x)$ be the distribution function of a random variable X on a probability space $(\mathcal{S}, \mathcal{E}, \mu)$. Then $F(x)$ is a right continuous, increasing function on $\mathbb{R}$ which satisfies $F(-\infty) = 0$ and $F(\infty) = 1$, defined as limits. The associated Borel probability measure μ_F is defined on $\mathcal{A}'$, the semi-algebra of right semi-closed intervals, in (6.8) of Proposition 6.3:

$$\mu_F((a, b]) = F(b) - F(a),$$

and so $\mu_F((-\infty, x]) = F(x)$. As noted following this result, this obtains that $\mu_F(\{b\}) = F(b) - F(b^-)$ where $F(b^-)$ denotes the left limit at b. Thus $F(x)$ is continuous at b if and only if $\mu_F(\{b\}) = 0$.

Conversely, given a probability measure μ on $(\mathbb{R}, \mathcal{B}(\mathbb{R}))$, the associated distribution function $F_\mu(x)$ is defined by:

$$F_\mu(x) \equiv \mu((-\infty, x]),$$

and thus $F_\mu(b) - F_\mu(a) = \mu((a, b])$. This function is then right continuous and increasing, and satisfies $F(-\infty) = 0$ and $F(\infty) = 1$, defined as limits.

Consistency of these constructions is addressed in Chapter 6, reflecting results from Sections I.5.3 and I.8.3.1.

Example 8.1 (On convergence of distributions) *Given a distribution function $F(x)$ of a random variable or a probability measure:*

1. *The sequence of distribution functions:*

$$F_n(x) \equiv F(x + 1/n),$$

   ***converges pointwise** to $F(x)$ by right continuity.*

   For the associated measures:

$$\mu_{F_n}((a, b]) \equiv F(b + 1/n) - F(a + 1/n),$$

   converges by right continuity to $\mu_F((a, b])$ for all $(a, b] \in \mathcal{A}'$, the semi-algebra of right semi-closed intervals.

   Here and below, $\mu_{F_n}((a, b])$ is defined as a limit for unbounded intervals.

2. *Conversely:*

$$F_n(x) \equiv F(x - 1/n),$$

*converges pointwise to $F(x^-)$, the left limit of $F(x)$, and hence **converges almost everywhere** to $F(x)$. Put another way, $F_n(x)$ converges to $F(x)$ at every continuity point of $F(x)$. By Proposition 6.1, $F(x)$ has at most countably many discontinuities, a set of measure zero.*

Now:

$$\mu_{F_n}((a,b]) \equiv F(b-1/n) - F(a-1/n),$$

converges to $\mu_F((a,b])$ for such intervals for which a and b are continuity points of $F(x)$. In general:,

$$
\begin{aligned}
\mu_{F_n}((a,b]) \quad &\rightarrow \quad F(b^-) - F(a^-) \\
&= \quad \mu_F([a,b)) \\
&= \quad \mu_F((a,b]) + \mu_F(a) - \mu_F(b).
\end{aligned}
$$

3. *Defining:*

$$F_n(x) \equiv F(x+n),$$

*then $F_n(x)$ **converges for all** x to $F_0(x) \equiv 1$, but F_0 is not a distribution function (of a random variable or probability measure) since $\lim_{x\to-\infty} F_0(x) \neq 0$.*

In addition,

$$\mu_{F_n}((a,b]) \equiv F(b+n) - F(a+n),$$

converges to 0 for all bounded $(a,b] \in \mathcal{A}'$, consistent with the measure induced by $F_0(x)$. But $\mu_{F_n}(\mathbb{R}) \equiv 1$ for all n, and this is inconsistent with the limiting function $F_0(x)$, since $\mu_{F_0}(\mathbb{R}) \equiv 0$.

Analogous observations apply to $F_n(x) \equiv F(x-n)$, but now $F_n(x)$ converges to $\widetilde{F}_0(x) \equiv 0$ for all x.

4. *Defining:*

$$F_n(x) \equiv F(x + (-1)^n n),$$

*then for any x, $F_{2n}(x)$ converges to $F_0(x) \equiv 1$, while $F_{2n+1}(x)$ converges to $\widetilde{F}_0(x) \equiv 0$, so $F_n(x)$ does **not converge in any sense**.*

However, for any bounded $(a,b] \in \mathcal{A}'$, $\mu_{F_n}((a,b])$ converges to 0.

8.1 Definitions of Weak Convergence

The goal of this section is to study the notion of **weak convergence** of a sequence of distribution functions or probability measures. In effect, weak convergence is a version of convergence almost everywhere, but where the exceptional set is restricted to be a subset of the at most countable set of discontinuities of $F(x)$. This section will focus largely on the theory of weak convergence of distribution functions and measures defined on $\mathbb{R}$. See Book VI for extensions of these results and generalizations to the case of measures defined on $\mathbb{R}^n$.

For the following definition, it is not assumed that there is a common probability space $(\mathcal{S}, \mathcal{E}, \mu)$ on which all random variables are defined. In other words, it is not required that these distribution functions or probability measures originate from random variables defined on a common probability space, though in many applications this will be the natural setting.

Definition 8.2 (Weak convergence - Distributions and measures on $\mathbb{R}$)

a. *A sequence of **distribution functions** $\{F_n(x)\}_{n=1}^{\infty}$ on $\mathbb{R}$ will be said to **converge weakly** to a distribution function $F(x)$, denoted:*

$$F_n \Rightarrow F,$$

if $F_n(x) \to F(x)$ for every continuity point of $F(x)$.

b. *A sequence of **probability measures** $\{\mu_n\}_{n=1}^{\infty}$ on $\mathbb{R}$ will be said to **converge weakly** to a probability measure μ, denoted:*

$$\mu_n \Rightarrow \mu,$$

if $\mu_n((-\infty, x])$ converges to $\mu((-\infty, x])$ for all x for which $\mu(\{x\}) = 0$.

Note that this definition **explicitly** assumes that $F(x)$ is a distribution function and μ is a probability measure. This raises a question: Can $F_n(x)$ or μ_n converge by the given criterion but converge to $F(x)$ which is not a distribution function, or to μ which is not a probability measure? In short "yes," but we will see below that we can characterize the situations for which the resulting $F(x)$ must be a distribution function, and resulting μ must be a probability measure.

Remark 8.3 (Convergence in distribution)

1. ***Random variables:*** *We have seen weak convergence earlier, but in a different context. Restating Definition 5.19 in the above terminology, a sequence of random variables $\{X_n\}_{n=1}^{\infty}$ **converges in distribution** or **converges in law** to a random variable X, denoted:*

$$X_n \to_d X, \ or, \ X_n \Rightarrow X,$$

if $F_n \Rightarrow F$ for the associated distribution functions.

2. ***Random vectors:*** *Though not yet studied, a sequence of random vectors $\{X_n\}_{n=1}^{\infty}$ can also be defined to **converge in distribution** or **converge in law** to a random vector X, denoted $X_n \to_d X$ or $X_n \Rightarrow X$. The requirement is analogous, that for the associated joint distribution functions, $F_n(x) \to F(x)$ for every continuity point of $F(x)$. Looking ahead to Definition 8.6, $X_n \to_d X$ if $F_n \Rightarrow F$ for the associated distribution functions.*

Exercise 8.4 (Definitional consistency of weak convergence) *As noted in the introduction, a probability measure μ_F on $\mathbb{R}$ is induced by a given distribution function F by defining:*

$$\mu_F((-\infty, x]) \equiv F(x).$$

Similarly, a distribution function F_μ is induced by probability measure μ by this same identity.

Prove that if x is a continuity point of F, then $\mu_F(\{x\}) = 0$, and conversely, if $\mu(\{x\}) = 0$, then x is a continuity point of F_μ. Thus $\mu_n \Rightarrow \mu$ if and only if $F_n \Rightarrow F$ when distribution functions and probability measures are induced from one another.

Hint: Recall that measures are continuous from above, so $\mu(\{x\}) = \lim \mu[I_n]$ where $\{I_n\}$ is a nested collection of intervals (or general sets) with $\bigcap_{n=1}^{\infty} I_n = x$.

Remark 8.5 (Weak convergence - Increasing functions on $\mathbb{R}$) *It is often convenient to use the terminology of* **weak convergence** *in the context of increasing functions rather than only for distribution functions. Such functions have left and right limits at every point by the proof of Proposition I.5.8, and thus are again continuous on all but at most countably many points.*

As an example, in Section 8.3 we will investigate the relationship between $F_n \Rightarrow F$ and $F_n^* \Rightarrow F^*$, where these latter functions are the left continuous inverses of Definition 3.12. These functions are increasing by Proposition 3.16 but are not distribution functions.

Definition: *A sequence of* **increasing functions** *$\{F_n(x)\}_{n=1}^{\infty}$ will be said to* **converge weakly** *to an increasing function $F(x)$, denoted $F_n \Rightarrow F$, if $F_n(x) \to F(x)$ for every continuity point of $F(x)$.*

The notion of weak convergence also applies in the multivariate context. We state the definition for completeness, but defer our investigations to Book VI.

Definition 8.6 (Weak convergence - Distributions and measures on $\mathbb{R}^m$) *Generalizing the above definition:*

   a. *A sequence of joint distribution functions $\{F_n(x)\}_{n=1}^{\infty}$ defined on $\mathbb{R}^m$ will be said to* **converge weakly** *to a joint distribution function $F(x)$, denoted $F_n \Rightarrow F$, if $F_n(x) \to F(x)$ for every continuity point of $F(x)$.*

   b. *A sequence of probability measures $\{\mu_n\}_{n=1}^{\infty}$ defined on $\mathbb{R}^m$ will be said to* **converge weakly** *to a probability measure μ, denoted $\mu_n \Rightarrow \mu$, if:*

$$\mu_n\left[\prod\nolimits_{j=1}^{m}(-\infty, x_j]\right] \to \mu\left[\prod\nolimits_{j=1}^{m}(-\infty, x_j]\right]$$

   for all rectangles $A_x \equiv \prod_{j=1}^{m}(-\infty, x_j]$ with $\mu(\{x\}) = 0$.

Note that as in Exercise 8.4, it is also the case here that these definitions of weak convergence are consistent when applied to distribution functions and probability measures on $\mathbb{R}^m$ when one is induced from the other. As discussed in Section 6.2.1 and more fully developed in Chapter I.8, probability measures and distribution functions on $\mathbb{R}^m$ are linked by:

$$F(x_1, x_2, ..., x_m) \equiv \mu\left[\prod\nolimits_{j=1}^{m}(-\infty, x_j]\right].$$

That is, from F we can induce μ_F, and from μ we can induce F_μ. It then follows that $\mu_n \Rightarrow \mu$ if and only if $F_n \Rightarrow F$, when distribution functions and probability measures are induced from one another.

Returning to the setting of Definition 8.2, we next investigate the following question:

If $\mu_n \Rightarrow \mu$ and $\mu(\{a\}) = \mu(\{b\}) = 0$, does $\mu_n(I) \to \mu(I)$ for any interval $I = \langle a, b \rangle$, whether open, closed or semi-closed?

From Exercise 8.4 we know that $F_n \Rightarrow F$ for the induced distribution functions, and that a and b are continuity points for F. The subtlety here is that a continuity point x of F need not be a continuity point of the various F_n, and thus the μ_n-measure of such x need not equal 0. However, the following result proves that the μ_n-measure of such x converges to 0 as $n \to \infty$.

Proposition 8.7 (On $\mu_n(\{x\}) \to 0$) *If $F_n \Rightarrow F$ and x is a continuity point of F, then $\mu_n(\{x\}) \to 0$.*

Proof. *Let $\epsilon > 0$ be given. We show that there exists N so that $\mu_n(\{x\}) < 3\epsilon$ for $n \geq N$.*

Since $F_n(x) \to F(x)$, there is an N_0 so that $|F_n(x) - F(x)| < \epsilon$ for $n \geq N_0$. Also, *because F is continuous at x there is a δ so that $|F(x) - F(y)| < \epsilon$ for $|x - y| < \delta$. Let $\{y_j\}_{j=1}^{\infty}$ be continuity points of F with $y_j < x$, $|x - y_j| < \delta$ and $y_j \to x$. These points exist since F has at most countably many discontinuities and hence continuity points are dense in $\mathbb{R}$.*

Now $F_n(y_j) \to F(y_j)$ for all j, and thus for each j there is an N_j so that $|F_n(y_j) - F(y_j)| < \epsilon$ for $n \geq N_j$. Choosing any j, it follows that for $n \geq \max\{N_0, N_j\}$:

$$|F_n(x) - F_n(y_j)| \leq |F_n(x) - F(x)| + |F(x) - F(y_j)| + |F(y_j) - F_n(y_j)| < 3\epsilon.$$

Since

$$F_n(x) - F_n(y_j) = \mu_n\left((y_j, x]\right),$$

monotonicity of μ_n obtains that for $n \geq \max\{N_0, N_j\}$:

$$\mu_n\left(\{x\}\right) \leq \mu_n\left((y_j, x]\right) < 3\epsilon.$$

■

Corollary 8.8 (When $\mu_n(I) \to \mu(I)$) *If $\mu_n \Rightarrow \mu$ and $\mu\{a\} = \mu\{b\} = 0$, then $\mu_n(I) \to \mu(I)$ for any interval $I = \langle a, b \rangle$, whether open, closed or semi-closed.*

Proof. *For any such interval, since a and b are continuity points of the distribution function F induced by μ, it follows that*

$$\mu(I) = F(b) - F(a).$$

Given n, as a and b need not be continuity points of the distribution function F_n associated with μ_n, the value of $\mu_n(I)$ depends on whether I is open, closed or semi-closed.

Recall the formulas following Proposition 6.3:

$$\begin{aligned}
\mu_n((a, b]) &= F_n(b) - F_n(a), \\
\mu_n((a, b)) &= F_n(b) - F_n(a) - \mu_n\left(\{b\}\right), \\
\mu_n([a, b)) &= F_n(b) - F_n(a) - \mu_n\left(\{b\}\right) + \mu_n\left(\{a\}\right), \\
\mu_n([a, b]) &= F_n(b) - F_n(a) + \mu_n\left(\{a\}\right).
\end{aligned}$$

Since $\mu_n\left(\{a\}\right) \to 0$ and $\mu_n\left(\{b\}\right) \to 0$ by Proposition 8.7, and $F_n(x) \to F(x)$ for continuity points $x = a, b$, the result follows. ■

Remark 8.9 (A generalization previewed) *In Book VI we return to this discussion with the tools of a general integration theory from Book V. There it will be proved that if $\mu_n \Rightarrow \mu$, then $\mu_n(A) \to \mu(A)$ for any Borel set $A \subset \mathbb{R}$ for which $\mu(\partial A) = 0$. Here ∂A denotes the **boundary of** A and is defined:*

$$\partial A = \{x | x \text{ is a limit point of } A \text{ and } \widetilde{A}\}. \tag{8.1}$$

By limit point is meant that there exists $\{x_n\}_{n=1}^{\infty} \subset A$ and $\{x_n'\}_{n=1}^{\infty} \subset \widetilde{A}$ with $x_n \to x$ and $x_n' \to x$. For the intervals I above, $\partial A = \{a, b\}$.

This result will also hold in the context of measures on $\mathbb{R}^m$.

It is then natural to wonder if more is true. Perhaps $\mu_n \Rightarrow \mu$ implies that $\mu_n(A) \to \mu(A)$ for every Borel set $A \in \mathcal{B}(\mathbb{R})$. The following example shows that this is not the case.

Example 8.10 ($\mu_n \Rightarrow \mu \nRightarrow \mu_n(A) \to \mu(A)$, all A) *On the interval $[0, 1]$, let $\mu_n^R\left(j/n\right) = \frac{1}{n}$, $j = 1, 2, .., n$, denote the discrete uniform measure introduced in (1.5), and $\mu \equiv m$, Lebesgue measure on this interval. Then the associated distribution functions are defined by:*

$$F(x) = x,$$

$$F_n(x) = \begin{cases} 0, & x < 0, \\ \lfloor nx \rfloor / n, & 0 \leq x \leq 1, \\ 1, & x > 1. \end{cases}$$

Here $\lfloor nx \rfloor$ denotes the **greatest integer less than or equal to** nx:

$$\lfloor nx \rfloor = \max\{m \in \mathbb{N} | m \leq nx\}. \tag{8.2}$$

Then $F_n(x) \to F(x)$ for all x, and hence $\mu_n^R \Rightarrow m$. However, $\mu_n^R(A) \nrightarrow m(A)$ for every Borel set $A \in \mathcal{B}(\mathbb{R})$.

For example, if A is the set of all rationals in $[0,1]$, then $\mu_n^R(A) = 1$ for all n but $m(A) = 0$. Conversely for the irrationals, $\mu_n^R(A) = 0$ for all n but $m(A) = 1$.

These examples are not inconsistent with the above noted Book VI result. Here, the set of boundary points for A defined as the set of rationals or irrationals is $\partial A = [0,1]$, which does not have m-measure zero. But if A' is any **finite** set of rationals or irrationals, then $\partial A' = A'$. Hence since $m(A') = 0$, the Book VI obtains that $\mu_n(A') \to \mu(A') = 0$.

But this is also clear by definition here. If A' contains N rational numbers, then for $n > N$ it follows that $\mu_n(A') \leq N/n \to 0$. For A' defined as a finite set of irrationals, then $\mu_n(A') \equiv 0 = m(A')$.

8.2 Properties of Weak Convergence

Turning next to properties of weak convergence, the first result states that weak convergence is strong enough to preclude ambiguities about the limit function. In other words, while $F_n \Rightarrow F$ is a "weak" form of convergence, it is strong enough to ensure that F is unique. Of course F must be uniquely defined on each continuity point x by definition of $F_n(x) \to F(x)$, so the only surprise is that F is also uniquely defined on its discontinuities.

Remark 8.11 (Distributions vs. measures) *For the following result, and any result below on weak convergence of distribution functions $F_n \Rightarrow F$, the same results apply **by definition** to weak convergence of a sequence of Borel measures $\mu_n \Rightarrow \mu$. So we will resist the urge to state results in both contexts, and leave it to the reader to reformulate as needed. That said, we will not resist the urge to remind the reader of this point periodically.*

Proposition 8.12 (F is unique) *If $\{F_n(x)\}_{n=1}^{\infty}$ is a sequence of distribution functions and there exists distribution functions F and G for which $F_n \Rightarrow F$ and $F_n \Rightarrow G$, then $F(x) = G(x)$ **for all** x.*
Proof. *Both F and G are continuous except on at most countably many points, and $F(x) = G(x)$ for each common continuity point by definition. Since the combined exceptional set is countable, the set of common continuity points is dense in $\mathbb{R}$.*

Thus for any x in this exceptional set, there is a sequence of common continuity points, $\{x_n\}_{n=1}^{\infty}$ with $x_n > x$ and $x_n \to x$. But then $F(x_n) = G(x_n)$ for all n, and the right continuity of F and G then assures that $F(x) = G(x)$. ∎

The next results address questions related to a given sequence of distribution functions, $\{F_n\}_{n=1}^{\infty}$. For example, what can be said about the existence of a function $F(x)$ so that $F_n(x) \to F(x)$ at all continuity points of F? More generally, does there exist a subsequence $\{F_{n_k}\}_{k=1}^{\infty}$ and a function $F(x)$ so that $F_{n_k}(x) \to F(x)$ at all continuity points of F? In either case, when can it be assured that $F(x)$ is in fact a distribution function so that $F_n(x) \Rightarrow F(x)$ or $F_{n_k}(x) \Rightarrow F(x)$, respectively?

The question concerning such $F(x)$ being a distribution function is not vacuous.

Example 8.13 (Convergence to non-d.f. F) *Let $F_n(x)$ be the distribution function associated with a constant random variable X_n defined on some probability space by $X_n \equiv (-1)^n n$. The distribution function is then given by $F_n(x) = \chi_{[(-1)^n n, \infty)}(x)$, where $\chi_A(x)$ is the **characteristic function of** A, and defined to equal 1 for $x \in A$ and is 0 otherwise.*

For every x, $F_{2n}(x) \to 0$, and hence $F_{2n}(x) \to F(x)$ for all x with $F(x) \equiv 0$. But obviously, such F is not a distribution function.

Similarly, $F_{2n+1}(x) \to F(x) \equiv 1$ for all x, but again, this F is not a distribution function.

The first proposition is the **Helly selection theorem,** named for **Eduard Helly** (1884–1943). Given an **arbitrary** sequence of distribution functions $\{F_n\}_{n=1}^\infty$, it provides a surprisingly general existence result. It states that there exists a subsequence $\{F_{n_k}\}_{k=1}^\infty$, and a right continuous, increasing function F, so that $F_{n_k} \to F$ at all continuity points of F. Since F is increasing, this means convergence is everywhere except on perhaps a countable set.

Importantly, there is no uniqueness result in this theorem. In fact, looking at the construction in the proof it will be clear that one often has a lot of choice along the way, and potentially there are many such subsequences and many such increasing, right continuous functions that satisfy this conclusion. This observation is important in understanding results below that refer to "any subsequential limit function F," or, "any convergent subsequence $\{F_{n_k}\}_{k=1}^\infty$."

Also note that Helly's result does not assure that any such F is a distribution function, and indeed given Example 8.13 above, it most assuredly cannot. But this question is addressed in Proposition 8.20.

Helly's theorem is stated here in terms of distribution functions, but elsewhere the equivalent focus on probability measures is common.

Proposition 8.14 (Helly Selection theorem) *Given a sequence of distribution functions $\{F_n\}_{n=1}^\infty$, there exists a subsequence $\{F_{n_k}\}_{k=1}^\infty$, and a right continuous, increasing function $F(x)$, so that $F_{n_k}(x) \to F(x)$ on all continuity points of F.*

Proof. *Let $\{r_j\}_{j=1}^\infty$ denote an enumeration of the rationals. Since $\{F_n(r_1)\}_{n=1}^\infty$ is a bounded sequence, in fact bounded by 1, there is at least one accumulation point which we denote $A(r_1)$ and a subsequence $\{n_{1,k}\}_{k=1}^\infty$ so that $F_{n_{1,k}}(r_1) \to A(r_1)$. Next, since $\{F_{n_{1,k}}(r_2)\}_{k=1}^\infty$ is a bounded sequence, there is an accumulation point $A(r_2)$ and a subsequence $\{n_{2,k}\}_{k=1}^\infty \subset \{n_{1,k}\}_{k=1}^\infty$ so that $F_{n_{2,k}}(r_2) \to A(r_2)$.*

Continuing in this way for each j, $\{F_{n_{j,k}}(r_{j+1})\}_{k=1}^\infty$ is a bounded sequence, so there is an accumulation point $A(r_{j+1})$ and a subsequence $\{n_{j+1,k}\}_{k=1}^\infty \subset \{n_{j,k}\}_{k=1}^\infty$ so that $F_{n_{j+1,k}}(r_{j+1}) \to A(r_{j+1})$. Now define $n_k = n_{k,k}$.

By construction, $F_{n_k}(r) \to A(r)$ for all rationals, and $0 \le A(r) \le 1$ for all r. Also, $A(r)$ is increasing in r. Given r_m, r_n with $r_m < r_n$, then $F_{n_k}(r_m) \le F_{n_k}(r_n)$ for $k > \max(n,m)$, since each F_{n_k} is increasing, and so:

$$A(r_m) = \lim_{k \to \infty} F_{n_k}(r_m) \le \lim_{k \to \infty} F_{n_k}(r_n) = A(r_n).$$

We define $F(x) = \inf_{r > x} A(r)$, and note that $F(x)$ is an increasing function, with $0 \le F(x) \le 1$ for all x.

To see that F is continuous from the right, let x and ϵ be given. By definition of infimum there is an $r > x$ so that $A(r) < F(x) + \epsilon$. Also, for y with $x \le y < r$, we have $F(x) \le F(y) \le A(r)$, and so if $y \to x+$, then combining results:

$$F(x) \le \lim_{y \to x+} F(y) < F(x) + \epsilon.$$

As $\epsilon > 0$ is arbitrary, F is right continuous at all x. As a right continuous and increasing function, $F(x)$ is thus continuous except on at most a countable collection of points.

If x is a continuity point of F, it is left to show that $F_{n_k}(x) \to F(x)$. Given $\epsilon > 0$, choose rational $r > x$ so that $A(r) < F(x) + \epsilon$ as above. By continuity there is $y < x$ so that $F(x) - \epsilon < F(y)$, and choosing rational s with $y < s < x$ obtains that $F(y) \leq A(s)$. Now since A is increasing:

$$F(x) - \epsilon < A(s) \leq A(r) < F(x) + \epsilon,$$

and so:

$$F(x) - \epsilon < \lim_{k \to \infty} F_{n_k}(s) \leq \lim_{k \to \infty} F_{n_k}(r) < F(x) + \epsilon. \tag{1}$$

Now for all k:

$$F_{n_k}(s) \leq F_{n_k}(x) \leq F_{n_k}(r),$$

and this with (1) implies that:

$$F(x) - \epsilon < \liminf F_{n_k}(x) \leq \limsup F_{n_k}(x) < F(x) + \epsilon.$$

As $\epsilon > 0$ was arbitrary, we conclude that $\lim_{k \to \infty} F_{n_k}(x) = F(x)$. ∎

Exercise 8.15 (Helly and Example 8.13) *Show that the functions $F(x) \equiv 0$ and $F(x) \equiv 1$ in Example 8.13 are functions that could be constructed in the proof of the Helly selection theorem. What determines which function and subsequence is chosen?*

We next investigate when the $F(x)$ constructed by the Helly process is $F(x) \equiv 0$ or $F(x) \equiv 1$ as seen in Example 8.13, rather than a distribution function. Looking at the proof above, the limiting function F is defined by:

$$F(x) = \inf_{r > x} A(r),$$

where $A(r) = \lim_k F_{n_k}(r)$.

1. $F(x) \equiv 0$: This construction can produce $F(x) = 0$ for all x if and only if $\lim_k F_{n_k}(r) = 0$ for all rational r. The "if" statement is true by definition. Conversely, if $\lim_k F_{n_k}(r) \geq \epsilon > 0$ for some r, then since $F(x)$ is increasing, it would follow that $F(x) \geq \epsilon$, for all $x \geq r$. Thus if one obtains $F(x) \equiv 0$ in the Helly selection process, it must be because of a subsequence of distribution functions $\{F_{n_k}\}_{k=1}^{\infty} \subset \{F_n\}_{n=1}^{\infty}$ with $F_{n_k}(x) \to 0$ for any x.

   Stated in terms of the associated Borel probability measures $\{\mu_{n_k}\}_{k=1}^{\infty}$, this means that for any right semi-closed interval $(a, b]$:

   $$\mu_{n_k}((a, b]) \to 0.$$

2. $F(x) \equiv 1$: Similarly, this construction can produce $F(x) = 1$ for all x if and only if $\lim_k F_{n_k}(r) = 1$ for all rational r. Again, if $\lim_k F_{n_k}(r) \leq 1 - \epsilon < 1$ for some r, then since $F(x)$ is increasing, it would follow that $F(x) \leq 1 - \epsilon$, for all $x \leq r$. Thus if one obtains $F(x) \equiv 1$ in the Helly selection process, it must be because of a subsequence of distribution functions $\{F_{n_k}\}_{k=1}^{\infty} \subset \{F_n\}_{n=1}^{\infty}$ with $F_{n_k}(x) \to 1$ for any x.

   Stated in terms of the associated Borel probability measures $\{\mu_{n_k}\}_{k=1}^{\infty}$, this again implies that for any right semi-closed interval $(a, b]$:

   $$\mu_{n_k}((a, b]) \to 0.$$

The following definition identifies a property of a collection of distribution functions $\{F_n\}_{n=1}^\infty$ or probability measures $\{\mu_n\}_{n=1}^\infty$ called **tightness,** which avoids these problems. It will be apparent that the distribution functions in Example 8.13 are not tight. We will then see in Proposition 8.20 that when this property applies to a collection $\{F_n\}_{n=1}^\infty$, any function $F(x)$ constructed in the Helly selection theorem will be a distribution function.

Definition 8.16 (Tight sequence) *A sequence of probability measures $\{\mu_n\}_{n=1}^\infty$ is said to be **tight** if for any $\epsilon > 0$ there is a finite interval $(a, b]$ so that $\mu_n\left((a, b]\right) > 1 - \epsilon$ for all n.*

*A sequence of distribution functions $\{F_n\}_{n=1}^\infty$ is said to be **tight** if for any $\epsilon > 0$ there is a finite interval $(a, b]$ so that $F_n(b) - F_n(a) > 1 - \epsilon$ for all n, or equivalently, $F_n(b) > 1 - \epsilon$ and $F_n(a) < \epsilon$ for all n.*

Remark 8.17 (Equivalence of d.f. definitions) *Note that for distribution functions, the stated criteria are equivalent because ϵ is arbitrary.*

Given $(a, b]$ so that $F_n(b) - F_n(a) > 1 - \epsilon$ for all n, then $F_n(b') - F_n(a') > 1 - \epsilon$ for all n for arbitrary $b' \geq b$ and $a' \leq a$. Letting $b' \to \infty$ obtains $F_n(a) < \epsilon$ for all n, while letting $a' \to \infty$ obtains $F_n(b) > 1 - \epsilon$ for all n.

Conversely, given $(a, b]$ with $F_n(b) > 1 - \epsilon/2$ and $F_n(a) < \epsilon/2$ for all n assures that $F_n(b) - F_n(a) > 1 - \epsilon$ for all n.

One connection between weak convergence and tightness is given in the following proposition. It states that weak convergence assures that the distribution sequence is tight.

Proposition 8.18 ($F_n \Rightarrow F$ implies tightness) *If $\{F_n\}_{n=1}^\infty$ is a sequence of distribution functions with $F_n \Rightarrow F$ for a distribution function F, then $\{F_n\}_{n=1}^\infty$ is tight.*
Proof. *Given $\epsilon > 0$, let M_n be defined so that $1 - F_n(M_n) + F_n(-M_n) < \epsilon$. Similarly let M be so defined in terms of F, so $1 - F(M) + F(-M) < \epsilon$. These values exist since for any distribution function:*

$$1 - F(x) + F(-x) \to 0,$$

as $x \to \infty$. This also obtains that if $1 - F(M) + F(-M) < \epsilon$, then the same is true for all $M' \geq M$. We can thus assume that M and $-M$ are continuity points of F since there are at most countably many discontinuities.

Hence $F_n(\pm M) \to F(\pm M)$, and there exists N so that for $n \geq N$:

$$|F_n(M) - F(M)| < \epsilon, \quad |F_n(-M) - F(-M)| < \epsilon.$$

By the triangle inequality we obtain for $n \geq N$:

$$\begin{aligned}
&1 - F_n(M) + F_n(-M) \\
\leq\ & |1 - F(M) + F(-M)| + |F_n(M) - F(M)| + |F_n(-M) - F(-M)| \\
<\ & 3\epsilon.
\end{aligned}$$

Letting $M' = \max_{n \leq N}\{M, M_n\}$ obtains that $1 - F_n(M') + F_n(-M') < 3\epsilon$ for all n. ∎

Remark 8.19 *The above proof very much relied on the assumption that F is a distribution function. This assumption was needed to anchor the inequality sequence with $1 - F(M) + F(-M) < \epsilon$. Indeed, this assumption is necessary for the conclusion, as Example 8.13 demonstrates that $F_n(x) \to F(x)$ at all continuity points of F does not assure tightness of $\{F_n\}$.*

The next few results address the implications of tightness for the Helly selection theorem. The first states that when the sequence is tight, **every** Helly construction produces a distribution function.

Proposition 8.20 (Helly with tight $\{F_n\}_{n=1}^{\infty}$) *If the sequence $\{F_n\}_{n=1}^{\infty}$ is tight, every subsequential limit function F in the Helly selection theorem is a distribution function, and thus $F_{n_k} \Rightarrow F$.*

Proof. *Let $\{F_{n_k}\}_{k=1}^{\infty}$ be the subsequence constructed in the proof of the Helly selection theorem, and $F(x)$ the function so that $F_{n_k}(x) \to F(x)$ at every continuity point of F. Then since $\{F_{n_k}\}_{k=1}^{\infty}$ is tight, for any $\epsilon > 0$ there is a finite interval $(a, b]$ so that $F_{n_k}(b) - F_{n_k}(a) > 1 - \epsilon$ for all k. But then for any continuity points x, y of F with $x \leq a$ and $y > b$, it follows that:*

$$F(y) - F(x) \geq \lim_{k \to \infty} (F_{n_k}(b) - F_{n_k}(a)) > 1 - \epsilon.$$

As noted in Remark 8.17, this implies that $F(y) > 1 - \epsilon$ for $y > b$ and $F(x) < \epsilon$ for $x < a$. Since $F(x)$ is increasing and right continuous from the above proof, and ϵ is arbitrary, this assures that F is a distribution function. ∎

Exercise 8.21 (Helly with non-tight $\{F_n\}_{n=1}^{\infty}$) *Prove that if that the sequence of distribution functions $\{F_n\}_{n=1}^{\infty}$ is not tight, then there is a subsequence for which the limit function $F(x)$ is not a distribution function. Hint: Define the index sequence n_k so that $F_{n_k}(-k) \geq \epsilon$ and $F_{n_k}(k) \leq 1 - \epsilon$.*

Though a corollary, the next result is very important in providing a basis for using the Helly selection theorem to not only conclude that $F_{n_k} \Rightarrow F$, but indeed, that $F_n \Rightarrow F$ for the original sequence.

Corollary 8.22 (Helly and when $F_n \Rightarrow F$) *If the sequence $\{F_n\}_{n=1}^{\infty}$ is tight, and any weakly convergent subsequence $\{F_{n_k}\}_{k=1}^{\infty}$ has the property that $F_{n_k} \Rightarrow F$ for a unique distribution function F, then $F_n \Rightarrow F$.*

Proof. *Assume that for the given F that $F_n \not\Rightarrow F$, and thus there is a continuity point x of F so that $F_n(x) \not\to F(x)$. This implies that for any $\epsilon > 0$ there is a subsequence $\{n_k\}_{k=1}^{\infty}$ so that $|F_{n_k}(x) - F(x)| \geq \epsilon$ for all k. Applying the Helly selection theorem to the sequence $\{F_{n_k}\}_{k=1}^{\infty}$, there is a subsequence which converges, and hence must converge to F by hypothesis. Further, this subsequence convergence is valid at all continuity points of F. But this is impossible at the above continuity point x since there, $|F_{n_k}(x) - F(x)| \geq \epsilon$ for all k. This is a contradiction and thus $F_n \Rightarrow F$.* ∎

Remark 8.23 (On Prokhorov's theorem) *The results of the Helly selection theorem of Proposition 8.14 and Proposition 8.20 can be combined to produce a special case of **Prokhorov's theorem**, named for **Yuri Vasilyevich Prokhorov** (1929–2013).*

Proposition 8.24 (Prokhorov's theorem) *Given a tight sequence $\{F_n\}_{n=1}^{\infty}$ of distribution functions on $\mathbb{R}$, there exists a subsequence $\{F_{n_k}\}_{k=1}^{\infty}$ and a distribution function F so that $F_{n_k}(x) \to F(x)$ at all continuity points of F. In other words, $F_{n_k} \Rightarrow F$.*

Beyond measures on $\mathbb{R}$, Prokhorov's theorem is true for measures on $\mathbb{R}^n$, and for measures on any metric space that is both **separable** and **complete**:

1. **Separable** means that this metric space contains a countable, dense collection of points.

2. **Complete** means that every **Cauchy sequence** converges to a point in the space. A sequence $\{x_n\}_{n=1}^{\infty}$ is a Cauchy sequence if for any $\epsilon > 0$ there is an N so that $d(x_n, x_m) < \epsilon$ for $n, m > N$, with d the given metric. Such sequences are named for **Augustin-Louis Cauchy** (1789–1857).

The generalization of **Prokhorov's theorem** to measures on $\mathbb{R}^n$ will be addressed in Book VI. The proof of this result will begin with a derivation of the Helly selection theorem on this space.

8.3   Weak Convergence and Left Continuous Inverses

The next results on weak convergence will be useful in the coming sections on extreme value theory and in Book IV. The first result states that weak convergence of a sequence of distribution functions implies weak convergence of the associated sequence of left continuous inverse functions. For this, recall the discussion in Remark 8.5 of weak convergence in the context of increasing functions. More generally, we prove that weak convergence of a sequence of increasing, right continuous functions is preserved in the associated sequence of left continuous inverses. We also address the reverse conclusions.

Remark 8.25 (On continuity points) *In the following result, note that we prove that $F_n^*(y) \to F^*(y)$ for every $y \in (0, 1)$ that is a continuity point of F^*. We do not claim that $F_n^*(y) \to F^*(y)$ for every $y = F(x)$ for which x is a continuity point of F. Indeed, if x is a continuity point of F it need not be the case that $F(x)$ is a continuity point of F^*. Recall Example 3.15 and let $x = 1.0$, a continuity point of F. But $F(1.0) = 0.5$ is not a continuity point of F^*.*

Proposition 8.26 (Distribution Functions) *If $\{F_n(x)\}_{n=1}^{\infty}$ is a sequence of distribution functions for which $F_n \Rightarrow F$ for a distribution function $F(x)$, and $\{F_n^*(y)\}_{n=1}^{\infty}$ and $F^*(y)$ are the associated left continuous inverse functions, then $F_n^* \Rightarrow F^*$. Specifically, $F_n^*(y) \to F^*(y)$ for every $y \in (0, 1)$ that is a continuity point of F^*.*
Proof. *Let $y \in (0, 1)$ be fixed but arbitrary and let x be an arbitrary continuity point of F with $x < F^*(y)$. Then $F(x) < y$ by (3.9), and since $F_n(x) \to F(x)$, there is an N_1 so that $F_n(x) < y$ for $n \geq N_1$. Again by (3.9), $x < F_n^*(y)$ for $n \geq N_1$. Hence $x \leq \liminf_n F_n^*(y)$.*

Similarly, let x' be an arbitrary continuity point of F with $F^*(y^+) < x'$, where $F^*(y^+) = \lim_{\epsilon \to 0+} F^*(y + \epsilon)$. Then $F^*(y + \epsilon) < x'$ for $\epsilon < \epsilon'$ say, and then $F[F^*(y + \epsilon)] \leq F(x')$ by monotonicity, and by (3.10) we obtain $y + \epsilon \leq F(x')$ for $\epsilon < \epsilon'$. Since $F_n(x') \to F(x')$, there is an N_2 so that $y + \epsilon \leq F_n(x')$ for $n \geq N_2$. This implies that $F_n^*(y + \epsilon) \leq x'$ for $n \geq N_2$ and $\epsilon < \epsilon'$ by (3.8), and so $F_n^*(y) \leq x'$ for $n \geq N_2$. Hence $\limsup_n F_n^*(y) \leq x'$.

Next, let $\{x_m\}$ and $\{x_m'\}$ be sequences of continuity points of F so that $x_m \to F^*(y)$ and $x_m' \to F^*(y^+)$, with $x_m < F^*(y)$ and $x_m' > F^*(y^+)$ for all m. This is possible since discontinuities of F are at most countable, and hence continuity points are dense in $\mathbb{R}$. We then have from the prior analyses that for all m:

$$x_m \leq \liminf F_n^*(y) \leq \limsup F_n^*(y) \leq x_m',$$

and hence by construction:

$$F^*(y) \leq \liminf F_n^*(y) \leq \limsup F_n^*(y) \leq F^*(y^+).$$

If y is a continuity point of F^*, then $F^*(y) = F^*(y^+)$ and so $\lim_n F_n^*(y) = F^*(y)$. In other words, $F_n^* \Rightarrow F^*$. ∎

We next generalize this result to increasing functions, which will be applied in the corollary that follows. Note that if we additionally assumed below that such increasing functions were right continuous, then no proof would be needed. The above proof works in this case without modification, other than a reference to Corollary 3.20. But since our goal is to apply this result to left continuous inverse functions, the more general statement and proof are required.

Proposition 8.27 (Increasing Functions) *Let $\{F_n(x)\}_{n=1}^\infty$ be a sequence of increasing functions for which $F_n(x) \to F(x)$ for every $x \in (a, b)$ which is a continuity point of an increasing function $F(x)$. If $\{F_n^*(y)\}_{n=1}^\infty$ and $F^*(y)$ are the associated left continuous inverse functions, then $F_n^*(y) \to F^*(y)$ for every $y \in (F(a), F(b))$ which is a continuity point of F^*.*

Proof. *Fix $y \in (F(a), F(b))$, a continuity point of F^*, and consider $F^*(y)$. We claim that for any $\epsilon > 0$, there exists $\epsilon', \epsilon'' < \epsilon$ so that $F^*(y) - \epsilon'$ and $F^*(y) + \epsilon''$ are continuity points of F, and:*

$$F[F^*(y) - \epsilon'] < y < F[F^*(y) + \epsilon''] . \tag{1}$$

For this it is enough that for any such ϵ:

$$F[F^*(y) - \epsilon] < y < F[F^*(y) + \epsilon],$$

and then using the density of continuity points in the interval $(F^*(y) - \epsilon, F^*(y) + \epsilon)$. This last statement is derived in the proof of 4 of Proposition 3.16.

Given $F[F^*(y) - \epsilon'] < y$ from (1), choose $\delta > 0$ with $\delta < y - F[F^*(y) - \epsilon']$. Since $F^*(y) - \epsilon'$ is a continuity point of F and $F_n(x) \to F(x)$ for all such points, we conclude that for n large:

$$F_n[F^*(y) - \epsilon'] < F[F^*(y) - \epsilon'] + \delta < y.$$

By definition of $F_n^*(y)$, this inequality implies that $F^*(y) - \epsilon' \leq F_n^*(y)$, and hence for any such $\epsilon' > 0$ there is an N_1 so that for $n \geq N_1$:

$$F^*(y) \leq F_n^*(y) + \epsilon'. \tag{2}$$

Given $F[F^*(y) + \epsilon''] > y$, choose $\delta' > 0$ with $\delta' < F[F^*(y) - \epsilon''] - y$. Since $F^*(y) - \epsilon''$ is a continuity point of F and $F_n(x) \to F(x)$ for all such points, we conclude that for n large:

$$F_n[F^*(y) + \epsilon''] > F[F^*(y) + \epsilon''] - \delta' > y.$$

Thus $F^*(y) + \epsilon'' \geq F_n^*(y)$ and for any such $\epsilon'' > 0$ there is an N_2 so that for $n \geq N_2$:

$$F^*(y) \geq F_n^*(y) - \epsilon''. \tag{3}$$

Combining bounds (2) and (3), we conclude that $F_n^*(y) \to F^*(y)$ for each y that is a continuity point of F^*. ∎

Corollary 8.28 ($F_n^* \Rightarrow F^*$ implies $F_n \Rightarrow F$) *Let $\{F_n(x)\}_{n=1}^\infty$ be a sequence of distribution functions, $F(x)$ a distribution function, and $\{F_n^*(y)\}_{n=1}^\infty$ and $F^*(y)$ the associated left continuous inverse functions.*

If $F_n^* \Rightarrow F^*$, then $F_n \Rightarrow F$.

Proof. *If $F_n^* \Rightarrow F^*$, then by definition, $\lim_n F_n^*(y) = F^*(y)$ for every continuity point of F^*. Since left continuous inverses of distribution functions are increasing functions by Proposition 3.16, we have by Proposition 8.27 that $\lim_n F_n^{**}(x) = F^{**}(x)$ for every continuity point of F^{**}. Now by Proposition 3.26, $F_n^{**}(x) = F_n(x^-)$, $F^{**}(x) = F(x^-)$, and the continuity points of F and F^{**} agree. Hence $F_n(x^-) \to F(x)$ at every continuity point of F. This completes the proof in the special case of continuous $\{F_n(x)\}_{n=1}^\infty$.*

For the general case, it must be proved that if x is a continuity point of F, then $|F_n(x) - F_n(x^-)| \to 0$. But this result is assured by Proposition 8.7 since $F_n(x) - F_n(x^-) = \mu_n(\{x\})$ for the associated Borel measures. Thus $F_n(x) \to F(x)$ for every continuity point of F, and so $F_n \Rightarrow F$. ∎

Remark 8.29 *The utility of this corollary is that in attempting to demonstrate that $F_n \Rightarrow F$, one can proceed directly, or if easier, demonstrate that $F_n^* \Rightarrow F^*$ for the left continuous inverses. This approach will be seen in Section 9.2 below as well as the follow-up section in Book IV.*

8.4 Skorokhod's Representation Theorem

For the next result on weak convergence, recall Proposition 3.6. If $F(x)$ is an increasing function which is right continuous and satisfies $F(-\infty) = 0$ and $F(\infty) = 1$ defined as limits, then there exists a probability space $(\mathcal{S}, \mathcal{E}, \mu)$ and random variable X on $\mathcal{S}$ so that F is the distribution function of X. That is, for all x:

$$F(x) = \mu[X^{-1}(-\infty, x]].$$

Given distribution functions $\{F_n(x)\}_{n=1}^{\infty}$ and $F(x)$ with $F_n \Rightarrow F$, this theorem implies that there are probability spaces $\{(\mathcal{S}_n, \mathcal{E}_n, \mu_n)\}_{n=1}^{\infty}$ and $(\mathcal{S}, \mathcal{E}, \mu)$, and random variables $\{X_n\}_{n=1}^{\infty}$ and X, with the given distribution functions. That is, $F_n(x) = \mu_n[X_n^{-1}(-\infty, x]]$ and $F(x) = \mu[X^{-1}(-\infty, x]]$. Since the probability space $(\mathcal{S}, \mathcal{E}, \mu) \equiv ((0, 1), \mathcal{B}(0, 1), m)$ used for Proposition 3.6 was independent of the distribution function, it follows that all such random variables can be defined on this common probability space. Finally, we obtain convergence in distribution by Definition 5.19:

$$X_n \Rightarrow X.$$

The next result is called **Skorokhod's representation theorem,** named for **A. V. (Anatoliy Volodymyrovych) Skorokhod** (1930–2011). This theorem improves the above conclusion in one significant way. If $F_n \Rightarrow F$, convergence in distribution $X_n \Rightarrow X$ is improved to:

$$X_n(s) \to X(s), \text{ for all } s \in \mathcal{S}.$$

The existence of distribution functions $\{F_n(x)\}_{n=1}^{\infty}$ and $F(x)$ with $F_n \Rightarrow F$ is equivalent to the existence of probability measures $\{\mu_n\}_{n=1}^{\infty}$ and μ on $(\mathbb{R}, \mathcal{B}(\mathbb{R}))$ with $\mu_n \Rightarrow \mu$. This follows from Definition 8.2 and Exercise 8.4.

The following theorem is conventionally stated in terms of weak convergence of probability measures, and we maintain this convention. Thus by this identification, rather than say the random variables have the given distribution functions, we say that these random variables have the given probability measures. More specifically, these random variables have distribution functions which induce the given probability measures.

Proposition 8.30 (Skorokhod's Representation theorem) *Let $\{\mu_n\}_{n=1}^{\infty}$ and μ probability measures on $(\mathbb{R}, \mathcal{B}(\mathbb{R}))$ with $\mu_n \Rightarrow \mu$.*

Then there exist random variables $\{X_n\}_{n=1}^{\infty}$ and X defined on a common probability space $(\mathcal{S}, \mathcal{E}, \upsilon)$, which have the given probability measures, such that $X_n(s) \to X(s)$ for all $s \in \mathcal{S}$.

Proof. As in Proposition 3.6, let $(\mathcal{S}, \mathcal{E}, v) = ((0,1), \mathcal{B}(0,1), m)$. Define $\{F_n(x)\}_{n=1}^{\infty}$ and $F(x)$ as the distribution functions induced by $\{\mu_n\}_{n=1}^{\infty}$ and μ, meaning $F_n(x) = \mu_n[(-\infty, x]]$, and similarly for F. We now define $\{X_n\}_{n=1}^{\infty}$ and X as the left continuous inverses of these distribution functions, so for $y \in (0,1)$:

$$X_n(y) \equiv F_n^*(y), \qquad X(y) \equiv F^*(y).$$

From Proposition 4.8, X_n has distribution function F_n and X has distribution function F. By the above discussion, these random variables then have the associated probability measures.

We claim that $X_n(s) \to X(s)$ for all $s \in (0,1)$. By Proposition 8.26, $X_n \Rightarrow X$ and hence $X_n(s) \to X(s)$ for all $s \in (0,1)$ that are continuity points of $X = F^*$. As F^* is left continuous and increasing, it has at most countably many discontinuities by Proposition 3.16. Now redefine $X_n(s) = X(s) = 0$ at every such discontinuity. Then $X_n(s) \to X(s)$ for all $s \in (0,1)$, and since this discontinuity set has Lebesgue measure 0, this redefinition does not change the distribution functions or the probability measures associated with $\{X_n\}_{n=1}^{\infty}$ and X. ∎

8.4.1   Mapping Theorem on $\mathbb{R}$

The next result is called the **mapping theorem on** $\mathbb{R}$, and sometimes the **continuous mapping theorem on** $\mathbb{R}$, and requires the notion of a **measure induced by a measurable transformation**. Such induced measures will be investigated in more generality and from an integration theory point of view in Book V. When the range space is $\mathbb{R}$, a measurable transformation is simply a measurable function.

Let $(\mathcal{S}, \mathcal{E}, \mu)$ be a measure space and $h : (\mathcal{S}, \mathcal{E}, \mu) \to (\mathbb{R}, \mathcal{B}(\mathbb{R}))$ a measurable function. Thus $h^{-1}(A) \in \mathcal{E}$ for all $A \in \mathcal{B}(\mathbb{R})$. This function induces a Borel measure on the range space, denoted μ_h, that is defined by:

$$\mu_h(A) = \mu\left(h^{-1}(A)\right), \qquad A \in \mathcal{B}(\mathbb{R}). \tag{8.3}$$

Because h is measurable, $h^{-1}(A) \in \mathcal{E}$ and so μ_h is well defined on $\mathcal{B}(\mathbb{R})$. Further, μ_h is a measure because h^{-1} preserves unions:

$$h^{-1}\left(\bigcup_{j=1}^{\infty} A_j\right) = \bigcup_{j=1}^{\infty} h^{-1}(A_j),$$

and $\{h^{-1}(A_j)\}_{n=1}^{\infty}$ are disjoint if $\{A_j\}_{n=1}^{\infty}$ are disjoint.

The mapping theorem investigates when weak convergence $\mu_n \Rightarrow \mu$ is preserved in the respective induced measures.

If this idea of an induced measure seems familiar, there is a good reason.

Example 8.31 (Induced probability measures) *A familiar example of a measure induced by a transformation is a probability measure induced on $(\mathbb{R}, \mathcal{B}(\mathbb{R}))$ by a random variable defined on a probability space $(\mathcal{S}, \mathcal{E}, \mu)$. If $X : \mathcal{S} \to \mathbb{R}$ is such a random variable, the measure μ induces the distribution function F_X defined by $F_X(x) = \mu\left(X^{-1}((-\infty, x])\right)$. This in turn induces the probability measure μ_F on $(\mathbb{R}, \mathcal{B}(\mathbb{R}))$ by Proposition 6.3. The measures μ_F and μ are related by $\mu_F(A) = \mu\left(X^{-1}(A)\right)$, $A \in \mathcal{B}(\mathbb{R})$, by (6.9).*

Definition 8.32 (Discontinuity set D_h) *For measurable $h : (\mathbb{R}, \mathcal{B}(\mathbb{R})) \to (\mathbb{R}, \mathcal{B}(\mathbb{R}))$, let D_h denote the collection of **discontinuity points** of h.*

Example 8.33 (Cardinality of D_h) *Since a general measurable function h of Definition 8.32 need not be increasing, D_h need not be a countable set. Indeed, if $h(x) \equiv \chi_{\mathbb{Q}}(x)$, the characteristic function of the rationals and defined as 1 on the rationals and 0 on the irrationals, then $D_h = \mathbb{R}$.*

However, every D_h-set is a Borel set, and thus is measurable relative to every probability measure, since probability measures are Borel measures by definition.

Exercise 8.34 (Measurability of D_h) *Show that $D_h \in \mathcal{B}(\mathbb{R})$. Hint: Define:*

$$A(\epsilon, \delta) \equiv \{x| \text{ there exists } |x - y| < \delta \text{ and } |x - z| < \delta, \text{ with } |h(y) - h(z)| \geq \epsilon\}.$$

Show that A is open for any (ϵ, δ), meaning if $x \in A(\epsilon, \delta)$, then $(x - \epsilon', x + \epsilon') \subset A(\epsilon, \delta)$ for some $\epsilon' > 0$. Now express D_h in terms of unions and intersections of $A(\epsilon, \delta)$-sets for rational (ϵ, δ).

We next show that if $\mu(D_h) = 0$, then weak convergence $\mu_n \Rightarrow \mu$ is preserved in the induced measures, $(\mu_n)_h \Rightarrow \mu_h$. This is known as the **mapping theorem on $\mathbb{R}$**, and also the **continuous mapping theorem on $\mathbb{R}$**. In this context, "continuous" implies sequence preserving. Recall that if f is continuous at x and $x_n \to x$, then $f(x_n) \to f(x)$. Here, if h is measurable and $\mu(D_h) = 0$, then h is sequence preserving in the sense that if $\mu_n \Rightarrow \mu$ then $\mu_n \circ h^{-1} \Rightarrow \mu \circ h^{-1}$.

Continuity also appears in another context. A special but important corollary of this result is that if h is a continuous function, weak convergence $\mu_n \Rightarrow \mu$ is always preserved in the associated induced measures.

The proof of this result is greatly simplified by **Skorokhod's representation theorem.**

Proposition 8.35 (Mapping theorem on $\mathbb{R}$) *Let $\{\mu_n\}_{n=1}^{\infty}$ and μ be probability measures on $(\mathbb{R}, \mathcal{B}(\mathbb{R}))$ with $\mu_n \Rightarrow \mu$, and $h : (\mathbb{R}, \mathcal{B}(\mathbb{R})) \to (\mathbb{R}, \mathcal{B}(\mathbb{R}))$ a measurable function with $\mu(D_h) = 0$. Then $(\mu_n)_h \Rightarrow \mu_h$.*

In particular, $(\mu_n)_h \Rightarrow \mu_h$ for all continuous functions h.

Proof. *Let $\{X_n\}_{n=1}^{\infty}$ and X be the random variables defined on $((0, 1), \mathcal{B}(0, 1), m)$ as given in Skorokhod's representation theorem and associated with $\{\mu_n\}_{n=1}^{\infty}$ and μ. Since $X_n(s) \to X(s)$ for all $s \in (0, 1)$, it follows that $h(X_n(s)) \to h(X(s))$ for all $s \in (0, 1)$ with $X(s) \notin D_h$. Since $\mu(D_h) = 0$ it follows that $h(X_n(s)) \to h(X(s))$ with probability 1.*

By Proposition 5.21 this implies convergence in distribution of random variables $h(X_n) \Rightarrow h(X)$. By definition then, if μ_n' is the probability measure of $h(X_n)$, and μ' is the probability measure of $h(X)$, then $\mu_n' \Rightarrow \mu'$. But $\mu' = \mu_h$ for $A \in \mathcal{B}(\mathbb{R})$. With m Lebesgue measure:

$$\begin{aligned}
\mu'(A) &= m\{h(X(s)) \in A\} \\
&= m\{X(s) \in h^{-1}(A)\} \\
&= \mu(h^{-1}(A)).
\end{aligned}$$

Similarly, $\mu_n' = (\mu_n)_h$.

Thus $\mu_n' \Rightarrow \mu'$ is equivalent to $(\mu_n)_h \Rightarrow \mu_h$. ∎

Remark 8.36 (Generalizations of the mapping theorem) *In Book VI we investigate more general results on weak convergence of measures. There we will be able to derive a more general mapping theorem related to measurable transformations between $\mathbb{R}^j$ and $\mathbb{R}^k$, using a general result on weak convergence known as the **portmanteau theorem**.*

8.5   Convergence of Random Variables 2

This section continues the investigation of Section 5.2. We present two important results on convergence of random variables, where key parts of the proofs are greatly simplified by Skorokhod's representation theorem.

8.5.1   Mann-Wald Theorem on $\mathbb{R}$

The following version of the mapping theorem is called the **Mann-Wald theorem,** named for **Henry Mann** (1905–2000) and **Abraham Wald** (1902–1950), who published it in 1943. It generalizes Proposition 5.28, and as noted in Remark 8.36, this result will be generalized further in Book VI to random vectors.

See Section 5.2.1 for definitions of convergence.

Proposition 8.37 (Mann-Wald theorem on $\mathbb{R}$) *Let $\{X_n\}_{n=1}^{\infty}$ and X be random variables defined on $(\mathcal{S}, \mathcal{E}, \upsilon)$, and h a measurable function $h : \mathbb{R} \to \mathbb{R}$ with discontinuity set D_h with $\upsilon\{X^{-1}(D_h)\} = 0$.*

1.   *Convergence with probability 1:*

     If $X_n \to_{a.e.} X$, then $h(X_n) \to_{a.e.} h(X)$.

2.   *Convergence in probability:*

     If $X_n \to_P X$, then $h(X_n) \to_P h(X)$.

3.   *Convergence in distribution:*

     If $X_n \to_d X$, then $h(X_n) \to_d h(X)$.

In particular, if h is a continuous function, then $X_n \to_\bullet X$ implies that $h(X_n) \to_\bullet h(X)$ for all three modes of convergence.
Proof.

1.   *If $X_n \to_{a.e.} X$, then there is an exceptional set $E \in \mathcal{E}$ with $\upsilon(E) = 0$ and $X_n(s) \to X(s)$ for $s \in \widetilde{E}$. Let $D \equiv \{X^{-1}(D_h)\}$. Then for $s \in (D \bigcup E)^c$, the complement of $D \bigcup E$, $X(s)$ is a continuity point of h and $X_n(s) \to X(s)$. Thus $h(X_n(s)) \to h(X(s))$, and so $h(X_n) \to_{a.e.} h(X)$ since $\mu(D \bigcup E) = 0$.*

2.   *Arguing by contradiction, if $X_n \to_P X$ and $h(X_n) \not\to_P h(X)$, then given $\delta > 0$ and $\epsilon > 0$ there is a subsequence $\{X_{n_m}\}$ so that*

$$\upsilon\{|h(X_{n_m}) - h(X)| > \epsilon\} > \delta.$$

     Since $\upsilon\{X^{-1}(D_h)\} = 0$, it can be assumed that the sets $\{|h(X_{n_m}) - h(X)| > \epsilon\}$ include only continuity points of h.

     *From $X_{n_m} \to_P X$, an application of Proposition 5.25 obtains a subsequence of $\{X_{n_m}\}_{m=1}^{\infty}$, say $\{X_{n_{m_k}}\}_{k=1}^{\infty}$, so that $X_{n_{m_k}} \to_{a.e.} X$, and thus by part **1**, $h\left(X_{n_{m_k}}\right) \to_{a.e.} h(X)$. But as a subsequence, it is the case that again $h\left(X_{n_{m_k}}\right) \not\to_P h(X)$. This contradicts Proposition 5.21, that convergence with probability 1 assures convergence in probability. Hence $h(X_n) \to_P h(X)$.*

3. *By Definition 5.19 and Exercise 8.4, $X_n \to_d X$ means that $\mu_n \Rightarrow \mu$ for the associated probability measures on $(\mathbb{R}, \mathcal{B}(\mathbb{R}))$. Also, $\mu(D_h) \equiv \upsilon(X^{-1}(D_h)) = 0$, so by Proposition 8.35, $(\mu_n)_h \Rightarrow \mu_h$. Equivalently:*

$$\mu_n\left[h^{-1}((-\infty, x])\right] \to \mu\left[h^{-1}((-\infty, x])\right], \tag{1}$$

for all x with $\mu_h(\{x\}) \equiv \mu\left[h^{-1}(\{x\})\right] = 0$.

Let F_n and F denote the distribution functions of $h(X_n)$ and $h(X)$, respectively. To prove that $h(X_n) \to_d h(X)$ is to prove that $F_n(x) \to F(x)$ for every continuity point of F. Since $(h \circ X_n)^{-1} = X_n^{-1} \circ h^{-1}$:

$$\begin{aligned}
F_n(x) &\equiv \nu\left[(h \circ X_n)^{-1}((-\infty, x])\right] \\
&= \mu_n\left[h^{-1}((-\infty, x])\right],
\end{aligned}$$

and similarly $F(x) = \mu\left[h^{-1}((-\infty, x])\right]$. Thus by (1), $F_n(x) \to F(x)$ for every x with $\mu\left[h^{-1}(\{x\})\right] = 0$.

To complete the proof we show that if x is a continuity point of F then $\mu\left[h^{-1}(\{x\})\right] = 0$. Since F is right continuous, it is enough to prove that left continuity at x assures that $\mu\left[h^{-1}(\{x\})\right] = 0$. Given a sequence $\{x_j\}$ with $x_j < x$ and $x_j \to x$, left continuity implies that $\lim_{j \to \infty}[F(x) - F(x_j)] = 0$. Since:

$$F(x) - F(x_j) = \mu\left[h^{-1}((x_j, x])\right],$$

this obtains that $\lim_{j \to \infty} \mu\left[h^{-1}((x_j, x])\right] = 0$ for such points.

But by continuity from above of measures:

$$\lim_{j \to \infty} \mu\left[h^{-1}((x_j, x])\right] = \mu\left[h^{-1}\left(\bigcap_{j=1}^{\infty}(x_j, x]\right)\right] = \mu\left[h^{-1}(\{x\})\right].$$

Thus if x is a continuity point of F, then $\mu\left[h^{-1}(\{x\})\right] = 0$, and the proof is complete. ■

8.5.2 The Delta-Method

The **Delta method**, or **Δ-method,** is the name given to a result which provides a more detailed conclusion for the limiting distribution of a sequence of random variables transformed by a **differentiable function** g. Applications will be seen in Book IV which make the general result presented here more transparent. But as this result is related to weak convergence and utilizes Skorokhod's representation theorem, we develop it here. See Book VI for generalizations.

Let $\{X_n\}_{n=1}^{\infty}$ and X be random variables defined on a probability space $(\mathcal{S}, \mathcal{E}, \upsilon)$, $\{c_n\}_{n=1}^{\infty} \subset \mathbb{R}$ a sequence of positive reals with $c_n \to \infty$ as $n \to \infty$, and assume that there is a constant x_0 so that:

$$c_n(X_n - x_0) \to_d X.$$

In the more general setting discussed in Section 9.2, there will be sequences $\{c_n\}_{n=1}^{\infty}$ and $\{x_n\}_{n=1}^{\infty}$, called **normalizing sequences,** so that $c_n(X_n - x_n) \to_d X$. Here we take $x_n = x_0$ for all n.

If g is a function that is differentiable at x_0, the goal of the Δ-method is to identify the limiting distribution of $g(X_n)$.

Applying a Taylor series centered on x_0, where implicitly the Δ of this method is $\Delta \equiv X_n - x_0$:

$$g(X_n) = g(x_0) + g'(x_0)(X_n - x_0) + h(X_n)(X_n - x_0). \tag{1}$$

The error term $h(X_n)$ is **big-O** of $X_n - x_0$, denoted:

$$h(X_n) = O(X_n - x_0).$$

This means that:

$$h(X_n) \to 0, \text{ as } X_n \to x_0.$$

This follows because g is assumed differentiable at x_0 and:

$$h(X_n) \equiv \frac{g(X_n) - g(x_0)}{X_n - x_0} - g'(x_0).$$

To make this approximation meaningful for a random variable sequence requires a specification of how $X_n \to x_0$, and this will be defined in terms of convergence in distribution. Specifically, we now show that if $c_n (X_n - x_0) \to_d X$, then $X_n \to_d x_0$. Perhaps surprisingly, the conclusion that $X_n \to_d x_0$ makes no mention of X, the limit in distribution of $c_n (X_n - x_0)$. As will be seen below, the only role X plays is to assure that $\{F_n\}_{n=1}^{\infty}$, the distribution functions of $c_n (X_n - x_0)$, is tight by Proposition 8.18.

Exercise 8.38 *Prove that the following result can also be derived by Slutsky's theorem of Proposition 5.29. (Hint: Let $Y_n \equiv 1/c_n$ and use part 2.)*

Proposition 8.39 (If $c_n (X_n - x_0) \to_d X$ then $X_n \to_d x_0$) *Let $\{X_n\}_{n=1}^{\infty}$ and X be random variables defined on a probability space $(\mathcal{S}, \mathcal{E}, \upsilon)$, $\{c_n\}_{n=1}^{\infty} \subset \mathbb{R}$ a positive sequence with $c_n \to \infty$ as $n \to \infty$, and x_0 a constant.*

If $c_n (X_n - x_0) \to_d X$, then $X_n \to_d x_0$.
Proof. *If F_n denotes the distribution function of $c_n (X_n - x_0)$, then given $\epsilon > 0$:*

$$\upsilon \{|X_n(s) - x_0| \geq \epsilon\} = \upsilon\{c_n (X_n(s) - x_0) \geq c_n \epsilon\} + \upsilon\{c_n (X_n(s) - x_0) \leq -c_n \epsilon\}$$
$$= 1 - F_n(c_n \epsilon^-) + F_n(-c_n \epsilon),$$

where $F_n(c_n \epsilon^-)$ denotes the left limit of F_n at $c_n \epsilon$.

With F the distribution function of X, that $F_n \Rightarrow F$ assures by Proposition 8.18 that $\{F_n\}_{n=1}^{\infty}$ is tight. Thus for any $\delta > 0$ there exists an interval $(a, b]$ so that for all n:

$$1 - F_n(b) + F_n(a) < \delta.$$

Since $c_n \to \infty$ as $n \to \infty$, it follows that there exists N so that $(a, b] \subset (-c_n \epsilon, c_n \epsilon]$ for $n \geq N$, and thus:

$$1 - F_n(c_n \epsilon^-) + F_n(-c_n \epsilon) < \delta.$$

Hence, $(X_n - x_0) \to_P 0$, and this is equivalent to $X_n \to_P x_0$ by definition. Thus $X_n \to_d x_0$ by Proposition 5.21. ■

Returning to the Taylor series approximation in (1):

$$c_n [g(X_n) - g(x_0)] = g'(x_0) [c_n (X_n - x_0)] + c_n h(X_n)(X_n - x_0).$$

If it can be proved that:

$$c_n h(X_n)(X_n - x_0) \to_d 0,$$

or equivalently by Proposition 5.27:

$$c_n h(X_n)(X_n - x_0) \to_P 0,$$

then the assumption that $c_n (X_n - x_0) \to_d X$ plus Slutsky's theorem would yield:

$$c_n [g(X_n) - g(x_0)] \to_d g'(x_0)X.$$

This is the essence of the following proof, which is simplified by Skorokhod's representation theorem and several applications of Slutsky's theorem.

Proposition 8.40 (The Δ-Method) *Let $\{X_n\}_{n=1}^{\infty}$ and X be random variables defined on a probability space $(\mathcal{S}, \mathcal{E}, \upsilon)$, and assume that there exists a positive sequence $\{c_n\} \subset \mathbb{R}$ with $c_n \to \infty$ as $n \to \infty$, and a constant x_0, so that $c_n (X_n - x_0) \to_d X$.*

If g is a Borel measurable function that is differentiable at x_0, then:

$$c_n [g(X_n) - g(x_0)] \to_d g'(x_0)X. \tag{8.4}$$

Proof. *For notational simplicity let $Y_n \equiv c_n(X_n - x_0)$, so now $Y_n \to_d X$ and this implies that $\mu_n \to_d \mu$ for the associated probability measures. Let $\{\widetilde{Y}_n\}_{n=1}^{\infty}$ and $\widetilde{X}$ be random variables constructed on the probability space $((0,1), \mathcal{B}(0,1), m)$ by Skorokhod's representation theorem with the given respective probability measures $\{\mu_n\}_{n=1}^{\infty}$ and μ, and with the property that $\widetilde{Y}_n(s) \to \widetilde{X}(s)$ for all $s \in (0,1)$.*

Define $\widetilde{X}_n \equiv \widetilde{Y}_n/c_n + x_0$ on $((0,1), \mathcal{B}(0,1), m)$. Since $X_n \equiv Y_n/c_n + x_0$ and $\widetilde{Y}_n$ and Y_n have the same distribution functions by the Skorokhod construction, it follows that $\widetilde{X}_n$ and X_n have the same distribution functions. Specifically, and with more detailed set notation for clarity:

$$
\begin{aligned}
m\left\{s \in (0,1)|\widetilde{X}_n(s) \le x\right\} &= m\left\{s \in (0,1)|\widetilde{Y}_n(s) \le c_n (x - x_0)\right\} \\
&= \upsilon\left\{s \in \mathcal{S}|Y_n(s) \le c_n (x - x_0)\right\} \\
&= \upsilon\left\{s \in \mathcal{S}|X_n(s) \le x\right\}.
\end{aligned}
$$

As $X_n \to_P x_0$ by Proposition 8.39, it follows that $\widetilde{X}_n \to_P x_0$. Specifically, for any $\epsilon > 0$,

$$m\left\{s \in (0,1)| \left|\widetilde{X}_n(s) - x_0\right| \ge \epsilon\right\} = \upsilon\left\{s \in \mathcal{S}| |X_n(s) - x_0| \ge \epsilon\right\} \to 0.$$

Define a new random variable $\widetilde{G}_n$ on $(0,1)$ by:

$$
\begin{aligned}
\widetilde{G}_n &= c_n\left[g(\widetilde{X}_n) - g(x_0)\right] \\
&= c_n(\widetilde{X}_n - x_0)\left[\frac{g(\widetilde{X}_n) - g(x_0)}{\widetilde{X}_n - x_0}\right],
\end{aligned}
$$

where the expression in square brackets is defined to be $g'(x_0)$ when $\widetilde{X}_n(s) = x_0$. Since g is differentiable at x_0:

$$\frac{g(\widetilde{X}_n) - g(x_0)}{\widetilde{X}_n - x_0} = g'(x_0) + h(\widetilde{X}_n)(\widetilde{X}_n - x_0),$$

where as noted above $h(\widetilde{X}_n) \to 0$ as $\widetilde{X}_n \to x_0$. Substituting $\widetilde{Y}_n \equiv c_n(\widetilde{X}_n - x_0)$:

$$\widetilde{G}_n - g'(x_0)\widetilde{X} = \left[\widetilde{Y}_n - \widetilde{X}\right] g'(x_0) + \widetilde{Y}_n h(\widetilde{X}_n)(\widetilde{X}_n - x_0). \tag{1}$$

Now $\widetilde{Y}_n(s) \to \widetilde{X}(s)$ for all $s \in (0,1)$ remains true after multiplication by $g'(x_0)$, and this obtains that $\left[\widetilde{Y}_n - \widetilde{X}\right] g'(x_0) \to_d 0$ by Proposition 5.21.

For the second term in (1), again $\widetilde{Y}_n \to_d \widetilde{X}$ and $\widetilde{X}_n - x_0 \to_P 0$ as noted above, and we claim that $h(\widetilde{X}_n) \to_P 0$. If not, then given $\delta > 0$ and $\epsilon > 0$, there is a subsequence $\{X_{n_m}\}_{m=1}^{\infty}$ so that:

$$\upsilon\{\left|h\left(\widetilde{X}_{n_m}\right)\right| > \epsilon\} > \delta. \tag{2}$$

As $\widetilde{X}_n - x_0 \to_P 0$ assures that $\widetilde{X}_{n_m} - x_0 \to_P 0$, by Proposition 5.25 choose a further subsequence $\{\widetilde{X}_{n_{m_k}}\}_{m=1}^{\infty}$ so that $\widetilde{X}_{n_{m_k}} \to_{a.e.} x_0$. As noted above, this obtains $h\left(\widetilde{X}_{n_{m_k}}\right) \to_{a.e.} 0$. Then by Proposition 5.27 it follows that $h\left(\widetilde{X}_{n_{m_k}}\right) \to_d 0$, contradicting (2). This contradiction proves that $h(\widetilde{X}_n) \to_P 0$.

It is an exercise using Proposition 5.27 and Slutsky's theorem to verify that $\widetilde{X}_n - x_0 \to_P 0$ and $h(\widetilde{X}_n) \to_P 0$ assure that $h(\widetilde{X}_n)(\widetilde{X}_n - x_0) \to_P 0$. Since $\widetilde{Y}_n \to_d \widetilde{X}$, another application of Slutsky's theorem obtains that $\widetilde{Y}_n h(\widetilde{X}_n)(\widetilde{X}_n - x_0) \to_d 0$, and thus $\widetilde{Y}_n h(\widetilde{X}_n)(\widetilde{X}_n - x_0) \to_P 0$. Yet another application of Slutsky's theorem obtains:

$$\left[\widetilde{Y}_n - \widetilde{X}\right] g'(x_0) + \widetilde{Y}_n h(\widetilde{X}_n)(\widetilde{X}_n - x_0) \to_d 0,$$

and so $\widetilde{G}_n - g'(x_0)\widetilde{X} \to_d 0$. By Proposition 5.27, this is equivalent to $\widetilde{G}_n - g'(x_0)\widetilde{X} \to_P 0$, which is equivalent to $\widetilde{G}_n \to_P g'(x_0)\widetilde{X}$ by definition. Proposition 5.21 now yields $\widetilde{G}_n \to_d g'(x_0)\widetilde{X}$.

Finally, on S define $G_n = c_n [g(X_n) - g(x_0)]$. Then since $\widetilde{X}_n$ and X_n have the same distribution functions as derived above, so too does $\widetilde{G}_n$ and G_n. Hence $\widetilde{G}_n \to_d g'(x_0)\widetilde{X}$ implies that $G_n \to_d g'(x_0)\widetilde{X}$. But as $\widetilde{X}$ and X have the same distribution function by the Skorokhod construction, it follows that $G_n \to_d g'(x_0)X$, which is (8.4). ∎

Example 8.41 (De Moivre-Laplace theorem) *In Book IV we will study various limit theorems, but one result that is accessible now is the **de Moivre-Laplace theorem**, which is another result related to the binomial distribution. This theorem provides additional details on the limiting distribution of $M_n(y)$, defined on the infinite product probability space associated with general coin flips $(Y^{\mathbb{N}}, \sigma(Y^{\mathbb{N}}), \mu_{\mathbb{N}})$. Here $Y = \{H, T\}$ is represented by the numerical space $Y = \{1, 0\}$, and $\mu_{\mathbb{N}}$ is the measure induced by the probability measure μ^B on Y in (1.7):*

$$\mu_B(1) = p, \ \mu_B(0) = 1 - p.$$

The definition of $M_n(y)$ on $y = (y_1, y_2, ...) \in Y^{\mathbb{N}}$ is given in (5.2):

$$M_n(y) \equiv \frac{1}{n} \sum_{j=1}^{n} y_j.$$

This result is named for the work of **Abraham de Moivre** *(1667–1754) in the special case of $p = 1/2$, and many years later generalized to all p, $0 < p < 1$, by **Pierre-Simon Laplace** (1749–1827). This theorem adds to the results of **Bernoulli's weak law of large numbers** of Proposition 5.3, named for **Jacob Bernoulli** (1654–1705), which yielded **convergence in probability**:*

$$M_n(y) \to_P p,$$

*and **Borel's strong law of large numbers** of Proposition 5.9, named for **Émile Borel** (1871–1956), which obtained **almost sure convergence**:*

$$M_n(y) \to_{a.s.} p.$$

*By Proposition 5.21, either of these earlier results implies **convergence in distribution**:*

$$M_n(y) \to_d p.$$

*The de Moivre-Laplace theorem studies the **normalized random variable** Y_n^B, defined by:*

$$Y_n^B \equiv \frac{\sqrt{n}}{p(1-p)} (M_n - p).$$

The conclusion which is proved in Book IV, is that:

$$Y_n^B \to_d Z, \tag{1}$$

*where Z has the **unit normal probability density** $\phi(x)$ defined in (1.23):*

$$\phi(x) = \frac{1}{\sqrt{2\pi}} \exp\left(-x^2/2\right),$$

and associated distribution function $\Phi(x)$ defined as in (1.17) by the Riemann integral:

$$\Phi(x) = \int_{-\infty}^{x} \phi(y)dy.$$

By Definition 5.19, since $\Phi(x)$ is continuous, (1) implies that for all x:

$$F_{Y_n^B}(x) \to \Phi(x).$$

More explicitly, for all x:

$$\Pr\left[\frac{\sqrt{n}}{p(1-p)}\left(M_n - p\right) \le x\right] \to \Pr\left[Z \le x\right]. \tag{2}$$

For large n, this obtains:

$$M_n \approx \sigma Z + \mu,$$

and has a distribution function that approximates the normal distribution function as in (1.22), with parameters:

$$\mu = p, \qquad \sigma = p(1-p)/\sqrt{n}.$$

The Δ-method can now be applied to $g(M_n)$ for any differentiable function, and yields that:

$$\frac{\sqrt{n}}{p(1-p)}\left(g(M_n) - g(p)\right) \to_d g'(p)Z.$$

For fixed but large n, this implies that $g(M_n)$ has a distribution function that approximates the normal distribution function as in (1.22), with parameters:

$$\mu = g(p), \qquad \sigma = p(1-p)g'(p)/\sqrt{n}.$$

9

Estimating Tail Events 1

In this chapter we begin the development of an investigation into the probability of "tail" events. We assume that there is a probability space $(\mathcal{S}, \mathcal{E}, \mu)$ in the background, and a random variable X or sequence $\{X_n\}_{n=1}^{\infty}$ with associated distribution functions F and $\{F_n\}_{n=1}^{\infty}$. By a tail event of a distribution function F is meant:

1. **Left Tail Event:** Given x, the event is defined as $\{X \leq x\}$ with associated probability $\Pr\{X \leq x\} = \mu\left(X^{-1}(-\infty, x]\right)$:

$$\Pr\{X \leq x\} = F(x).$$

2. **Right Tail Event:** Given x, the event is defined as $\{X \geq x\}$ with associated probability $\Pr\{X \geq x\} \equiv \mu\left(X^{-1}[x, \infty)\right)$:

$$\Pr\{X \geq x\} = 1 - F\left(x^-\right).$$

Since we are typically interested in the behavior of these probabilities as $x \to \infty$, it rarely matters if one uses "<" or ">" in the above definitions, respectively. For the theoretical development, a convention must be established, and we use the above conventions here.

It is common for applications in finance to parametrize a given problem so that the "right tail" is of interest, and thus many results are stated in terms of properties of $1 - F\left(x^-\right)$, or properties of F as $x \to \infty$. For example, if $F_G(x)$ is the distribution function associated with gains and losses on an investment portfolio over a fixed period of time, then losses are in the left tail and gains are in the right. This is appropriate for investment analyses.

However, for many risk management applications it is the potential severity of losses that is of interest, and one commonly parametrizes the distribution to be the loss distribution, $F_L(x)$. Here $L \equiv -G$ and so:

$$F_L(x) = 1 - F_G(-x^-),$$

and the right tail of F_L is investigated.

Below we initiate two investigations into properties of the right tail, where in both cases the focus is on results as $n \to \infty$:

- **Large Deviation Theory** studies tail probabilities related to the **average** of n independent random variables;

- **Extreme Value Theory** studies the limiting distribution of the **maximum** of n independent random variables.

This investigation will be continued in Book IV.

9.1   Large Deviation Theory 1

Recalling Definition 4.1 and the subsequent constructions, let $\{X_j\}_{j=1}^{\infty}$ be a collection of **independent, identically distributed** random variables on a probability space $(\mathcal{S}, \mathcal{E}, \mu)$, abbreviated as **i.i.d.**. Define S_n on $(\mathcal{S}, \mathcal{E}, \mu)$ by:

$$S_n = \sum\nolimits_{j=1}^{n} X_j.$$

For t large, we will be interested in estimating $\Pr\{S_n \geq nt\}$ as a function of t, and in the limit of this probability as $n \to \infty$. This probability is equivalent to $\Pr\{S_n/n \geq t\}$, which is to say, the probability that the simple average of these random variables equals or exceeds t.

Remark 9.1 (On large deviation theory) *The theory initiated here is known as* **large deviation theory** *because given $s > 0$, it addresses probability estimates with $t \equiv E[X] + s$:*

$$\Pr\left\{\sum\nolimits_{j=1}^{n} X_j \geq n\left[E[X] + s\right]\right\},$$

as $n \to \infty$. Here $E[X]$ denotes the "mean" or the "expectation" of the random variable X, a concept with which the reader is undoubtedly familiar. Expectations will be introduced in Book IV, but not formally justified until Book VI, using the general integration theory of Book V. This probability statement can be equivalently expressed:

$$\Pr\left\{\sum\nolimits_{j=1}^{n} X_j - nE[X] \geq ns\right\}.$$

That $n\left[E[X] + s\right]$ is considered a "large" deviation reflects the presence of n on the right-hand side of this second inequality. This order of magnitude is in contrast with a result of the central limit theorem, the first version of which will appear in Book IV.

The reader may be familiar with this result, which provides estimates of the form:

$$\Pr\left\{\sum\nolimits_{j=1}^{n} X_j - nE[X] \geq \sqrt{n}s\right\},$$

as $n \to \infty$. The de Moivre-Laplace theorem of Example 8.41 is a left tail version of this type. This can be seen by rewriting (2) of that example:

$$\Pr\left[\sum\nolimits_{j=1}^{n} y_j - np \leq \sqrt{n}xp(1-p)\right] \to \Pr\left[Z \leq x\right].$$

As will be seen in Book IV if not already known, $E[Y] = p$ for this example.

Thus the central limit theorem provides a theoretical basis for estimating the probabilities of **small deviations** *from $nE[X]$, where by "small" is meant that they grow like $\sqrt{n}$. Large deviation theory provides a basis for estimating* **large deviations** *from $nE[X]$, which by definition grow like n. It will be seen that for any $s > 0$, the central limit theorem provides a limit $p(s)$ as $n \to \infty$ for the stated probability, whereas large deviation theory will identify the rate at which the associated probability converges to 0 as $n \to \infty$.*

For completeness, **moderate deviation theory** *studies questions about the limit as $n \to \infty$ of:*

$$\Pr\left\{\sum\nolimits_{j=1}^{n} X_j - nE[X] \geq c_n s\right\},$$

with $\sqrt{n} << c_n << n$. This notation means that $c_n/n \to 0$ and $\sqrt{n}/c_n \to 0$ as $n \to \infty$. For example, let $c_n = n^p$ for $.5 < p < 1$.

Below we initiate the study $\Pr\left\{\sum_{j=1}^{n} X_j \geq nt\right\}$, *where it will be assumed that $t > 0$. But we will come to understand in Book IV that this theory only provides useful right tail estimates when $t > E[X]$, and thus $s > 0$ with the notation above.*

In Book VI we will develop the distribution function of S_n from the distribution function of X, but this result is not needed for this preliminary investigation. It is enough to represent this distribution function here only notationally.

For $t > 0$ fixed, define:

$$p_n = \Pr\{S_n \geq nt\}, \qquad \pi_n = \ln p_n, \tag{9.1}$$

with the convention that $\pi_n = -\infty$ if $p_n = 0$. Then for all n:

$$0 \leq p_n \leq 1, \qquad -\infty \leq \pi_n \leq 0.$$

The one property of the distribution function of sums that is needed below is that if $\{X_j\}_{j=1}^{m+n}$ are independent random variables, then S_m and $T_{m,n}$ are independent random variables, where:

$$S_m = \sum_{j=1}^{m} X_j, \quad T_{m,n} \equiv S_{m+n} - S_m.$$

In other words, for any $A, B \in \mathcal{B}(\mathbb{R})$:

$$\mu\left(S_m^{-1}[A] \bigcap T_{m,n}^{-1}[B]\right) = \mu\left(S_m^{-1}[A]\right)\mu\left(T_{m,n}^{-1}[B]\right).$$

This follows from Proposition 3.56, and is an application of items 3 then 2 of Example 3.58.

For a statement on the associated distribution functions, define the transformation $T : \mathbb{R}^{m+n} \to \mathbb{R}^2$:

$$(X_1, X_2, ..., X_{m+n}) \to (Y_1, Y_2),$$

where:

$$Y_1 = \sum_{j=1}^{m} X_j, \quad Y_2 = \sum_{j=m+1}^{m+n} X_j.$$

Then by Proposition 3.53, independence of $\{X_j\}_{j=1}^{m+n}$ obtains that $F(X_1, X_2, ..., X_{m+n}) = \prod_{j=1}^{m+n} F(X_j)$, and similarly $F(Y_1, Y_2) = F(Y_1)F(Y_2)$. For this and similar statements, we recall Notation 3.54.

Proposition 9.2 (On $\lim_{n\to\infty}(\pi_n/n)$) *Let π_n be defined in (9.1) above. Then for any $n, m \in \mathbb{N}$:*

$$\pi_{n+m} \geq \pi_n + \pi_m, \tag{9.2}$$

and by iteration, this is true for any finite index sum. Further, as $n \to \infty$:

$$\lim_{n\to\infty}(\pi_n/n) = \sup_m(\pi_m/m). \tag{9.3}$$

Proof. *As measurable subsets of $\mathcal{S}$, for $t > 0$:*

$$\left\{\sum_{j=1}^{m} X_j(s) \geq mt\right\} \cap \left\{\sum_{j=m+1}^{m+n} X_j(s) \geq nt\right\}$$
$$\subset \left\{\sum_{j=1}^{n+m} X_j(s) \geq (n+m)t\right\}.$$

By monotonicity of μ and independence:

$$\mu\left\{\sum_{j=1}^{n+m} X_j(s) \geq (n+m)\,t\right\}$$

$$\geq \mu\left\{\sum_{j=1}^{m} X_j(s) \geq mt\right\}\mu\left\{\sum_{j=m+1}^{m+n} X_j(s) \geq nt\right\}.$$

This is equivalent to $p_{m+n} \geq p_m\,p_n$, which is (9.2) after taking logarithms.

For (9.3), it first must be verified that this limit exists, since all that is certain is that $-\infty \leq \pi_n \leq 0$. If $\pi_n \to -\infty$ as $n \to \infty$, the existence of this limit is a statement about the speed of this divergence.

If $\sup_m (\pi_m/m) = -\infty$, then $\pi_m = -\infty$ for all m and hence $\lim_{n\to\infty}(\pi_n/n) = -\infty$. If $-\infty < \sup_m(\pi_m/m) \leq 0$, it follows by I.(3.14) that $\limsup_n(\pi_n/n) \leq \sup_m(\pi_m/m)$. The result in (9.3) will follow from a proof that:

$$\liminf_n (\pi_n/n) \geq \sup_m (\pi_m/m), \tag{1}$$

since then:

$$\limsup_n (\pi_n/n) \leq \sup_m (\pi_m/m) \leq \liminf_n (\pi_n/n).$$

This implies equality of these three expressions, and so $\lim_{n\to\infty}(\pi_n/n)$ exists and equals this common value by Corollary I.3.46.

To prove (1), let m be given and write $n = am + b$ with nonnegative integers a, b and $0 \leq b < m$. By an iterative application of (9.2), $\pi_n \geq a\pi_m + \pi_b$, and so:

$$\pi_n/n \geq \left(1 - \frac{b}{n}\right)\pi_m/m + \pi_b/n.$$

Letting $n \to \infty$ and noting that b and π_b are bounded, it follows that $\liminf_n(\pi_n/n) \geq \pi_m/m$. This is then true for every m, and (1) is proved. ∎

Remark 9.3 (Significance of π_1) *The value of $\pi_1 \equiv \ln[\Pr\{X \geq t\}]$ can be used to derive inferences about the probabilities of tail events, though ultimately they are not useful to the investigation at hand.*

1. **If** *$\pi_1 = 0$, then $p_1 \equiv \Pr\{X_1 \geq t\} = 1$. Thus $\pi_n = 0$ for all n by (9.2). This conclusion also follows directly from the observation that $\Pr\{S_n \geq nt\} = 1$ in this case by definition that $\{X_j\}_{j=1}^{\infty}$ are i.i.d.*

2. **If** *$\pi_1 = -\infty$, then $p_1 = 0$ and (9.2) provides no additional insights on π_n. However, this can be handled by direct methods. By monotonicity of measures and independence of $\{X_j\}_{j=1}^{\infty}$:*

    $$\Pr\{S_n < nt\} \geq \Pr\{X_j < t, \text{ all } 1 \leq j \leq n\}\} = 1.$$

    Thus $\pi_n = -\infty$ for all n as expected.

3. **If** *$-\infty < \pi_1 < 0$, then $0 < p_1 < 1$ and again by monotonicity and independence:*

    $$\Pr\{S_n < nt\} \geq \Pr\{X_j < t, \text{ all } 1 \leq j \leq n\}\} = (1 - p_1)^n,$$

    and so for all n:

    $$p_n \leq 1 - (1 - p_1)^n.$$

    By (9.2) and the Taylor series approximation $\ln(1 - x) \leq -x$ for $|x| < 1$:

    $$n\pi_1 \leq \pi_n \leq -(1 - p_1)^n.$$

Thus:

$$e^{n\pi_1} \leq \Pr\{S_n \geq nt\} \leq e^{-(1-p_1)^n}.$$

As $e^{n\pi_1} \to 0$ and $e^{-(1-p_1)^n} \to 1$, this does not provide useful estimates of $\Pr\{S_n \geq nt\}$ as $n \to \infty$.

The following proposition is an important first result in the theory of large deviations.

Proposition 9.4 (Large Deviation Estimates) *Let $\pi_n = \ln[\Pr\{S_n \geq nt\}]$ and for any $t > 0$, define:*

$$\pi(t) \equiv \lim_{n \to \infty}(\pi_n/n),$$

which exists by Proposition 9.2.
Then:

$$\pi_1 \leq \pi(t) \leq 0,$$

and for all n:

$$\Pr\{S_n \geq nt\} \leq e^{n\pi(t)}. \tag{9.4}$$

Proof. *By the estimate in part 3 of Remark 9.3:*

$$\pi_1 \leq \pi_n/n \leq -(1-p_1)^n/n,$$

which proves the bounds for $\pi(t)$.
As $\lim_{n \to \infty}(\pi_n/n) = \sup_m(\pi_m/m)$ by (9.3), it follows that $\pi_n/n \leq \pi(t)$ for all n, which is (9.4). ∎

Remark 9.5 (Significance of $\pi(t)$) *From (9.4) and the observation that $\pi_1 \leq \pi(t) \leq 0$, we can draw several conclusions:*

1. *If $\pi(t) = 0$, then (9.4) provides only the uninformative conclusion that $\Pr\{S_n \geq nt\} \leq 1$. But if $\pi(t) = 0$ because $\pi_1 = 0$, then as in case 1 of Remark 9.3, $\Pr\{S_n \geq nt\} = 1$ for all n.*

2. *If $\pi(t) = -\infty$, then (9.4) formally implies the result that $\Pr\{S_n \geq nt\} = 0$ for all n. This is the same conclusion as case 2 in Remark 9.3 since then $\pi_1 = -\infty$.*

3. *If $-\infty < \pi(t) < 0$, then (9.4) provides the important estimate that the tail probability **decreases exponentially** with n, and $\pi(t)$ identifies the speed of convergence. This is far better than the result in case 3 of Remark 9.3.*

The following result translates the above result to the context of a weak law of large numbers, and states that convergence could be quite fast.

Proposition 9.6 (A General Weak Law of Large Numbers) *Let $\{X_j\}_{j=1}^{\infty}$ be a collection of **independent, identically distributed** random variables, and define S_n by:*

$$S_n = \sum_{j=1}^{n} X_j.$$

If $t > 0$ and $\pi(t) \neq 0$, then:

1. *$\Pr\{S_n/n \geq t\} \to 0$ exponentially fast as $n \to \infty$, or,*
2. *$\Pr\{S_n/n \geq t\} \equiv 0$ for all n.*

Proof. *This follows directly from (9.4).* ∎

Essential to the usefulness of this kind of result is the ability to estimate $\pi(t)$ given $\{X_j\}_{j=1}^{\infty}$, which is to say, given the associated distribution function $F(x)$ of X. We return to large deviation theory in Book IV. There, with an additional assumption on $F(x)$, it will be possible to evaluate $\pi(t)$ explicitly. For now, we return to the weak law of large numbers of Bernoulli's theorem of Proposition 5.3.

Example 9.7 (Bernoulli's Theorem) *Recall (5.4) of Bernoulli's theorem in the current notation, that if $\epsilon > 0$:*

$$\mu_{\mathbb{N}}\left[\{|S_n(y)/n - p| \geq \epsilon\}\right] \to 0, \ \text{as } n \to \infty,$$

where $S_n(y) \equiv \sum_{j=1}^{n} y_j$ and is defined on $(Y^{\mathbb{N}}, \sigma(Y^{\mathbb{N}}), \mu_{\mathbb{N}})$, the infinite product probability space associated with general coin flips. Here $Y = \{H, T\}$ is represented by the numerical space $Y = \{1, 0\}$, and $\mu_{\mathbb{N}}$ is the measure induced by the probability measure μ_B on Y in (1.7).

Proposition 9.6 improves the statement of Bernoulli's theorem in terms of generality, in that it is applicable to any sum of independent, identically distributed random variables with $\pi(t) \neq 0$. In addition, this general weak law provides information on the rate of convergence when $\pi(t) \neq 0$.

That said, this general law is not immediately applicable to Bernoulli's result. While $0 < p_1 < 1$ for any $t > 0$ by definition, we do not currently have the tools to estimate $\pi(t)$.

*However **Bernstein's inequalities**, derived in the proof of Proposition 5.3, provided explicit upper bounds. For the right tail, if $t = p + \epsilon$ for $\epsilon > 0$:*

$$\mu_{\mathbb{N}}\left[\{S_n(y)/n \geq t\}\right] \leq e^{-n\epsilon^2/4},$$

and for the left tail, if $t = p - \epsilon$:

$$\mu_{\mathbb{N}}\left[\{S_n(y)/n \leq t\}\right] \leq e^{-n\epsilon^2/4}.$$

9.2 Extreme Value Theory 1

In this section we begin an investigation into what is known as **extreme value theory**, or **EVT**, a branch of probability theory which studies properties of certain types of distribution functions $F(x)$ as $x \to \infty$, or as $x \to -\infty$. As noted in the previous section, it is common to parametrize models in finance so that the area of interest is in the right tail, and we follow this convention here. In that context, EVT studies the limiting distribution of the maximum of n independent random variables.

9.2.1 Introduction and Examples

If $\{X_m\}_{m=1}^{\infty}$ is a collection of independent, identically distributed random variables with common distribution function $F(x)$, define the **maximum random variables**, $\{M_n\}_{n=1}^{\infty}$, by:

$$M_n = \max_{m \leq n}\{X_m\}. \tag{9.5}$$

If $\{X_m\}_{m=1}^{\infty}$ are defined on a probability space $(\mathcal{S}, \mathcal{E}, \mu)$, then M_n is measurable for all n by Proposition I.3.47, and hence these are random variables on this space.

If $F_n(x)$ denotes the distribution function of M_n, then $M_n \leq x$ if and only if $X_m \leq x$ for all $m \leq n$. It then follows by the independence and Proposition 3.53 that:

$$F_n(x) = F^n(x). \tag{9.6}$$

It is natural to wonder if there is some distribution function $G(x)$ so that $F_n \Rightarrow G$ as $n \to \infty$. But this question needs refinement to make sense. If $0 \leq F(x) < 1$, then $F^n(x) \to 0$ as $n \to \infty$. If X is bounded and $\max X \leq x$, then $F^n(x) \equiv 1$. To formalize this discussion, define:

$$x^* = \inf\{x | F(x) = 1\}, \tag{9.7}$$

where we define $x^* \equiv \infty$ if $F(x) < 1$ for all x.

Then for all $x < x^*$, $F^n(x) \to 0$ as $n \to \infty$. Hence:

- If $x^* < \infty$, then $F^n(x) \to 0$ for $x < x^*$ and $F^n(x) = 1$ for $x \geq x^*$. In this case $\{F^n(x)\}_{n=1}^{\infty}$ is tight and converges weakly to a distribution function: $F^n \Rightarrow \chi_{[x^*, \infty)}$.

- If $x^* = \infty$, then $F^n(x) \to 0$ for all x. In this case, $\{F^n(x)\}_{n=1}^{\infty}$ is not tight and there can be no distribution function for which $F^n \Rightarrow F$. Indeed, in this latter case, $F^n \Rightarrow 0$, understood in the sense of weak convergence of increasing functions.

Remark 9.8 (On $M_n \to x^*$) *In Book IV it will be proved that when $F(x)$ is continuous, that $M_n \to x^*$ with probability 1.*

To ensure a tight sequence of distribution functions in all cases, it is common to **normalize** the sequence of maximum random variables. This is accomplished by defining a new random variable sequence $\{M_n^N\}_{n=1}^{\infty}$:

$$M_n^N = \frac{M_n - b_n}{a_n}, \tag{9.8}$$

for some sequences of real numbers $\{a_n\}_{n=1}^{\infty}$ and $\{b_n\}_{n=1}^{\infty}$ where $a_n > 0$ for all n. The role of these sequences is to provide tightness and hence ensure that $F^n(x)$ converges weakly to a proper distribution function.

Because $M_n^N \leq x$ if and only if $M_n \leq a_n x + b_n$, it follows that the distribution function of M_n^N is given:

$$F_n^N(x) = F^n(a_n x + b_n). \tag{9.9}$$

Definition 9.9 (Extreme value distribution) *If there exists **normalizing sequences** $\{a_n\}_{n=1}^{\infty}$ and $\{b_n\}_{n=1}^{\infty}$ where $a_n > 0$ for all n, and a distribution function $G(x)$ so that:*

$$F^n(a_n x + b_n) \Rightarrow G(x),$$

*meaning $F^n(a_n x + b_n) \to G(x)$ for every continuity point x of $G(x)$, then $G(x)$ is called an **extreme value distribution**, or an **EV distribution**.*

The following questions come to mind:

1. Which distribution functions $F(x)$ have the property that there exist sequences $\{a_n\}_{n=1}^{\infty}$ and $\{b_n\}_{n=1}^{\infty}$ and an associated EV distributions $G(x)$?
2. Which distribution functions $G(x)$ can arise from such $F(x)$ given appropriate sequences $\{a_n\}_{n=1}^{\infty}$ and $\{b_n\}_{n=1}^{\infty}$?
3. How does $G(x)$ depend on the sequences $\{a_n\}_{n=1}^{\infty}$ and $\{b_n\}_{n=1}^{\infty}$ used?
4. How can these sequences be determined?

Below and in Book IV we will investigate such questions. Perhaps surprisingly, the first question will be the most elusive for this book. That is, it is quite difficult to characterize properties of $F(x)$ which assure the existence of $G(x)$. In many applications, a given distribution function can be shown to answer this question by explicitly deriving the associated function $G(x)$. More general results on this first question will be developed in Book IV. In this book we will investigate the other questions.

We begin with an analysis of 3 well-known examples.

Example 9.10 (Well-known EV distributions) *The following distribution functions have associated EV distributions.*

1. *Let $F_E(x)$ be the **exponential distribution function** given in (1.21), but with parameter here denoted $\alpha > 0$ for consistency below:*

$$F_E(x) = \begin{cases} 0, & x \leq 0, \\ 1 - \exp(-\alpha x), & x \geq 0. \end{cases}$$

Then with $a_n \equiv 1$ and $b_n = \alpha^{-1} \ln n$, $F_n^E(x + \alpha^{-1} \ln n) \Rightarrow G_E(x)$ where:

$$G_E(x) \equiv \exp\left[-e^{-\alpha x}\right], \ x \in \mathbb{R}. \tag{9.10}$$

*The distribution function $G_E(x)$ is in the **double-exponential** or **Gumbel** class of distributions, named for **Emil Julius Gumbel** (1891–1966).*

2. *Let $F_P(x)$ be the **Pareto distribution function**, named for **Vilfredo Pareto** (1848–1923), with parameter $\alpha > 0$:*

$$F_P(x) = \begin{cases} 0, & x \leq 1, \\ 1 - x^{-\alpha}, & x \geq 1. \end{cases} \tag{9.11}$$

Then with $a_n \equiv n^{1/\alpha}$ and $b_n = 0$, $F_n^P(n^{1/\alpha}x) \Rightarrow G_P(x)$ where:

$$G_P(x) \equiv \begin{cases} 0, & x \leq 0, \\ \exp(-x^{-\alpha}), & x > 0. \end{cases} \tag{9.12}$$

*The distribution function $G_P(x)$ is in the **Fréchet** class of distributions, named for **Maurice Fréchet** (1878–1973).*

3. *Let $F_\beta(x)$ be a one-parameter version of the **Beta distribution function**, which usually has two parameters. With a single parameter $\alpha > 0$:*

$$F_\beta(x) = \begin{cases} 0, & x \leq 0, \\ 1 - (1 - x)^\alpha, & 0 \leq x \leq 1. \\ 1, & x \geq 1. \end{cases} \tag{9.13}$$

Then with $a_n \equiv n^{-1/\alpha}$ and $b_n = 1$, $F_n^\beta(n^{-1/\alpha}x + 1) \Rightarrow G_\beta(x)$ where:

$$G_\beta(x) \equiv \begin{cases} \exp\left[-(-x)^\alpha\right], & x < 0, \\ 1, & x \geq 0. \end{cases} \tag{9.14}$$

*The distribution function $G_\beta(x)$ is in the **reversed-Weibull** (or **reverse-Weibull**) class of distributions, named for **Waloddi Weibull** (1887–1979). This distribution is "reversed" in the sense that the standard Weibull distribution is parametrized to have all its probability mass on $\{x > 0\}$.*

Proof. *Recall that for any real number a:*

$$\lim_{n \to \infty} (1 + a/n)^n = e^a. \tag{9.15}$$

This is proved by taking logarithms and noting that by the Taylor series expansion in (1.15):

$$n \ln(1 + a/n) = a + O(1/n).$$

To prove that $F_n \Rightarrow G$, we show that $F_n(x) \to G(x)$ for all continuity points of G, which in the above examples means for all x.

1. *For $x \geq -\alpha^{-1} \ln n$:*

$$F_n^E(x + \alpha^{-1} \ln n) = \left[1 - e^{-\alpha x}/n\right]^n,$$

and hence with $a = -e^{-\alpha x}$, (9.10) is proved for $x \in \mathbb{R}$.

2. *For $x \geq n^{-1/\alpha}$:*

$$F_n^P(n^{1/\alpha} x) = \left[1 - x^{-\alpha}/n\right]^n,$$

and hence with $a = -x^{-\alpha}$, (9.12) is proved for $x > 0$. If $x \leq 0$, then $F_n^P(n^{1/\alpha} x) = 0$, completing the proof.

3. *For $-n^{1/\alpha} \leq x < 0$:*

$$F_n^\beta(n^{-1/\alpha} x + 1) = \left[1 - (-x)^\alpha/n\right]^n,$$

and hence with $a = -(-x)^\alpha$, (9.14) is proved for $x < 0$. If $x \geq 0$, then $F_n^\beta(n^{-1/\alpha} x + 1) = 1$, completing the proof.

■

It turns out that each of the three limiting distributions above can be obtained by a transformation of a single family of distributions.

Definition 9.11 (Generalized extreme value distributions) *For $\gamma \in \mathbb{R}$, the distribution function $G_\gamma(x)$ is said to be a member of the **extreme value class of distributions**, where:*

$$G_\gamma(x) \equiv \begin{cases} \exp\left(-(1 + \gamma x)^{-1/\gamma}\right), & 1 + \gamma x \geq 0, \\ 0 & 1 + \gamma x < 0. \end{cases} \qquad (9.16)$$

When $\gamma = 0$, this distribution is interpreted consistently with the limit obtained with (9.15) as $\gamma \to 0$:

$$G_0(x) \equiv \exp\left(-e^{-x}\right), \qquad x \in \mathbb{R}. \qquad (9.17)$$

*Given $a, b \in \mathbb{R}$, the distribution function $G_\gamma(ax + b)$ is called a **generalized extreme value distribution**, abbreviated **GEV distribution**.*

Remark 9.12 (Extreme value index) *The parameter $\gamma \in \mathbb{R}$ is called the **extreme value index**.*

- *When $\gamma = 0$, $G_\gamma(x)$ is defined for all x, and hence has long left and right tails.*

- *When $\gamma > 0$, $G_\gamma(x)$ is defined for $x \geq -1/\gamma$, and hence has a short left tail and a long right tail.*

- *When $\gamma < 0$, $G_\gamma(x)$ is defined for $x \leq -1/\gamma$, and hence has a short right tail and a long left tail.*

Example 9.13 (Well-known EV distributions) *Relating $G_\gamma(x)$ to the three distributions in Example 9.10:*

1. **Gumbel** *class ($\gamma = 0$):*

$$G_E(x) = G_0(\alpha x), \qquad x \in \mathbb{R}. \qquad (9.18)$$

*This is also called a **Type I extreme value distribution**.*

2. **Fréchet** *class* $(\gamma > 0)$:

$$G_P(x) = G_\gamma \left((x-1)/\gamma\right), \; with \; \gamma = 1/\alpha, \qquad x \geq 0. \qquad (9.19)$$

This is also called a **Type II extreme value distribution.**

3. **Reversed-Weibull** *class* $(\gamma < 0)$:

$$G_\beta(x) = G_\gamma \left(-(1+x)/\gamma\right), \; with \; \gamma = -1/\alpha \qquad x \leq 0. \qquad (9.20)$$

This is also called a **Type III extreme value distribution.**

Exercise 9.14 *Confirm the statements in Remark 9.12 and Example 9.13.*

9.2.2 Extreme Value Distributions

We now return to the following question:

Given $F(x)$, *if there exist* **normalizing sequences** $\{a_n\}_{n=1}^\infty$ *and* $\{b_n\}_{n=1}^\infty$ *where* $a_n > 0$ *for all* n, *and a distribution function* $G(x)$ *so that* $F^n(a_n x + b_n) \Rightarrow G(x)$, *meaning* $F^n(a_n x + b_n) \to G(x)$ *for every continuity point* x *of* $G(x)$, *what can be said about the distribution functions* $G(x)$ *that can arise?*

Such distribution functions are called **extreme value distributions,** abbreviated **EV distributions** as noted above.

The reader will undoubtedly have observed that "extreme value distribution" has now been used in two contexts:

- in Definition 9.9, in the general context of being the weak limit of a specified sequence of distribution functions;

- in Definition 9.11, as a specific collection of distributions.

It is fair to say that this observation gives the punchline of this section away.

We now proceed toward this anticipated goal, to show that the distribution functions in Definition 9.11 are the only weak limits possible in Definition 9.9. In Book IV we will return to the important reverse question, and that is, given such a G, how can we determine which if any distribution functions F have the property that $F^n(a_n x + b_n) \Rightarrow G(x)$ for some $\{a_n\}_{n=1}^\infty$ and $\{b_n\}_{n=1}^\infty$?

If $F^n(x)$ denotes the distribution function in (9.6) of $M_n = \max_{m \leq n}\{X_m\}$, we defined x^* in (9.7):

$$x^* = \inf\{x | F(x) = 1\},$$

where $x^* \equiv \infty$ if $F(x) < 1$ for all x. If $x^* < \infty$, $\{F^n(x)\}_{n=1}^\infty$ was seen to be tight, whereas if $x^* = \infty$, a common result in finance applications, then $\{F^n(x)\}_{n=1}^\infty$ was not tight.

However tightness is a requirement here. Using current notation:

- By Proposition 8.18, if $F^n \Rightarrow G$, with G a distribution function by definition, then $\{F^n\}_{n=1}^\infty$ is tight.

- Conversely by Exercise 8.21, if $\{F^n\}_{n=1}^\infty$ is not tight, there is a subsequence with a limiting function that is not a distribution function.

- By Proposition 8.20, if $\{F^n\}_{n=1}^\infty$ is tight, then every subsequence $\{F^{n_k}\}_{k=1}^\infty$ in Helly's selection theorem has limit function G that is a distribution function, so $F^{n_k} \Rightarrow G$.

- If all such subsequences have the same limit function G, then $F^n \Rightarrow G$ by Corollary 8.22.

Thus for the current investigation into when $F^n \Rightarrow G$, it is necessary to modify the sequence $\{F^n(x)\}_{n=1}^\infty$ to assure tightness in all cases, meaning whether $x^* < \infty$ or $x^* = \infty$.

To this end we "normalized" the sequence of maximum random variables in (9.8) by defining:

$$M_n^N = \frac{M_n - b_n}{a_n},$$

for sequences of real numbers $\{a_n\}_{n=1}^\infty$ and $\{b_n\}_{n=1}^\infty$, where $a_n > 0$ for all n. The goal of this normalization is to provide tightness to $\{F_n^{(N)}\}_{n=1}^\infty$, the distribution functions of M_n^N, and hence ensure that $F_n^{(N)}(x)$ has a chance of converging to a proper distribution function.

Because $M_n^N \le x$ if and only if $M_n \le a_n x + b_n$, it followed that:

$$F_n^{(N)}(x) = F^n(a_n x + b_n),$$

where F^n is the distribution function of M_n. Thus if identified, G will be the weak limit of the normalized sequence of distribution functions $\{F_n^{(N)}\}_{n=1}^\infty$, meaning $F_n^{(N)}(x) \Rightarrow G(x)$, and this is equivalent to $F^n(a_n x + b_n) \Rightarrow G(x)$.

Notation 9.15 *There seems to be no elegant way to avoid the notational confusion between the name of the distribution function, say F, and the value of this distribution function at x, which is $F(x)$. In mathematics it is almost universally accepted that $F(x)$ can notationally stand in for either the function or the function's value at x. This convention is employed throughout these books.*

In the case of the normalized distribution function $F_n^{(N)}$ which can be defined in terms of F^n, one can define this relationship in terms of the respective values at x, or to avoid the valuation notation, introduce a transformation, T say. Here $T_n(x) = a_n x + b_n$, and then define the translated distribution function as a composition of functions:

$$F_n^{(N)} = F^n \circ T_n.$$

Rather than further complicate our work with this unnecessarily general notation, we resort to the well-worn approach of denoting the function $F_n^{(N)}$ in terms of its values at x, for which $F_n^{(N)}(x) \equiv F^n(a_n x + b_n)$.

It is important to recognize at the start that if normalizing sequences $\{a_n\}_{n=1}^\infty$ and $\{b_n\}_{n=1}^\infty$ exist, they cannot be unique. The following proposition provides the necessary technical result.

Proposition 9.16 (Weak convergence under transformations) *If $F_n(x) \Rightarrow G(x)$, and $\{c_n\}_{n=1}^\infty$ and $\{d_n\}_{n=1}^\infty$ are sequences with $c_n \to c$ and $d_n \to d$, then:*

$$F_n(c_n x + d_n) \Rightarrow G(cx + d). \tag{9.21}$$

Proof. *If $cx + d$ is a continuity point of G, then for any $\epsilon > 0$ there are continuity points y, z with $y < cx + d < z$ and $G(z) - G(y) < \epsilon$. This follows because G has at most countably many discontinuities. As $c_n x + d_n \to cx + d$, this obtains that $y < c_n x + d_n < z$ for $n \ge N_1$. Further $F_n(u) \to G(u)$ for all continuity points of G such as y, z, and so there exists N_2 so that $|F_n(u) - G(u)| < \epsilon$ for $u = y, z$ and $n \ge N_2$.*

Combining obtains that for any $\epsilon > 0$ there exists $N = \max\{N_1, N_2\}$, so that for $n \ge N$:

$$y < c_n x + d_n < z, \quad G(z) - G(y) < \epsilon,$$
$$|F_n(y) - G(y)| < \epsilon, \quad |F_n(z) - G(z)| < \epsilon.$$

These inequalities and monotonicity of distribution functions provide three sets of inequalities:

$$G(cx + d) - 2\epsilon < G(y) - \epsilon < F_n(y),$$
$$F_n(y) \leq F_n(c_n x + d_n) \leq F_n(z),$$
$$F_n(z) < G(z) + \epsilon < G(cx + d) + 2\epsilon.$$

Hence:

$$G(cx + d) - 2\epsilon \leq F_n(c_n x + d_n) \leq G(cx + d) + 2\epsilon,$$

and the result follows. ∎

Corollary 9.17 (On nonuniqueness of $\{a_n\}_{n=1}^{\infty}$ and $\{b_n\}_{n=1}^{\infty}$) *Assume that $F^n(a_n x + b_n) \Rightarrow G(x)$, where $a_n > 0$ for all n. If $\{a_n'\}_{n=1}^{\infty}$ and $\{b_n'\}_{n=1}^{\infty}$ are sequences, $a_n' > 0$ for all n, so that:*

$$\frac{a_n'}{a_n} \to a > 0, \qquad \frac{b_n' - b_n}{a_n} \to b, \tag{9.22}$$

then:

$$F^n(a_n' x + b_n') \Rightarrow G(ax + b). \tag{9.23}$$

Proof. *Define $c_n = a_n'/a_n$ and $d_n = (b_n' - b_n)/a_n$, so $c_n \to a$, $d_n \to b$. By Proposition 9.16 applied to $F_n(x) \equiv F^n(a_n x + b_n)$:*

$$
\begin{aligned}
F^n(a_n' x + b_n') &\equiv F^n(a_n(c_n x + d_n) + b_n) \\
&= F_n(c_n x + d_n) \\
&\Rightarrow G(ax + b).
\end{aligned}
$$

∎

Thus if normalizing sequences $\{a_n\}_{n=1}^{\infty}$ and $\{b_n\}_{n=1}^{\infty}$ exist, they can always be altered when desired or convenient to $\{a_n'\}_{n=1}^{\infty}$ and $\{b_n'\}_{n=1}^{\infty}$ where the latter sequences satisfy (9.22). Such a substitution does not change the limiting distribution other than by a rescaling and recentering. In addition, if $a = 1$ and $b = 0$, the new normalizing sequences produce the same limiting distribution.

Remark 9.18 (Converse of Corollary 9.17) *The converse of the above corollary is also true. If $F_n(a_n x + b_n) \Rightarrow G_1(x)$ and $F_n(a_n' x + b_n') \Rightarrow G_2(x)$ for G_j non-degenerate distribution functions, where $a_n > 0$ and $a_n > 0$ for all n, then there exists a, b satisfying (9.22) and $G_2(x) = G_1(ax + b)$. In the special case of $F_n(a_n x + b_n) = F^n(a_n x + b_n)$, which is the case of the extreme value distributions of this section, we prove this result in Corollary 9.33. See Section 14 of Billingsley (1995) for the general result.*

9.2.3　The Fisher-Tippett-Gnedenko Theorem

Remarkably, though already revealed at the beginning of the prior section, the only distribution functions G that can arise as weak limits as above are members of the class of extreme value distributions $G_\gamma(x)$ of Definition 9.11, appropriately scaled and centered. Specifically, we will show that if $F^n(a_n x + b_n) \Rightarrow G(x)$ for some sequences $\{a_n\}_{n=1}^{\infty}$ and $\{b_n\}_{n=1}^{\infty}$ with $a_n > 0$ for all n, then $G(x) = G_\gamma(ax + b)$ for some real parameters $a > 0, b,$ and γ.

This result is called the **Fisher-Tippett theorem**, named for the earliest developers of this theory, **Ronald Fisher** (1890–1962) and **Leonard (L. H. C.) Tippett** (1902–1985). This result is also called the **Fisher-Tippett-Gnedenko theorem**, recognizing the later contributions of **Boris Gnedenko** (1912–1995).

Remark 9.19 *The **Fisher-Tippett-Gnedenko theorem** is one of the two centerpiece results of extreme value theory, the other being the **Pickands-Balkema-de Haan theorem**, named for 1974-5 papers of **A. A. Balkema** and **Laurens de Haan**, and, **James Pickands III**. It is also called the **Gnedenko-Pickands-Balkema-de Haan theorem**, and even the **Gnedenko theorem** in recognition of the earlier 1943 work of **Boris Gnedenko**. This latter result will be introduced below, and then studied again in Book IV.*

In order to prove the Fisher-Tippett-Gnedenko theorem, we will first transform the requirement underlying $F^n(a_n x + b_n) \Rightarrow G(x)$, of convergence at all continuity points of G, to a more accessible functional formulation. Following **de Haan and Ferreira** (2006), the proof will then proceed using associated left continuous inverses and the results in Sections 3.2 and 8.3.

The transformation will be undertaken in the next proposition. The major innovation underlying this transformation and needed for the later proof is that we need only prove such convergence at continuity points x with $0 < G(x) < 1$. This introduces a question.

If $F_n(x) \to G(x)$ for all continuity points of $G(x)$ with $0 < G(x) < 1$, does this imply that $F_n(x) \Rightarrow G(x)$?

In general, the answer is negative.

Example 9.20 ($F_n(x) \to G(x)$ *for continuity points with* $0 < G(x) < 1$) *Let $G(x)$ be the distribution function of the standard binomial:*

$$G(x) = \begin{cases} 0, & x < 0, \\ \frac{1}{2}, & 0 \le x < 1, \\ 1, & 1 \le x. \end{cases}$$

Define:

$$F_n(x) = \begin{cases} \frac{n+1}{2n+1} e^x, & x < 0, \\ \frac{n+1}{2n+1}, & 0 \le x < 1, \\ \frac{n}{2n+1}\left(1 - e^{-(x-1)}\right) + \frac{n+1}{2n+1}, & 1 \le x. \end{cases}$$

Then $F_n(x) \to G(x)$ for all continuity points x with $0 < G(x) < 1$, meaning $0 < x < 1$. But $F_n(x) \nRightarrow G(x)$. We see that $F_n(x) \to \frac{1}{2}e^x \ne G(x)$ for $x < 0$, all of which are continuity points, and similarly for $x > 1$ where $F_n(x) \to 1 - \frac{1}{2}e^{-(x-1)}$ for $x > 1$.

Hence $F_n(x) \Rightarrow \widetilde{G}(x)$, the continuous distribution function given by:

$$\widetilde{G}(x) = \begin{cases} \frac{1}{2}e^x, & x < 0, \\ \frac{1}{2}, & 0 \le x < 1, \\ 1 - \frac{1}{2}e^{-(x-1)}, & 1 \le x. \end{cases}$$

In order to assure a favorable result, we will require that $G(x)$ satisfy the condition of **endpoint continuity**. Any continuous distribution function satisfies this condition, but continuity is a much stronger condition. Since all distribution functions are right continuous, endpoint continuity is really an assumption about left continuity at the selected points. It will be observed that the binomial distribution function is not endpoint continuous.

Definition 9.21 (Endpoint continuity) *For a given distribution function $G(x)$, let:*

$$x_{(0)} \equiv \sup\{x | G(x) = 0\}, \quad x_{(1)} \equiv \inf\{x | G(x) = 1\}, \tag{9.24}$$

with the convention that if either of the respective sets is empty:

$$x_{(0)} \equiv \sup \emptyset = -\infty, \quad x_{(1)} \equiv \inf \emptyset = \infty.$$

*Then $G(x)$ will be said to be **endpoint continuous** if each of $x_{(0)}$ and $x_{(1)}$ is either unbounded or a continuity point of $G(x)$.*

Notation 9.22 *(CP(G)) Let $CP(G)$ denote the set of **all continuity points of** G.*

We now provide the above noted transformation of the convergence assumption underlying the definition of weak convergence for an endpoint continuous distribution function. The proof of item 1 below is perfectly general in that for such functions, $F_n \Rightarrow G$ if and only if $F_n(x) \to G(x)$ for all $x \in CP(G)$ with $0 < G(x) < 1$. The other statements in this proposition are specifically related to $F_n(x) \equiv F^n(a_n x + b_n)$, the distribution functions for the **normalized** sequence of maximum random variables in (9.9).

Proposition 9.23 (Equivalent formulations for $F^n \Rightarrow G$) *Let $G(x)$ be an endpoint continuous distribution function, and $\{F^n(a_n x + b_n)\}_{n=1}^{\infty}$ a collection of distribution functions as in Definition 9.9.*

Then $F^n(a_n x + b_n) \Rightarrow G(x)$ if and only if any of the following conditions are satisfied:

1. For all $x \in CP(G)$ with $0 < G(x) < 1$:

$$F^n(a_n x + b_n) \to G(x). \tag{1}$$

2. For all $x \in CP(G)$ with $0 < G(x) < 1$:

$$n \ln F(a_n x + b_n) \to \ln G(x). \tag{2}$$

3. For all $x \in CP(G)$ with $0 < G(x) < 1$:

$$n\left(1 - F(a_n x + b_n)\right) \to -\ln G(x). \tag{3}$$

4. For all $x \in CP(G)$ with $0 < G(x) < 1$:

$$\frac{1}{n\left(1 - F(a_n x + b_n)\right)} \to -\frac{1}{\ln G(x)} \tag{9.25}$$

Proof. *Since $G(x) > 0$, (2) is equivalent to statement 1 by taking logarithms in (1).*

For statement 3, recall the Taylor series expansion of $\ln y$ about $y = 1$:

$$\ln y = y - 1 + O\left[(y-1)^2\right].$$

Noting that $\ln G(x)$ is finite, $n \ln F(a_n x + b_n) \to \ln G(x)$ if and only if $\ln F(a_n x + b_n) \to 0$. It then follows that $F(a_n x + b_n) \to 1$, and thus by the above series:

$$\frac{-\ln F(a_n x + b_n)}{1 - F(a_n x + b_n)} \to 1.$$

This and (2) obtains (3).

Finally, statement 4 follows from statement 3 since $0 < G(x) < 1$ implies $-\ln G(x) > 0$, and then by convergence in (3) there exists N so that $n\left(1 - F(a_n x + b_n)\right) > 0$ for $n \geq N$. Thus reciprocals converge and (9.25) follows from (3).

The proof is completed once statement 1 is established, and only the "if" statement requires discussion by definition of weak convergence. Thus assume convergence for $x \in CP(G)$ with $0 < G(x) < 1$.

To prove convergence for continuity points with $G(x) = 1$, consider $\{x | G(x) = 1\}$. If this set is empty, then there is nothing to prove, and otherwise define $x_{(1)} \equiv \inf\{x | G(x) = 1\}$ as in Definition 9.21. By endpoint continuity $x_{(1)} \in CP(G)$, $G(x_{(1)}) = 1$, and $G(x) < 1$ for $x < x_{(1)}$. Denoting $F_n(x) \equiv F^n(a_n x + b_n)$, let $\{x_m\} \subset CP(G)$ with $x_m < x_{(1)}$ be a sequence *of points with* $x_m \to x_{(1)}$. *Since each* F_n *is increasing,* $F_n(x_{(1)}) \geq F_n(x_m)$ *for all* m, *and hence:*

$$\liminf_n F_n(x_{(1)}) \geq \liminf_n F_n(x_m) = G(x_m).$$

As $x_{(1)} \in CP(G)$, $G(x_m) \to G(x_{(1)}) = 1$ *and so* $\liminf_n F_n(x_{(1)}) \geq 1$. *But* $F_n(x_{(1)}) \leq 1$ *for all* n, *so* $\limsup_n F_n(x_{(1)}) \leq 1$ *and it follows that* $F_n(x_{(1)}) \to 1 = G(x_{(1)})$.

If $x' > x_{(1)}$, *then* $x' \in CP(G)$. *As* $F_n(x_{(1)}) \leq F_n(x') \leq 1$, *the prior result assures that* $F_n(x') \to 1 = G(x')$.

To prove convergence for continuity points with $G(x) = 0$, *consider* $\{x|G(x) = 0\}$. *Again, there is nothing to prove if this set is empty, and otherwise define* $x_{(0)} \equiv \sup\{x|G(x) = 0\}$. *By endpoint continuity* $x_{(0)} \in CP(G)$, $G(x_{(0)}) = 0$, *and* $G(x) > 0$ *for* $x > x_{(0)}$. *Let* $\{x_m\} \subset CP(G)$ *with* $x_m > x_{(0)}$ *be a sequence of points with* $x_m \to x_{(0)}$. *Since each* F_n *is increasing,* $F_n(x_{(0)}) \leq F_n(x_m)$ *for all* m, *and hence:*

$$\limsup_n F_n(x_{(0)}) \leq \limsup_n F_n(x_m) = G(x_m).$$

As $x_{(0)} \in CP(G)$, $G(x_m) \to G(x_{(0)}) = 0$ *and so* $\limsup_n F_n(x_{(0)}) \leq 0$. *But* $F_n(x_{(0)}) \geq 0$ *for all* n, *so* $\liminf_n F_n(x_{(0)}) \geq 0$, *and it follows that* $F_n(x_{(0)}) \to 0 = G(x_{(0)})$.

If $x' < x_{(0)}$, *then* $x' \in CP(G)$. *As* $F_n(x_{(0)}) \geq F_n(x') \geq 0$, *the prior result assures that* $F_n(x') \to 0 = G(x')$. ∎

The next step of the development is to evaluate the left continuous inverses of the functions in (9.25). But first:

Exercise 9.24 *Prove that* $H(x) \equiv -\frac{1}{\ln G(x)}$ *is an increasing function over the given domain defined above, meaning all* x *with* $0 < G(x) < 1$. *Prove that* $H(x) \equiv \frac{1}{n(1-F(a_n x+b_n))}$ *is an increasing function for all* n. *Finally, show that all such functions are right continuous. Hint: Recall* $a_n > 0$.

As the functions in (9.25) are increasing, the associated left continuous inverses are well defined and increasing by Proposition 3.16. Further, the sequence of left continuous inverses will converge at all continuity points of $\left(\frac{1}{-\ln G}\right)^*(y)$ by Proposition 8.27.

The major result of this section then proceeds by analyzing the implication of convergence of these left continuous inverses for the parametric form of G. Once that analysis is complete, we can then conclude that if these inverses converge at all such continuity points, then so too do their left continuous inverses, which will bring us back to the statement in (9.25) with a parametrized G term. From there we can reverse the steps in the above proposition to obtain the parametrization for G.

The first result is to derive these left continuous inverses. Recall that these functions are restricted to x with $0 < G(x) < 1$.

Proposition 9.25 ($H^*(y)$ **for** H **in (9.25)**) *With the notation above and* $H^*(y)$ *denoting the left continuous inverse of the given function* $H(x)$:

$$\left(\frac{1}{n\left(1 - F(a_n x + b_n)\right)}\right)^*(y) = \frac{U(ny) - b_n}{a_n},$$

and

$$\left(\frac{1}{-\ln G}\right)^*(y) = D(y),$$

where:

$$U(y) \equiv \left(\frac{1}{1-F}\right)^*(y), \qquad D(y) \equiv G^*(e^{-1/y}), \tag{9.26}$$

with $U(y)$ *defined for* $y > 1$, *and* $D(y)$ *defined for* $y > 0$.

Proof. *These left continuous inverses are well defined by Proposition 3.16 since the respective functions are increasing in x by Exercise 9.24. By definition:*

$$\left(\frac{1}{n\left(1-F(a_n x + b_n)\right)}\right)^* (y) \equiv \inf\left\{x \,\middle|\, \frac{1}{n\left(1-F(a_n x + b_n)\right)} \geq y\right\}.$$

If $z = a_n x + b_n$, then $x = (z - b_n)/a_n$ and so:

$$\inf\left\{x \,\middle|\, \frac{1}{n\left(1-F(a_n x + b_n)\right)} \geq y\right\} = \left[\inf\left\{z \,\middle|\, \frac{1}{1-F(z)} \geq ny\right\} - b_n\right]\bigg/ a_n$$

$$= \frac{U(ny) - b_n}{a_n}.$$

Similarly, from $-\ln G(x) > 0$:

$$\left(\frac{1}{-\ln G}\right)^* (y) \equiv \inf\left\{x \,\middle|\, \frac{-1}{\ln G(x)} \geq y\right\}$$

$$= \inf\left\{x \,\middle|\, G(x) \geq e^{-1/y}\right\}$$

$$= G^*(e^{-1/y}).$$

Since $G^*(z)$ is defined on the range of $G(x)$, which is $(0, 1)$ by Proposition 9.23, it follows that $G^*(e^{-1/y})$ is defined on $(0, \infty)$. By (9.25), the range of $G(x)$ restricts the range of $F(x)$ to $(0, 1)$ by convergence, and thus $1/(1 - F(x))$ has range $(1, \infty)$, which is the domain of $U(y)$. ∎

Exercise 9.26 (Alternative $U(y)$) *Derive an alternative expression for $U(y)$, that for $y > 1$:*

$$U(y) = F^*(1 - 1/y). \tag{9.27}$$

Corollary 9.27 (Final equivalent formulation for $F^n \Rightarrow G$) *Let $G(x)$ be an endpoint continuous distribution function, $\{F^n(a_n x + b_n)\}_{n=1}^{\infty}$ a collection of distribution functions as in Definition 9.9, and U and D as defined in (9.26).*

Then $F^n(a_n x + b_n) \Rightarrow G(x)$ if and only if for all $y \in CP(D)$ with $y > 0$:

$$\frac{U(ty) - b(t)}{a(t)} \to D(y), \qquad as\ t \to \infty. \tag{9.28}$$

*Here $a(t) \equiv a_{\lfloor t \rfloor}$, and $b(t) \equiv b_{\lfloor t \rfloor}$ where $\lfloor t \rfloor$ is the **greatest integer function** defined by $\lfloor t \rfloor = \max\{n \,|\, n \leq t\}$.*

Proof. *If $F^n(a_n x + b_n) \Rightarrow G(x)$, then with n in place of t, (9.28) follows from (9.25) by Propositions 8.27 and 9.25. That is, weak convergence of increasing functions implies weak convergence of left continuous inverses. To express this latter convergence with an assumption in real t, we need to check that $U(ty)$, which is well defined for real $t > 0$, does not vary much from $U(\lfloor t \rfloor y)$ as $t \to \infty$.*

If y is a continuity point of $D(y)$, then as U is increasing we obtain that for any $t \geq 1$ with $n \leq t < n+1$:

$$U(ny) \leq U(ty) \leq U((n+1)y_m),$$

for any $y_m \in CP(D)$ with $y < y_m$. As $a(t) = a_n$ and $b(t) = b_n$ by definition for $n \leq t < n+1$, for any such y_m:

$$\frac{U(ny) - b_n}{a_n} \leq \frac{U(ty) - b(t)}{a(t)} \leq \frac{U((n+1)y_m) - b_n}{a_n}. \tag{1}$$

Letting $t, n \to \infty$ obtains from (1) that for any such y_m :

$$D(y) \leq \lim_{t \to \infty} \frac{U(ty) - b(t)}{a(t)} \leq D(y_m). \qquad (2)$$

Only the upper limit needs comment, and this is proved as follows. Let $y_m^{(N)} \in CP(D)$ with $y_m^{(N)} > \left(\frac{n+1}{n}\right) y_m$ for $n \geq N$. As U is increasing:

$$\lim_{n \to \infty} \frac{U((n+1) y_m) - b_n}{a_n} \leq \lim_{n \to \infty} \frac{U(n y_m^{(N)}) - b_n}{a_n} = D(y_m^{(N)}).$$

Letting $N \to \infty$, choose $y_m^{(N)} \to y_m$ as is possible by density of continuity points, and then $D(y_m^{(N)}) \to D(y_m)$ since $y_m \in CP(D)$.

Since we can choose $y_m \to y$ by construction, and D is continuous at y, (9.28) follows from (2).

Conversely, assume that (9.28) is satisfied, and thus by Proposition 9.25, for all $y \in CP\left[\left(\frac{-1}{\ln G}\right)^*\right]$:

$$\left(\frac{1}{n\left(1 - F(a_n x + b_n)\right)}\right)^* (y) \to \left(\frac{-1}{\ln G}\right)^* (y). \qquad (3)$$

The functions inside the left continuous inverse transformations are increasing and right continuous by Exercise 9.24. All these functions have limits of 0 and 1 at $-\infty$ and ∞, respectively, and so by Proposition 6.6 these are distribution functions.

Corollary 8.28 now obtains that for all $x \in CP\left(\frac{-1}{\ln G}\right)$:

$$\frac{1}{n\left(1 - F(a_n x + b_n)\right)} \to \frac{-1}{\ln G(x)}.$$

As $CP\left(\frac{-1}{\ln G}\right) = CP(G)$, it follows from item 4 of Proposition 9.23 that $F^n(a_n x + b_n) \Rightarrow G(x)$. $\blacksquare$

We are now ready for the statement and proof of the **Fisher-Tippett-Gnedenko theorem**. Its statement makes the assumption that G is endpoint continuous, and also **not a degenerate distribution** function. Recall that **degenerate** means that $G = \chi_{[a,\infty)}$ for some a. Equivalently, G is the distribution function of a random variable X defined on $(\mathcal{S}, \mathcal{E}, \mu)$ with:

$$\mu\left[X^{-1}(a)\right] = 1.$$

The proof is made relatively easy because all the hard work needed for the key technical result is deferred to Proposition 9.34.

Remark 9.28 (Extreme value index) *As noted above, the family of distributions identified in this theorem and parametrized by γ is called the **generalized extreme value class of distributions**, and the distribution function $G_\gamma(ax+b)$ is called a **generalized extreme value distribution**, abbreviated **GEV distribution**. The parameter $\gamma \in \mathbb{R}$ is called the **extreme value index**. Finally, $G_0(x)$ equals the pointwise limit of $G_\gamma(x)$ as $\gamma \to 0$.*

Definition 9.29 (Domain of attraction) *When $F^n(a_n x + b_n) \Rightarrow G_\gamma(Ax + B)$, we say that the distribution function F is in the **domain of attraction of G_γ**, denoted $F \in \mathcal{D}(G_\gamma)$.*

Proposition 9.30 (Fisher-Tippett-Gnedenko theorem) *Let F be the distribution function of a random variable X and F^n the distribution function of $M_n = \max_{m \leq n}\{X_m\}$ for independent $\{X_m\}_{m=1}^n$. Assume that there exists sequences $\{a_n\}_{n=1}^\infty$ and $\{b_n\}_{n=1}^\infty$ with $a_n > 0$ for all n, so that $F^n(a_n x + b_n) \Rightarrow G(x)$ for an endpoint continuous and nondegenerate distribution function G.*

Then there are real constants $A > 0$, B, and γ so that:

$$G(x) = G_\gamma\left(Ax + B\right),$$

with:

$$G_\gamma(x) = \exp\left(-(1+\gamma x)^{-1/\gamma}\right), \qquad 1 + \gamma x > 0. \tag{9.29}$$

When $\gamma = 0$, $G_0(x)$ is defined on $\mathbb{R}$ by the limiting function:

$$G_0(x) \equiv \exp\left(-e^{-x}\right). \tag{9.30}$$

Proof. *By Corollary 9.27, for every continuity point $y > 0$ of D:*

$$\frac{U(ty) - b(t)}{a(t)} \to D(y), \tag{1}$$

with U and D given as in (9.26). Then by Proposition 9.34, this obtains that there are real constants d, $c > 0$, and γ so that:

$$D(y) = \begin{cases} d + c\left(y^\gamma - 1\right)/\gamma, & \gamma \neq 0, \\ d + c\ln y, & \gamma = 0. \end{cases} \tag{2}$$

This proposition applies because if $D(y)$ is constant, then by definition:

$$\inf\{x | G(x) \geq e^{-1/y}\} = x_0,$$

for all y. This implies that $G \equiv \chi_{[x_0,\infty)}$ and is degenerate, a contradiction.

Given (2), D is monotonic and continuous, and so $D^* = D^{-1}$ by Proposition 3.22. If $\gamma \neq 0$ then:

$$\begin{aligned} D^*(x) &= D^{-1}(x) \\ &= \left(1 + \gamma\frac{x-d}{c}\right)^{1/\gamma}. \end{aligned}$$

If $\gamma = 0$ then:

$$D^*(x) = \exp\left(\frac{x-d}{c}\right),$$

which equals the limit of the previous formula as $\gamma \to 0$ by (9.15).

Since $D(y) \equiv G^*(e^{-1/y})$, we calculate directly that $D^*(x) = -1/\ln G(x)$. Equating the two expressions for D^* obtains:

$$\begin{aligned} G(x) &= \exp\left[-\left(1 + \gamma\frac{x-d}{c}\right)^{-1/\gamma}\right] \\ &= G_\gamma\left(\frac{x-d}{c}\right). \end{aligned}$$

Hence $A = 1/c$ and $B = -d/c$.

Note that since in this case the resulting $G(x)$ is continuous, and thus endpoint continuous, this justifies the equivalent formulations derived in Proposition 9.23 and Corollary 9.27. ∎

Corollary 9.31 (Fisher-Tippett-Gnedenko theorem; $A = 1$, $B = 0$**)** *The functions* $a(t)$ *and* $b(t)$ *in* (1) *above can be chosen so that the resulting* $D(y)$ *is defined with* $c = 1$ *and* $d = 0$, *and thus* $A = 1$, $B = 0$ *and:*

$$G(x) = G_\gamma(x).$$

Proof. *Corollary 9.35 states that if such* $a(t)$ *and* $b(t)$ *exists to produce* (1), *then in fact one can choose these functions so that the expression for* $D(y)$ *is defined with* $c = 1$ *and* $d = 0$. *The result now follows from the last line of the above proof.* ∎

Example 9.32 ($G_\gamma \in \mathcal{D}(G_\gamma)$) *It is natural to wonder if a generalized extreme value distribution function* G_γ *is a member of* $\mathcal{D}(G_\gamma)$, *and thus is in its own domain of attraction.*
    Let:

$$a_n \equiv \begin{cases} n^\gamma & \gamma \neq 0, \\ 1, & \gamma = 0, \end{cases} \qquad b_n \equiv \begin{cases} (n^\gamma - 1)/\gamma & \gamma \neq 0, \\ \ln n, & \gamma = 0. \end{cases}$$

Then $G_\gamma^n(a_n x + b_n) = G_\gamma(x)$, *so by definition* $G_\gamma^n(a_n x + b_n) \Rightarrow G_\gamma(x)$.
    Thus for all γ:

$$G_\gamma \in \mathcal{D}(G_\gamma). \tag{9.31}$$

That is, every **generalized extreme value distribution** *is in its own domain of attraction.*

The next result proves that the limiting distribution in the **Fisher-Tippett-Gnedenko theorem,** under different pairs of normalizing sequences, is unique in the sense that γ is uniquely defined. In other words, when it exists, the extreme value index associated with F is unique.

This result does not assume that the pairs of normalizing sequences in the statement satisfy (9.22), since in that case the conclusion follows from Corollary 9.17. Recalling Remark 9.18, the following result is the converse of Corollary 9.17 in the special case where the limiting distribution function $G(x)$ is a GEV.

Corollary 9.33 (Fisher-Tippett-Gnedenko theorem; Uniqueness of γ**)** *Let* F *be the distribution function of a random variable* X *and* F^n *the distribution function of* $M_n = \max_{m \leq n}\{X_m\}$ *for independent* $\{X_m\}_{m=1}^n$. *Assume that there exist normalizing sequences* $\{a_n\}_{n=1}^\infty$, $\{b_n\}_{n=1}^\infty$, *and* $\{a_n'\}_{n=1}^\infty$, $\{b_n'\}_{n=1}^\infty$, *where* $a_n > 0$ *and* $a_n' > 0$ *for all* n, *so that:*

$$F^n(a_n x + b_n) \Rightarrow G_\gamma(Ax + B), \quad F^n(a_n' x + b_n') \Rightarrow G_{\gamma'}(A'x + B').$$

Then $\gamma = \gamma'$.
Proof. *By Corollary 9.27 and subscripting* D *for clarity, the above weak convergences imply that as* $t \to \infty$:

$$\frac{U(ty) - b(t)}{a(t)} \to D_\gamma(y), \qquad \frac{U(ty) - b'(t)}{a'(t)} \to D_{\gamma'}'(y), \qquad y > 0.$$

As in the proof of Proposition 9.30, $D_\gamma(y)$ *and* $D_{\gamma'}'(y)$ *are given as in* (2) *of that proof.*
    Addition obtains:

$$\frac{U(ty) - b''(t)}{a''(t)} \to D_\gamma(y) + D_{\gamma'}'(y),$$

where:

$$a''(t) = \frac{a(t)a'(t)}{a'(t) + a(t)}, \qquad b''(t) = \frac{a(t)b'(t) + a'(t)b(t)}{a'(t) + a(t)}.$$

Since $a''(t) > 0$, the same derivation applies to obtain that:

$$D_\gamma(y) + D'_{\gamma'}(y) = D''_{\gamma''}(y).$$

But $c > 0$ and $c' > 0$ in (9.32), and so $D_\gamma(y) + D'_{\gamma'}(y)$ is increasing and nonconstant. The same is true for $D''_{\gamma''}(y)$.

This is only possible given (2) in the proof of Proposition 9.30 if $\gamma = \gamma'$. ∎

We now turn to the critical technical detail needed in the Fisher-Tippett-Gnedenko theorem, related to the necessary functional form of $D(y)$. The proof is somewhat long, and so is divided into several steps.

Proposition 9.34 (Parametric form for $D(y)$) *Let $U(y)$ be an increasing function on $y > 1$, and assume there are functions, $a(t) > 0$ and $b(t)$ so that as $t \to \infty$:*

$$\frac{U(ty) - b(t)}{a(t)} \to D(y),$$

for every continuity point $y > 0$ of an increasing, non-constant function D. Then there exist real constants d, $c > 0$, and γ, so that:

$$D(y) = \begin{cases} d + c\frac{y^\gamma - 1}{\gamma}, & \gamma \neq 0, \\ d + c \ln y, & \gamma = 0. \end{cases} \tag{9.32}$$

Proof. *Since D is increasing, it is continuous almost everywhere by Proposition I.5.8, so let $x_0 > 0$ be such a continuity point. Define $D_0(x)$ on $x > 0$ by $D_0(x) \equiv D(x_0 x)$, and so $x = 1$ is a continuity point of $D_0(x)$. If $E(x) \equiv D_0(x) - D_0(1)$, then for any continuity point $x > 0$ of D_0:*

$$E(x) = \lim_{t \to \infty} \frac{U_0(tx) - U_0(t)}{a(t)}, \tag{1}$$

where $U_0(x) \equiv U(x_0 x)$ for $x > 0$.

We now work with $E(x)$, which is simplified due to the absence of $b(t)$.

1. *There is a function $A(y)$ on $y > 0$ so that for $x, y > 0$:*

$$E(xy) = E(x)A(y) + E(y). \tag{2}$$

To see this fix $y > 0$, then note:

$$\frac{U_0(txy) - U_0(t)}{a(t)} = \frac{U_0(txy) - U_0(ty)}{a(ty)}\frac{a(ty)}{a(t)} + \frac{U_0(ty) - U_0(t)}{a(t)}.$$

If $x > 0$ is a continuity point of D_0, then $(U_0(txy) - U_0(ty))/a(ty) \to E(x)$ as $ty \to \infty$ by (1).

For given $y > 0$, we now show that $(U_0(txy) - U_0(t))/a(t)$ also converges as $t \to \infty$ by showing that:

$$a(ty)/a(t) \to A, \quad (U_0(ty) - U_0(t))/a(t) \to B.$$

If not convergent, let $A_1 \neq A_2$ be two accumulation points of $a(ty)/a(t)$ as $t \to \infty$, and $B_1 \neq B_2$ be two accumulation points of $(U_0(ty) - U_0(t))/a(t)$ as $t \to \infty$. Then for any x for which both x and xy are continuity points of D_0, it follows that:

$$E(yx) = E(x)A_1 + B_1 = E(x)A_2 + B_2. \tag{3}$$

For any other x, we can find continuity points $x_n \to x$ and $x_n y \to xy$, both as left limits, so that this identity holds with x_n and $x_n y$, and by the left continuity of E from (9.26), it follows that for all x:

$$E(x)(A_1 - A_2) = B_2 - B_1.$$

But $E(x)$ cannot be a constant function since this would imply that D is constant.

Hence this identity implies that $A_1 = A_2 = A$ and $B_2 = B_1 = B$. Since $y > 0$ was arbitrary, it follows from (3) with $x = 1$ and $E(1) = 0$ that $B = E(y)$. As A exists for all $y > 0$ define:

$$A(y) = \lim_{t \to \infty} a(ty)/a(t). \tag{4}$$

This proves (2) for all $x > 0$ and $y > 0$.

2. *If $H(x) \equiv E(e^x)$, then H is differentiable everywhere with $H'(t) = H'(0)A(e^x)$.*

   For this, let $s \equiv \ln x$ and $t \equiv \ln y$. Then with $H(x) \equiv E(e^x)$, the functional equation in (2) obtains:

   $$H(s + t) = H(s)A(e^t) + H(t). \tag{5}$$

   Since $H(0) = E(1) = 0$, for $s \neq 0$ this equation can be expressed as:

   $$\frac{H(s + t) - H(t)}{s} = \frac{H(s) - H(0)}{s} A(e^t).$$

   This implies that if H is differentiable at any t, then it is differentiable at $t = 0$, and this then implies it is differentiable everywhere. As an increasing function, we can assume that H has one such differentiable point. In Book III, it will in fact be shown that monotonic functions are differentiable almost everywhere.

   So the existence of one differentiable point implies that H is differentiable everywhere, with:

   $$H'(t) = H'(0)A(e^t), \tag{6}$$

   and $H'(0) \neq 0$. If $H'(0) = 0$, then $H'(t) = 0$ for all t, and constant H implies E and hence D are constant, a contradiction.

3. *With $Q(t) \equiv H(t)/H'(0)$, there exists $\gamma \in \mathbb{R}$ so that:*

   $$Q(t) = \begin{cases} (e^{\gamma t} - 1)/\gamma, & \gamma \neq 0, \\ t & \gamma = 0. \end{cases} \tag{7}$$

   To prove this, note that $Q(0) = E(1)/H'(0) = 0$ from part 2, $Q'(t) = A(e^t)$ from (6) and $Q'(0) = 1$ from (4). Then from the functional equation for H in (5):

   $$Q(s + t) - Q(t) = Q(s)Q'(t),$$

   and by interchanging s and t:

   $$Q(s + t) - Q(s) = Q(t)Q'(s).$$

   Subtracting expressions obtains:

   $$Q(t)\frac{Q'(s) - Q'(0)}{s} = \frac{Q(s) - Q(0)}{s}\left(Q'(t) - 1\right).$$

Letting $s \to 0$ proves that $Q''(0)$ exists and $Q(t)Q''(0) = Q'(t) - 1$. This now implies $Q''(t)$ exists for all t:

$$Q''(t) = Q'(t)Q''(0). \tag{8}$$

Now let $\gamma \equiv Q''(0)$, and so $(\ln Q')'(t) \equiv \gamma$ from (8). Integrating from 0 to s obtains $\ln Q'(s) = \gamma s$ since $Q'(0) = 1$, and this can be exponentiated and integrated from 0 to t. Since $Q(0) = 0$, this obtains $Q(t)$ as specified above in the cases $\gamma \neq 0$ and $\gamma = 0$.

4. *The final steps are to reverse these transformations to obtain $D(y)$.*

   From the Q formula in (7) for $\gamma \neq 0$:

$$H(t) = H'(0)\frac{e^{\gamma t} - 1}{\gamma}.$$

   Now $H(t) \equiv E(e^t) \equiv D_0(e^t) - D_0(1)$ produces for $x \equiv e^t > 0$:

$$D_0(x) = D_0(1) + H'(0)\frac{x^\gamma - 1}{\gamma},$$

   or with $y = xx_0$,

$$D(y) = D(x_0) + H'(0)\frac{\left(\frac{y}{x_0}\right)^\gamma - 1}{\gamma}.$$

   A little algebra obtains:

$$D(y) = d + c\frac{y^\gamma - 1}{\gamma},$$

   with

$$d = D(x_0) + \frac{(1 - x_0^\gamma)H'(0)}{\gamma x_0^\gamma}, \qquad c = \frac{H'(0)}{x_0^\gamma}.$$

   The case $\gamma = 0$ is similar and left as an exercise.

5. *Finally, note that $c > 0$. First, $H'(0) \neq 0$ from part 2, while $H'(0) < 0$ implies that $H'(t) < 0$ for all t, since $H'(t) = H'(0)A(e^t)$ by (6) and $A(e^t) > 0$. But then with $H(0) = 0$, this implies by integration that $H(t) < 0$ for $t > 0$. This then contradicts that $H(t) \equiv E(e^t) > 0$ since $E(e^t) = D(x_0 e^t) - D(x_0)$ and D is increasing.*

■

Corollary 9.35 (Replacing $b(t)$) *Let $U(y)$ be an increasing function on $y > 1$, and assume there are functions, $a(t) > 0$ and $b(t)$ so that as $t \to \infty$:*

$$\frac{U(ty) - b(t)}{a(t)} \to D(y),$$

for every continuity point $y > 0$ of an increasing, non-constant function D.

   Then $b(t)$ can be defined to equal $U(t)$, and with $a_c(t) \equiv ca(t)$:

$$\frac{U(ty) - U(t)}{a_c(t)} \to \begin{cases} (y^\gamma - 1)/\gamma, & \gamma \neq 0, \\ \ln y, & \gamma = 0, \end{cases} \tag{9.33}$$

as $t \to \infty$.

Proof. *From Proposition 9.34:*

$$\frac{U(ty) - U(t)}{ca(t)} = \frac{U(ty) - b(t)}{ca(t)} - \frac{U(t) - b(t)}{ca(t)}$$

$$\rightarrow \frac{D(y) - D(1)}{c}.$$

So for $y > 0$*, (9.33) follows from (9.32).* ∎

Remark 9.36 (On $\gamma > 0$**)** *In the case of* $\gamma > 0$*, which is of critical importance in finance, a result in Book IV provides an even simpler parametrization of the normalizing functions* $a(t)$ *and* $b(t)$*, as well as the normalizing sequences* a_n *and* b_n*.*

9.3   The Pickands-Balkema-de Haan Theorem

In this section we investigate two applications of the above theory to a typical finance question in which we are given a sample of data points parametrized in increasing order, $\{x_i\}_{i=1}^N$. Our goal is to study properties of the right tail of the unknown underlying distribution function F. In the most typical situation the data are sparse in the far right tail, and one is often interested in one of the following related questions:

1. **Quantile Estimation:** Given α with $0 < \alpha < 1$ but usually $\alpha \approx 1$, estimate x_α defined so that $F(x_\alpha) = \alpha$, or $x_\alpha = F^*(\alpha)$. These are equivalent when F is strictly increasing and continuous by Proposition 3.22, while otherwise $F^*(\alpha)$ is taken as the definition of x_α.

2. **Tail Probability Estimation:** Given x large, estimate $1 - F(x)$.

In the development of the second question we will derive insights into the **Pickands-Balkema-de Haan theorem**, which will be further studied in Book IV.

9.3.1   Quantile Estimation

Recall that while $U(y) \equiv \left(\frac{1}{1-F}\right)^*(y)$ as is derived in (9.26), an alternative representation was provided in Exercise 9.26:

$$U(y) = F^*(1 - 1/y).$$

Letting $x_\alpha \equiv F^*(\alpha)$, (9.33) obtains that for fixed $y > 0$:

$$\frac{x_{1-1/ty} - x_{1-1/t}}{a(t)} \rightarrow \frac{y^\gamma - 1}{\gamma}, \text{ as } t \rightarrow \infty,$$

relabeling $a_c(t)$ as simply $a(t)$.

   Given $\alpha' > \alpha$, it follows that $x_{\alpha'} \geq x_\alpha$ since F^* is increasing. Transforming $\alpha \rightarrow 1 - 1/t$ and $\alpha' \rightarrow 1 - 1/ty$ for $y > 0$ produces:

$$t = 1/(1 - \alpha), \qquad y = (1 - \alpha)/(1 - \alpha'),$$

and hence $U(t) = x_\alpha$ and $U(ty) = x_{\alpha'}$.

Under the assumption that $t = (1 - \alpha)^{-1}$ is large enough for the above limit to be used as an approximation, meaning α is close enough to 1, obtains:

$$x_{\alpha'} \approx x_\alpha + k \left[\left(\frac{1 - \alpha}{1 - \alpha'} \right)^\gamma - 1 \right], \qquad k = \frac{a((1 - \alpha)^{-1})}{\gamma}. \tag{9.34}$$

Example 9.37 (Model tail estimates) *We will investigate methodologies for estimating* γ *in theory below, and empirically in Book IV. Given such an estimate* $\widehat{\gamma}$, *an example of an application of (9.34) is to choose* $\alpha \approx 0.90$, *say, and estimate* $x_{0.90} = \widehat{x}_{0.90}$ *based on the ordered data set* $\{x_i\}_{i=1}^N$. *Then by regression on the remaining 10% of the data,* $x_i > \widehat{x}_{0.90}$, *estimate* k *so that:*

$$x_{\alpha'} \approx \widehat{x}_{0.90} + k \left[(0.1)^{\widehat{\gamma}} (1 - \alpha')^{-\widehat{\gamma}} - 1 \right].$$

Specifically, if $\{x_i\}_{i=0.9N+1}^N$ *represents the subsample of the original data for which* $x_i > \widehat{x}_{0.90}$, *then the regression identifies:*

$$x_{\alpha'} = x_i, \text{ for } \alpha' = 0.9 + (i - .9N)/N.$$

Once k *is estimated, the resulting formula can then be used to estimate* $x_{\alpha'}$ *for* α' *large. More generally, one could simultaneously estimate* $\widehat{\gamma}$ *and* k *from such tail subsamples.*

9.3.2 Tail Probability Estimation

Under the assumption of continuity in the tail of $F(x)$, $x_\alpha \equiv F^*(\alpha)$ obtains that $\alpha = F(x_\alpha)$ by Proposition 3.22, while in the general case Proposition 3.19 provides:

$$F(x_\alpha^-) \leq \alpha \leq F(x_\alpha).$$

Assuming continuity for simplicity, $1 - F(x_\alpha) \equiv 1 - \alpha$ and similarly $1 - F(x_{\alpha'}) \equiv 1 - \alpha'$, and the formula in (9.34) produces for $x \equiv x_{\alpha'} > x_\alpha$:

$$1 - F(x) \approx (1 - F(x_\alpha)) \left(1 + \frac{1}{k} (x - x_\alpha) \right)^{-1/\gamma}, \tag{9.35}$$

where $k = a((1 - \alpha)^{-1})/\gamma$ as in (9.34).

To justify the merits of this manipulation, we now derive this formula within the theoretical framework of the limit theorem above. This result is related to the **Pickands-Balkema-de Haan theorem**, named for the 1974-5 papers of **A. A. Balkema** and **Laurens de Haan**, and, **James Pickands III**, and sometimes called the **Gnedenko-Pickands-Balkema-de Haan theorem**, and even the **Gnedenko theorem** in recognition of the earlier 1943 work of **Boris Gnedenko** (1912–1995). We will return to a more detailed development of this result in Book IV in the case where $\gamma > 0$.

For the statement of this result, recall that as defined in (9.7), $x^* \equiv \inf\{x | F(x) = 1\}$, with $x^* \equiv \infty$ if $F(x) < 1$ for all x.

Proposition 9.38 (Pickands-Balkema-de Haan theorem I) *Assume that* F *is in the domain of attraction of* G_γ, $F \in \mathcal{D}(G_\gamma)$. *Then:*

- *For all* $x > -1/\gamma$ *when* $\gamma > 0$, *or,*

- *For* $0 \leq x < -1/\gamma$ *when* $\gamma < 0$:

$$\lim_{t \to x^*} \frac{1 - F(t + xh_a(t))}{1 - F(t)} = (1 + \gamma x)^{-1/\gamma}. \tag{9.36}$$

- *For all x when $\gamma = 0$:*

$$\lim_{t \to x^*} \frac{1 - F(t + x h_a(t))}{1 - F(t)} = e^{-x}. \tag{9.37}$$

Here:

$$h_a(t) \equiv a \left(\frac{1}{1 - F(t)} \right),$$

with $a(t) \equiv a_c(t)$ defined in terms of the normalizing function in (9.33).

Proof. *Consider $U \left(\frac{1}{1 - F(t)} \right)$. As derived in Exercise 9.26, $U(y) = F^*(1 - 1/y)$ and so $U \left(\frac{1}{1 - F(t)} \right) = F^*(F(t))$. As F is a distribution function and thus increasing, it follows from (3.7) that:*

$$F^*(F(t)) \leq t \leq F^* \left(F(t)^+ \right).$$

Hence for any $\epsilon > 0$,

$$U \left(\frac{1}{1 - F(t)} \right) \leq t \leq U \left(\frac{1 + \epsilon}{1 - F(t)} \right),$$

and so:

$$0 \leq \frac{t - U \left(\frac{1}{1 - F(t)} \right)}{a \left(\frac{1}{1 - F(t)} \right)} \leq \frac{U \left(\frac{1 + \epsilon}{1 - F(t)} \right) - U \left(\frac{1}{1 - F(t)} \right)}{a \left(\frac{1}{1 - F(t)} \right)}.$$

By (9.33), as $1/(1 - F(t)) \to \infty$, which is to say as $t \to x^*$, the upper bounding expression has limit:

$$\frac{U \left(\frac{1 + \epsilon}{1 - F(t)} \right) - U \left(\frac{1}{1 - F(t)} \right)}{a \left(\frac{1}{1 - F(t)} \right)} \to \begin{cases} [(1 + \epsilon)^\gamma - 1]/\gamma, & \gamma \neq 0, \\ \ln(1 + \epsilon), & \gamma = 0. \end{cases}$$

As these bounding expressions converge to 0 as $\epsilon \to 0$, it follows that as $t \to x^*$:

$$\frac{t - U \left(\frac{1}{1 - F(t)} \right)}{a \left(\frac{1}{1 - F(t)} \right)} \to 0.$$

This and (9.33) obtain that for all $x > 0$:

$$\frac{U \left(\frac{x}{1 - F(t)} \right) - t}{a \left(\frac{1}{1 - F(t)} \right)} \to \begin{cases} [x^\gamma - 1]/\gamma, & \gamma \neq 0, \\ \ln x, & \gamma = 0, \end{cases}$$

as $t \to x^*$.

Convergence of this sequence of increasing functions of x implies convergence of the left continuous inverses by Proposition 8.27. We now calculate these inverses.

Denoting $\frac{1}{1 - F(t)} = k$, a constant, and recalling that $a(k) > 0$ and $U(y) = F^*(1 - 1/y)$:

$$
\begin{aligned}
\left(\frac{U(kx) - t}{a(k)} \right)^* (y) &= \inf \left\{ x \left| \frac{U(kx) - t}{a(k)} \geq y \right. \right\} \\
&= \inf \left\{ x | F^*(1 - 1/x) \geq t + a(k) y \right\} / k \\
&= \inf \left\{ \frac{1}{1 - y} \left| F^*(y) \geq t + a(k) y \right. \right\} / k \\
&= \frac{k}{1 - \inf \{ y | F^*(y) \geq t + a(k) y \}} \\
&= \frac{k}{1 - F(t + a(k) y)}.
\end{aligned}
$$

The last step follows from Proposition 3.26 by right continuity of F:

$$\inf\{y|F^*(y) \geq t + a(k)y\} \equiv F^{**}(t + a(k)y) = F(t + a(k)y).$$

Similarly,

$$
\begin{aligned}
\left(\frac{x^\gamma - 1}{\gamma}\right)^*(y) &= \inf\left\{x \,\middle|\, \frac{x^\gamma - 1}{\gamma} \geq y\right\} \\
&= \inf\left\{x | x \geq (1 + \gamma y)^{1/\gamma}\right\} \\
&= (1 + \gamma y)^{1/\gamma}.
\end{aligned}
$$

While:

$$(\ln x)^*(y) = \inf\{x \,|\ln x \geq y\} = e^y.$$

The results now follow by taking reciprocals. ∎

The limits in (9.36) and (9.37) reflect the conditional probability distribution of **exceedances over the threshold** t, and can be expressed as in (3.21):

$$\lim_{t \to x^*} \Pr[X \leq t + xh_a(t)|X > t] = 1 - (1 + \gamma x)^{-1/\gamma},$$

when $\gamma \neq 0$, and with limit $1 - e^{-x}$ when $\gamma = 0$.

Alternatively, for $u = t$ "close" to x^*:

$$\Pr[X \leq u + y|X > u] \approx 1 - \left(1 + \frac{\gamma}{h_a(u)}y\right)^{-1/\gamma}, \tag{9.38}$$

when $\gamma \neq 0$, and when $\gamma = 0$:

$$\Pr[X \leq u + y|X > u] \approx 1 - \exp[y/h_a(u)]. \tag{9.39}$$

The analysis underlying this result is often referred to as the **peak over threshold method**, and a more user-friendly result will be derived in Book IV for $\gamma > 0$ as noted above.

The limiting and approximating distributions above are members of an important class of distributions.

Definition 9.39 (Generalized Pareto distribution) *The distribution function:*

$$H_{\gamma,u,\beta}(x) \equiv 1 - \left(1 + \gamma\left(\frac{x - u}{\beta}\right)\right)^{-1/\gamma}, \tag{9.40}$$

*is a **generalized Pareto distribution**, abbreviated **GPD**. When $\gamma = 0$, this distribution is defined as the limit in (9.40):*

$$H_{0,u,\beta}(x) \equiv 1 - \exp\left(-\frac{x - u}{\beta}\right). \tag{9.41}$$

The function $H_{\gamma,u,\beta}(x)$ is defined for $x \geq u$ when $\gamma \geq 0$, and for $u \leq x \leq u - \beta/\gamma$ when $\gamma < 0$.

*The parameter $\gamma \in \mathbb{R}$ is called the **shape** parameter, $u \in \mathbb{R}$ the **location** parameter, and $\beta \in (0, \infty)$ the **scale** parameter.*

Remark 9.40 *The generalized Pareto distribution function is often denoted $G_{\gamma,\mu,\sigma}(x)$, but we use $H_{\gamma,u,\beta}(x)$ to avoid notational confusion with the closely related extreme value distribution $G_\gamma(x)$ defined in (9.29) and (9.30), and also to avoid confusion that the parameters μ and σ in this notation, or u and β in (9.40), are not the mean and standard deviation of this distribution when these exist.*

*This distribution generalizes the **Pareto distribution,** parametrized by $\alpha > 0$, $c > 0$ and defined for $x \geq c$ by:*

$$P_{\alpha,c}(x) = 1 - \left(\frac{x}{c}\right)^{-\alpha}. \tag{9.42}$$

*This distribution is named for **Vilfredo Pareto** (1848–1923), an economist who discovered that income was approximately distributed according to this law.*

In terms of this distribution function, (9.36) and (9.37) can be restated for all γ, assuming the limit function is defined at x:

$$\lim_{t \to x^*} \Pr\left[X \leq t + x h_a(t) | X > t\right] = H_{\gamma,0,1}(x),$$

and the approximations in (9.38) and (9.39) can be restated with $\beta(u) \equiv h_a(u)$:

$$\Pr\left[X \leq u + y | X > u\right] \approx H_{\gamma,0,\beta(u)}(y).$$

Proposition 9.44 below will address the quality of this approximation.

That the shape parameter β of the generalized Pareto distribution can be a function of the threshold u is better appreciated with an example.

Example 9.41 ($\beta = \beta(u)$, $\gamma \neq 0$) *Let $F(x) = H_{\gamma,0,\beta}(x)$ be defined as in (9.40) with $\gamma \neq 0$:*

$$F(x) = 1 - \left(1 + \gamma\left(\frac{x}{\beta}\right)\right)^{-1/\gamma}$$

Denoting:

$$F_u(y) \equiv \Pr\left[X \leq u + y | X > u\right], \tag{9.43}$$

then for y and $u + y$ in the domain of $F(x)$:

$$
\begin{aligned}
F_u(y) &= \frac{F(u+y) - F(u)}{1 - F(u)} \\
&= 1 - \left(1 + \frac{\gamma}{\beta + \gamma u}y\right)^{-1/\gamma}.
\end{aligned}
$$

Thus if $F(x) = H_{\gamma,0,\beta}(x)$, then $F_u(x) = H_{\gamma,0,\beta+\gamma u}(x)$.

In other words, the conditional distribution function $F_u(y)$ is again generalized Pareto, but with shape parameter $\beta(u) = \beta + \gamma u$.

Exercise 9.42 ($\beta = \beta(u)$, $\gamma = 0$) *Show that the same result does not apply with $F(x) = H_{0,0,\beta}(x)$ defined as in (9.41):*

$$F(x) = 1 - \exp\left(-\frac{x}{\beta}\right),$$

and now $F_u(x) = F(x)$.

Example 9.43 (Example 9.37 cont'd) *Recall we are given a sample of data points parametrized in increasing order, $\{x_i\}_{i=1}^N$, and our goal is to study properties of the right tail of the unknown underlying distribution function F. Assume $\gamma \neq 0$, and t is large enough for the limit in (9.36) to be approximated, meaning α is large enough relative to 1. Then for $x > x_\alpha$:*

$$1 - F(x) \approx (1 - F(x_\alpha)) \left(1 + \gamma \frac{x - x_\alpha}{k'}\right)^{-1/\gamma}, \tag{9.44}$$

where $k' = h_a(x_\alpha)$. If $\gamma = 0$:

$$1 - F(x) \approx (1 - F(x_\alpha)) \exp\left[\frac{x - x_\alpha}{k'}\right].$$

The first formula with $\gamma \neq 0$ is functionally identical to the approximation in (9.35) derived by inverting the percentile approximation formula in (9.35), since:

$$\frac{\gamma}{k'} = \frac{\gamma}{h_a(x_\alpha)} = \frac{\gamma}{a\left(1/\left(1 - F(x_\alpha)\right)\right)} = \frac{\gamma}{a\left(1/\left(1 - \alpha\right)\right)} = \frac{1}{k}.$$

Any application requires the selection of α and hence x_α, as well as γ and k'. Estimation of γ is addressed in Book IV, so assume for example that the estimate $\widehat{\gamma} \neq 0$ is given, and then choose $\alpha = 0.9$ and $x_{0.90} = \widehat{x}_{0.90}$ from the data set $\{x_i\}_{i=1}^N$. Then for $x > \widehat{x}_{0.90}$:

$$1 - F(x) \approx 0.10 \left(1 + \frac{\widehat{\gamma}}{k'} (x - \widehat{x}_{0.90})\right)^{-1/\widehat{\gamma}},$$

and k' can be estimated from the remaining 10% of the data, $x_i > \widehat{x}_{0.90}$.

We next develop another, more powerful representation of the **Pickands-Balkema-de Haan theorem**, which is often used as the statement of this named result. It addresses the quality of the approximation in (9.38), restated following Remark 9.40. We do not prove this result here, but only discuss the subtleties involved in a formal demonstration.

In Book IV we return to this result and provide a proof in the case $\gamma > 0$ using estimates based on **Karamata's representation theorem,** named for **Jovan Karamata** (1902–1967).

Proposition 9.44 (Pickands-Balkema-de Haan theorem II) *Assume that $F \in \mathcal{D}(G_\gamma)$ is in the **domain of attraction of** G_γ, $\gamma > 0$, and denote $F_u(y)$ as in (9.43).*

Then the approximation in (9.38) is uniform in y in the sense that with $\beta(u) \equiv h_a(u)$:

$$\lim_{u \to x^*} \sup_{0 \leq y < x^* - u} |F_u(y) - H_{\gamma, 0, \beta(u)}(y)| = 0. \tag{9.45}$$

Further, (9.45) is true with $\beta(u) \equiv \gamma u$.

Discussion. *As noted following Remark 9.40, (9.36) of Proposition 9.38 can be restated: For each x for which the limit function is defined:*

$$\lim_{t \to x^*} \Pr[X \leq t + x h_a(t) | X > t] = H_{\gamma, 0, 1}(x).$$

The approximation in (9.38) can also be restated: For $u \equiv t$ large and $\beta(u) \equiv h_a(u)$:

$$\Pr[X \leq u + y | X > u] \approx H_{\gamma, 0, \beta(u)}(y).$$

Now $y \equiv x h_a(u)$ in this latter statement, and for fixed u this variable transformation is one-to-one.

But the limit in the first statement is pointwise for each x, and we cannot seek to prove (9.45) without better estimates on the rate of convergence.

For example, if the convergence in t in (9.36) is uniform in x, this would imply that given $\epsilon > 0$ there exists u so that for $t \geq u$ and for all x:

$$|\Pr[X \leq t + xh_a(t)|X > t] - H_{\gamma,0,1}(x)| < \epsilon.$$

Then with $y \equiv xh_a(u)$, this implies that for all y:

$$\left| F_u(y) - H_{\gamma,0,\beta(u)}(y) \right| < \epsilon,$$

and thus:

$$\sup_{0 \leq y < x^* - u} \left| F_u(y) - H_{\gamma,0,\beta(u)}(y) \right| < \epsilon.$$

Hence the proof of (9.45) would follow from an estimate that assures that the limit in Proposition 9.38 is uniform in x. This will be proved in Book IV under the assumption that $\gamma > 0$, by proving that $\beta(u) \equiv \gamma u$ provides the result in (9.45).

It will further be seen that any other $\beta(u)$ that satisfies this limit has the property that $\beta(u)/\gamma u \to 1$ as $u \to x^*$. ∎

9.4 γ in Theory: von Mises' Condition

The question of estimating γ based on sample data for a presumably unknown distribution function F will be addressed in Book IV. Here we address the following question:

If F is given, what can be said about the value of γ for which $F \in D(G_\gamma)$, meaning that F is in the domain of attraction of G_γ?

This question was addressed in 1936 by **Richard von Mises** (1883–1953) who is also credited with the parametrization of the **extreme value class of distributions** in (9.16) underlying the Fisher-Tippett-Gnedenko theorem. The resulting sufficient condition on F to ensure that $F \in \mathcal{D}(G_\gamma)$ has come to be known as **von Mises' condition.**

We state this result under the assumption that F is twice **continuously** differentiable, which is stronger than von Mises' assumption of twice differentiable. But the continuity assumption assures the existence of the Riemann integral of various second derivatives below, whereas the more general assumption requires Lebesgue integration theory which will be developed in Book III. However, for many applications the added assumption of continuity is true in any case.

That said, the reader is encouraged to reproduce this proof within the Lebesgue integration theory of Book III for the general case.

Proposition 9.45 (von Mises' Condition) *Let F be a twice continuously differentiable distribution function with $F'(x) > 0$ for some interval (x_0, x^*), where x^* is defined in (9.7). If:*

$$\lim_{x \to x^*} \left(\frac{1-F}{F'} \right)'(x) = \gamma, \tag{9.46}$$

*then $F \in \mathcal{D}(G_\gamma)$. That is, F is in the **domain of attraction of** G_γ.*

Proof. *As F is by assumption continuous everywhere, and with $U(t)$ defined in (9.27), then by (3.10):*

$$F(U(t)) = F(F^*(1 - 1/t)) = 1 - 1/t,$$

and so for t > 0:

$$t = \frac{1}{1 - F(U(t))}. \tag{1}$$

That F is twice differentiable and $F'(x) > 0$ for some interval (x_0, x^*) ensures that F is strictly increasing for $x > x_0$. Thus $U(t) = F^{-1}(1 - 1/t)$ for $t > 1/(1 - F(x_0))$ by Proposition 3.22, and $U(t)$ is also twice differentiable on this domain. Taking two derivatives in (1) obtains:

$$U'(t) = \frac{(1 - F(U(t)))^2}{F'(U(t))},$$

and so:

$$
\begin{aligned}
t\frac{U''(t)}{U'(t)} &= -2 - \frac{F''(U(t))\,[1 - F(U(t))]}{[F'(U(t))]^2} \\
&= -1 + \left(\frac{1-F}{F'}\right)'(U(t)).
\end{aligned}
$$

We claim that if:

$$\lim_{t \to \infty} t\frac{U''(t)}{U'(t)} = \gamma - 1, \tag{2}$$

then $F \in \mathcal{D}(G_\gamma)$. This limit is equivalent to that in (9.46) since $t \to \infty$ is equivalent to $U(t) \equiv x \to x^*$.

To this end, continuity of derivatives and the assumption that $F'(x) > 0$ for (x_0, x^*) obtains that $U''(t)/U'(t)$ is continuous and hence Riemann integrable on bounded intervals. Letting $1 < t < tx$ for $x > 1$:

$$
\begin{aligned}
\ln\left[\frac{U'(tx)}{U'(t)}\right] &= \int_t^{tx} \frac{U''(s)}{U'(s)}ds \\
&= \int_1^x (st)\frac{U''(st)}{U'(st)}\frac{ds}{s}.
\end{aligned}
$$

Since $\ln x^{\gamma-1} = \int_1^x (\gamma - 1)\frac{ds}{s}$:

$$\left|\ln\left[\frac{U'(tx)}{U'(t)}\right] - \ln x^{\gamma-1}\right| \le \int_1^x \left|(st)\frac{U''(st)}{U'(st)} - (\gamma - 1)\right|\frac{ds}{s}. \tag{3}$$

Now given (2), then for any $\epsilon > 0$ there is a T so that $|tU''(t)/U'(t) - (\gamma - 1)| < \epsilon$ for $t \ge T$. This implies that for $t \ge T/s$:

$$\left|(st)\frac{U''(st)}{U'(st)} - (\gamma - 1)\right| < \epsilon,$$

and since $s \ge 1$ this obtains from (3) that for $t \ge T$:

$$\left|\ln\left[\frac{U'(tx)}{U'(t)}\right] - \ln x^{\gamma-1}\right| \le \epsilon \ln x.$$

Thus for any closed interval $[a, b]$:

$$\sup_{x \in [a,b]} \left|\ln\left[\frac{U'(tx)}{U'(t)}\right] - \ln x^{\gamma-1}\right| \le \epsilon \ln b. \tag{4}$$

As $\epsilon > 0$ is arbitrary, the limit of this supremum is 0 as $t \to \infty$.

Now $|e^u - e^w| \leq c|u - w|$ for u, w in a compact interval, and equivalently $|u - w| \leq c|\ln u - \ln w|$, so it follows from (4) that for any compact interval $[a, b]$:

$$\lim_{t \to \infty} \sup_{x \in [a,b]} \left| \frac{U'(tx)}{U'(t)} - x^{\gamma - 1} \right| \to 0.$$

This assures that the limit of the integral over such intervals also converges to zero, since with $f(x) \equiv \frac{U'(tx)}{U'(t)} - x^{\gamma - 1}$:

$$\left| \int_a^b f(x)dx \right| \leq (b - a) \sup_{x \in [a,b]} |f(x)|.$$

Now:

$$\int_1^x \left(\frac{U'(ts)}{U'(t)} - s^{\gamma - 1} \right) ds = \frac{U(tx) - U(t)}{tU'(t)} - \frac{x^\gamma - 1}{\gamma},$$

and hence for any compact interval $[1, x]$, this integral converges to 0 as $t \to \infty$, and hence:

$$\frac{U(tx) - U(t)}{tU'(t)} \to \frac{x^\gamma - 1}{\gamma}. \tag{5}$$

With $a(t) \equiv tU'(t) > 0$ and $b(t) \equiv U(t)$, this implies that $F \in \mathcal{D}(G_\gamma)$ by the proof of the Fisher-Tippett-Gnedenko theorem. ∎

Corollary 9.46 (von Mises' normalizing sequences $\{a_n\}_{n=1}^\infty$, $\{b_n\}_{n=1}^\infty$) *Under the assumptions of the above proposition:*

$$F^n(a_n x + b_n) \Rightarrow G_\gamma(x),$$

with:

$$b_n \equiv U(n) = F^{-1}(1 - 1/n), \tag{9.47}$$

and:

$$a_n \equiv nU'(n) = 1/(nF'(b_n)). \tag{9.48}$$

Proof. *Equation (5) of the last proof showed that for all $x > 0$:*

$$\frac{U(nx) - U(n)}{nU'(n)} \to D(x) = \frac{x^\gamma - 1}{\gamma},$$

and so $b_n = U(n)$ and $a_n = nU'(n)$ follow by Propositions 9.23 and 9.25.

For a continuous distribution function with $F'(x) > 0$ for the interval (x_0, x^*), it follows that $F^*(y) = F^{-1}(y)$ for $y > F(x_0)$. Hence:

$$U(y) \equiv F^*(1 - 1/y) = F^{-1}(1 - 1/y),$$

and as derived above:

$$U'(y) = \frac{(1 - F(U(y)))^2}{F'(U(y))} = \frac{1}{y^2 F'(U(y))}.$$

The results for a_n and b_n now follow. ∎

Remark 9.47 (Other $\{a_n'\}_{n=1}^\infty$, $\{b_n'\}_{n=1}^\infty$) *The above corollary provides one pair of normalizing sequences $\{a_n\}_{n=1}^\infty$ and $\{b_n\}_{n=1}^\infty$ given the assumptions underlying von Mises' result. In general, there are infinitely many such pairs $\{a_n'\}_{n=1}^\infty$, $\{b_n'\}_{n=1}^\infty$ as given by (9.22) of Corollary 9.17. All that is required is that:*

$$\frac{a_n'}{a_n} \to a > 0, \qquad \frac{b_n' - b_n}{a_n} \to b.$$

Example 9.48 (The normal distribution) *Let the random variable X have the standard normal distribution function defined in (1.23):*

$$\Phi(x) = \int_{-\infty}^{x} \phi(y)dy \equiv \frac{1}{\sqrt{2\pi}} \int_{-\infty}^{x} \exp\left(-y^2/2\right) dy.$$

In this example we prove that $\gamma = 0$ for this distribution and thus $\Phi \in \mathcal{D}(G_0)$, and also derive normalizing sequences.

1. $\gamma = 0$:

   First, $\Phi(x)$ is certainly twice continuously differentiable with:

   $$\frac{1 - \Phi}{\Phi'} = \frac{\int_x^{\infty} \phi(y)dy}{\phi(x)}.$$

   Hence:

   $$\begin{aligned}
   \left(\frac{1-\Phi}{\Phi'}\right)'(x) &= \frac{-\phi^2(x) - \phi'(x)\int_x^{\infty}\phi(y)dy}{\phi^2(x)} \\
   &= \frac{-\phi^2(x) + x\phi(x)\int_x^{\infty}\phi(y)dy}{\phi^2(x)} \\
   &= \frac{x\int_x^{\infty}\phi(y)dy}{\phi(x)} - 1.
   \end{aligned} \tag{1}$$

   By integration by parts:

   $$x\int_x^{\infty}\phi(y)dy \le \int_x^{\infty} y\phi(y)dy = \phi(x). \tag{2}$$

   With:

   $$e(x) \equiv \int_x^{\infty}\phi(y)dy - \frac{x}{x^2+1}\phi(x),$$

   then $e(0) = 0.5, \lim_{x\to\infty} e(x) = 0$, and:

   $$\begin{aligned}
   e'(x) &= -\phi(x) + \frac{x^2}{x^2+1}\phi(x) + \frac{x^2-1}{(x^2+1)^2}\phi(x) \\
   &= \left[\frac{-2}{(x^2+1)^2}\right]\phi(x) < 0.
   \end{aligned}$$

   Thus $e(x)$ is a strictly decreasing positive function with minimum value of 0 at infinity. Combining with the inequality in (2):

   $$\frac{x}{x^2+1}\phi(x) \le \int_x^{\infty}\phi(y)dy \le \frac{1}{x}\phi(x), \tag{9.49}$$

   which obtains from the estimate in (1):

   $$\frac{x^2}{x^2+1} - 1 \le \left(\frac{1-\Phi}{\Phi'}\right)'(x) \le 0.$$

   Letting $x \to \infty$ it follows that $\left(\frac{1-\Phi}{\Phi'}\right)'(x) \to 0$.

   In other words, von Mises' condition yields $\gamma = 0$ and $\Phi \in \mathcal{D}(G_0)$ for the normal distribution.

2. *Normalizing sequences:*

One set of normalizing sequences for the standard normal distribution is given in (9.47) and (9.48):

$$b_n = U(n) = \Phi^{-1}(1 - 1/n),$$

$$a_n = nU'(n) = \frac{1}{n\Phi'(b_n)} = \frac{1}{n\phi(b_n)}.$$

Since $\Phi^{-1}(1-1/n) \equiv z_{1-1/n}$, b_n so defined equals the quantile value z_α for $\alpha = 1-1/n$:

$$b_n = z_{1-1/n}.$$

Thus $\int_{b_n}^\infty \phi(y)dy = 1/n$ and so from (9.49):

$$\frac{\phi(b_n)}{b_n + 1/b_n} \leq \frac{1}{n} \leq \frac{\phi(b_n)}{b_n}. \tag{3}$$

These inequalities can also be used to numerically estimate b_n for large n because the ratio of the upper and lower bounds converges quickly to 1 as $b_n \to \infty$.

The inequalities in (3) also imply that:

$$\frac{1}{b_n + 1/b_n} \leq a_n \leq \frac{1}{b_n}. \tag{4}$$

Because the ratio of the bounds converges quickly to 1 as $b_n \to \infty$, these inequalities can also be used to numerically estimate a_n for large n.

The inequalities in (3) and (4) also provide alternative normalizing sequences by using estimates. As a numerical example, define a_n' and b_n' by:

$$\frac{1}{n} = \frac{\phi(b_n')}{b_n'}, \quad a_n' = \frac{1}{b_n'}.$$

The equation for b_n' is equivalent to $ye^y = n^2/2\pi$ with $b_n' = \sqrt{y}$. The solution y satisfies $y < y_0 \equiv \ln(n^2/2\pi)$, the solution to $e^y = n^2/2\pi$. It thus seems logical to set $y = \ln(n^2/c_n)$, and determine a sequence $\{c_n\}_{n=1}^\infty$, which works well as $n \to \infty$. Substitution leads to:

$$(n^2/c_n) \ln(n^2/c_n) = n^2/2\pi,$$

and thus the goal is to have:

$$\frac{2\ln n - \ln c_n}{c_n/2\pi} \approx 1. \tag{5}$$

Since any sequence with $c_n \to \infty$ assures that $\ln c_n/c_n \to 0$, this goal can be achieved if $c_n \approx 4\pi \ln n$. Choosing $c_n = 4\pi \ln n$, then $b_n' = \sqrt{y} = \sqrt{\ln(n^2/c_n)}$ and $a_n' = 1/b_n'$ produce:

$$b_n' = (2\ln n - \ln\ln n - \ln 4\pi)^{0.5}, \quad a_n' = (2\ln n - \ln\ln n - \ln 4\pi)^{-0.5}.$$

By Corollary 9.17 these sequences can be simplified to:

$$b_n'' = (2\ln n)^{0.5}, \quad a_n'' = (2\ln n)^{-0.5},$$

if:

$$\frac{a_n''}{a_n'} \to 1, \quad \frac{b_n'' - b_n'}{a_n'} \to 0. \tag{6}$$

The proof of (6) is left as an exercise.

9.5 Independence vs. Tail Independence

With the aid of Example 9.48, we are now able to complete our investigation into tail independence of variates $\{X_j\}_{j=1}^n$ by Definitions 7.42 and 7.47, versus independence of these variates by Definition 3.47. As noted in Remark 7.45, if $\{X_j\}_{j=1}^n$ are independent, then they are both upper and lower tail independent. It was also noted that the converse is not in general true: tail independence does not imply independence.

A result noted there and developed below comes from a 1960 paper by **Masaaki Sibuya**, who showed that normal variates that were not independent could be tail independent. The derivation requires some tools from later books to formally justify, but readers should be able to follow this proof based on prior knowledge, the development in Example 9.48, and by referencing later materials.

We develop the result for two random variables, but with much the same approach this can be generalized to $(X_1, ..., X_n)$ with a joint normal distribution. For this generalization we would assume a fixed correlation ρ between each pair (X_i, X_{i+1}) and also (X_1, X_n), with all other correlations 0. The matrix R then has a ρ in the diagonals above and below the main diagonal of 1s, as well as in the $(1, n)$ and $(n, 1)$ positions, and zeros elsewhere. Details are left to the interested reader.

Example 9.49 (M. Sibuya) *Let X and Y have standard normal distribution functions as defined in (1.23):*

$$\Phi(x) = \int_{-\infty}^{x} \phi(z)dz = \frac{1}{\sqrt{2\pi}} \int_{-\infty}^{x} \exp\left(-z^2/2\right) dz,$$

*so each has mean 0 and variance 1. Assume that (X, Y) has a **joint normal distribution function** as in (7.13):*

$$\Phi(x, y) = \frac{1}{2\pi(1-\rho^2)^{1/2}} \int_{-\infty}^{x} \int_{-\infty}^{y} \exp\left[-\frac{1}{2(1-\rho^2)}\left(z_1^2 - 2\rho z_1 z_2 + z_2^2\right)\right] dz_1 dz_2.$$

Here the matrix R of correlation coefficients in (7.13) is given:

$$R = \begin{pmatrix} 1 & \rho \\ \rho & 1 \end{pmatrix}, \qquad R^{-1} = \begin{pmatrix} 1/(1-\rho^2) & -\rho/(1-\rho^2) \\ -\rho/(1-\rho^2) & 1/(1-\rho^2) \end{pmatrix},$$

and so det $R = 1 - \rho^2$.

*The parameter ρ is the **correlation** between X and Y, to be formally defined in Book IV. It is there shown that $|\rho| \leq 1$ in general, and for normal variates, X and Y are independent if and only if $\rho = 0$.*

We now show that if $0 < |\rho| < 1$, then X and Y are upper tail independent.

Since continuous and strictly increasing, $\Phi^* = \Phi^{-1}$ by Proposition 3.22, and the conditional probability in (7.31) is well defined:

$$\Pr\left[X > \Phi^{-1}(t) \,|\, Y > \Phi^{-1}(t)\right] = \frac{\Pr\left[\left(X > \Phi^{-1}(t)\right) \bigcap \left(Y > \Phi^{-1}(t)\right)\right]}{\Pr\left[Y > \Phi^{-1}(t)\right]}.$$

As always, we use Pr as shorthand for the μ-measures of these sets in the probability space on which X and Y are defined. Now:

$$\left\{(X > \Phi^{-1}(t)) \bigcap (Y > \Phi^{-1}(t))\right\} \subset \left\{X + Y > 2\Phi^{-1}(t)\right\},$$

and so by monotonicity of measures:

$$\Pr\left[\left(X > \Phi^{-1}(t)\right) \bigcap \left(Y > \Phi^{-1}(t)\right)\right] \leq \Pr\left[X + Y > 2\Phi^{-1}(t)\right].$$

It will be proved in the Book VI that $W \equiv X + Y$ *is normally distributed with mean* 0 *and variance* $\sigma^2 = 2(1 + \rho)$. *Thus:*

$$\Pr\left[X > \Phi^{-1}(t) \,|\, Y > \Phi^{-1}(t)\right] \leq \frac{\Pr\left[W > 2\Phi_W^{-1}(t)\right]}{\Pr\left[Y > \Phi^{-1}(t)\right]}, \tag{1}$$

where the subscript on $\Phi_W^{-1}(t)$ *is to denote that unlike* X *and* Y, *the variance of* W *is not equal to* 1. *As* $Z \equiv W/\sqrt{2(1 + \rho)}$ *is standard normal:*

$$\Pr\left[W > 2\Phi_W^{-1}(t)\right] = \Pr\left[Z > \alpha \Phi^{-1}(t)\right],$$

where $\alpha = \sqrt{\frac{2}{1+\rho}} > 1$. *Hence with* $T \equiv \Phi^{-1}(t)$, *the expression in* (1) *becomes:*

$$\Pr\left[X > \Phi^{-1}(t) \,|\, Y > \Phi^{-1}(t)\right] \leq \frac{\Pr\left[Z > \alpha \Phi^{-1}(t)\right]}{\Pr\left[Y > \Phi^{-1}(t)\right]} = \frac{1 - \Phi[\alpha T]}{1 - \Phi[T]}. \tag{2}$$

The upper tail index λ_U *is the limit of the expression in* (2) *as* $T \to \infty$. *Applying the estimates in* (9.49) *of Example* 9.48:

$$\Pr\left[X > \Phi^{-1}(t) \,|\, Y > \Phi^{-1}(t)\right] \leq \frac{T^2 + 1}{\alpha T^2} \frac{\phi(\alpha T)}{\phi(T)}.$$

Because $\alpha > 1$, *it follows by the definition of* ϕ *that the limit is* 0 *as* $T \to \infty$, *and thus* $\lambda_U = 0$.

That $\lambda_L = 0$ *is left as an exercise in symmetry.*

9.6 Multivariate Extreme Value Theory

The notion of an extreme value joint distribution function was introduced in Chapter 7 in the discussion on extreme value copulas. With the aid of the **Fisher-Tippett-Gnedenko theorem**, we can now provide additional details.

Recall the setup there, that $F(x_1, x_2, ..., x_n)$ is a joint distribution function with marginal distributions $\{F_j(x_j)\}_{j=1}^n$, and $\{(X_{1k}, X_{2k}, ..., X_{nk})\}_{k=1}^m$ a random sample from this joint distribution function. The random vector of component-wise maximum variates is defined consistent with (9.5) by:

$$\left(X_1^{(m)}, X_2^{(m)}, ..., X_n^{(m)}\right) \equiv \left(\max_{k \leq m} X_{1k}, \max_{k \leq m} X_{2k}, ..., \max_{k \leq m} X_{nk}\right). \tag{9.50}$$

If $F^{(m)}(x_1, x_2, ..., x_n)$ denotes the joint distribution function of these maximal random vectors, then since $\max_{k \leq m} X_{ik} \leq x_i$ if and only if $X_{ik} \leq x_i$ for all k:

$$F^{(m)}(x_1, x_2, ..., x_n) = F^m(x_1, x_2, ..., x_n).$$

Similarly, if $F_j^{(m)}(x_j)$ is the distribution function of $\max_{k \leq m} X_{jk}$, then:

$$F_j^{(m)}(x_j) = F_j^m(x_j).$$

Finally, $\{F_j^{(m)}(x_j)\}_{j=1}^n$ so defined are the marginal distributions of $F^{(m)}(x_1, x_2, ..., x_n)$ by Definition 3.34.

9.6.1 Multivariate Fisher-Tippett-Gnedenko Theorem

As in the one variable case we "normalize" the maximum variates in (9.50) with vector sequences $\left\{\left(a_1^{(m)}, a_2^{(m)}, ..., a_n^{(m)}\right)\right\}_{m=1}^{\infty}$ with $a_j^{(m)} > 0$ for all j, m, and

$\left\{\left(b_1^{(m)}, b_2^{(m)}, ..., b_n^{(m)}\right)\right\}_{m=1}^{\infty}$, and investigate the normalized joint distribution function:

$$F_N^{(m)}(x_1, x_2, ..., x_n) \equiv \Pr\left[\left(X_j^{(m)} - b_j^{(m)}\right) / a_j^{(m)} \leq x_j, \text{ all } j\right].$$

Since $\left(X_j^{(m)} - b_j^{(m)}\right) / a_j^{(m)} \leq x_j$ if and only if $X_j^{(m)} \leq a_j^{(m)} x_j + b_j^{(m)}$, it follows from the above discussion that:

$$F_N^{(m)}(x_1, x_2, ..., x_n) = F^m\left(a_1^{(m)} x_1 + b_1^{(m)}, a_2^{(m)} x_2 + b_2^{(m)}, ..., a_n^{(m)} x_n + b_n^{(m)}\right), \qquad (9.51)$$

and $\{F_j^{(m)}(a_j^{(m)} x_j + b_j^{(m)})\}_{j=1}^{n}$ are the marginal distributions of $F_N^{(m)}(x_1, x_2, ..., x_n)$.

Definition 9.50 (Multivariate extreme value distribution) *If for some vector seque-nces* $\left\{\left(a_1^{(m)}, a_2^{(m)}, ..., a_n^{(m)}\right)\right\}_{m=1}^{\infty}$ *with* $a_j^{(m)} > 0$ *for all* j, m, *and* $\left\{\left(b_1^{(m)}, b_2^{(m)}, ..., b_n^{(m)}\right)\right\}_{m=1}^{\infty}$, *there exists a distribution function* $G(x_1, x_2, ..., x_n)$ *with nondegenerate marginal distribu-tions so that* $F_N^{(m)} \Rightarrow G$, *which is to say that for every continuity point of* G:

$$F^m\left(a_1^{(m)} x_1 + b_1^{(m)}, a_2^{(m)} x_2 + b_2^{(m)}, ..., a_n^{(m)} x_n + b_n^{(m)}\right) \to G(x_1, x_2, ..., x_n), \qquad (9.52)$$

then G *is called a **multivariate extreme value distribution**, abbreviated, **MEV dis-tribution**.*

In this case, the distribution function F *is said to be in the **domain of attraction** of the distribution function* G, *and denoted* $F \in \mathcal{D}(G)$.

Remark 9.51 *To prove the following result, we need part 1 of the Cramér-Wold theorem, which will not be proved until Book VI as part of the study of weak convergence of mea-sures. Its proof requires the mapping theorem on* $\mathbb{R}^k / \mathbb{R}^j$, *which substantially generalizes the mapping theorem on* $\mathbb{R}$ *of Proposition 8.35.*

Fortunately, the needed result is quite plausible intuitively, and thus should provide no obstacle to the comprehension of the following proof.

Proposition 9.52 (Multivariate Fisher-Tippett-Gnedenko theorem) *If the multi-variate extreme value distribution* $G(x_1, x_2, ..., x_n)$ *exists, so that:*

$$F_N^{(m)} \Rightarrow G,$$

as defined in (9.52), then:

1. *There exist real constants* $\{(A_j, B_j, \gamma_j)\}_{j=1}^{n}$ *with* $A_j > 0$ *for all* j, *so that the marginal distribution functions* $\{G_j(x_j)\}_{j=1}^{n}$ *are given by:*

$$G_j(x_j) = G_{\gamma_j}(A_j x_j + B_j), \qquad (9.53)$$

   with $G_{\gamma_j}(x_j)$ *defined in (9.29) for* $\gamma_j \neq 0$ *or (9.30) for* $\gamma_j = 0$.

2. $G(x_1, x_2, ..., x_n)$ *is continuous.*

3. *There exists an extreme value copula $C_G(u_1, u_2, ..., u_n)$ that is a copula for G:*

$$G(x_1, x_2, ..., x_n) = C_G\left(G_1\left(x_1\right), ..., G_n\left(x_n\right)\right), \tag{9.54}$$

and C_F, the copula associated with F, is in the domain of attraction of C_G.

Proof.

1. *By a corollary to the Cramér-Wold theorem in Book VI, if $F_N^{(m)} \Rightarrow G$ then the jth marginal distribution of $F_N^{(m)}$ converges weakly to the jth marginal distribution of G for all j. By the above discussion, the marginals of $F_N^{(m)}$ are $\left\{F_j^m\left(a_j^{(m)}x_j + b_j^{(m)}\right)\right\}_{j=1}^n$ where $\{F_j\left(x_j\right)\}_{j=1}^n$ are the marginals of F, and thus the Cramér-Wold corollary implies that for all j:*

$$F_j^m\left(a_j^{(m)}x_j + b_j^{(m)}\right) \Rightarrow G_j\left(x_j\right).$$

By Definition 9.50, each $G_j\left(x_j\right)$ is nondegenerate and so the Fisher-Tippett-Gnedenko theorem of Proposition 9.30 is applicable. This result obtains the existence of $\{(A_j, B_j, \gamma_j)\}_{j=1}^n$ with $A_j > 0$ so that (9.53) is satisfied with $G_{\gamma_j}\left(x_j\right)$ defined in (9.29) for $\gamma_j \neq 0$, or (9.30) for $\gamma_j = 0$.

2. *Since G has continuous marginal distributions, a copula C_G for G exists and satisfies (9.54) by Sklar's theorem of Proposition 7.18. Continuity of G follows from Corollary 7.4.*

3. *We prove that C_G from part 2 is an extreme value copula using (7.24) of Definition 7.28. Let $C_F(u_1, u_2, ..., u_n)$ be a copula associated with F:*

$$F(x_1, x_2, ..., x_n) = C_F\left(F_1\left(x_1\right), F_2\left(x_2\right), ..., F_n\left(x_n\right)\right).$$

As noted above:

$$F_N^{(m)}(x_1, ..., x_n) = F^m\left(a_1^{(m)}x_1 + b_1^{(m)}, ..., a_n^{(m)}x_n + b_n^{(m)}\right),$$

and so:

$$F_N^{(m)}(x_1, x_2, ..., x_n) = C_F^m\left(F_1\left(a_1^{(m)}x_1 + b_1^{(m)}\right), ..., F_n\left(a_n^{(m)}x_n + b_n^{(m)}\right)\right). \tag{1}$$

Since $\left\{F_j^m\left(a_j^{(m)}x_j + b_j^{(m)}\right)\right\}_{j=1}^n$ are the marginal distributions for $F_N^{(m)}$ as noted above, it follows from this last identity that a copula $C^{(m)}$ associated with $F_N^{(m)}$ is given by:

$$C^{(m)}(u_1, u_2, ..., u_n) = C_F^m(u_1^{1/m}, u_2^{1/m}, ..., u_n^{1/m}).$$

As G is continuous, weak converge $F_N^{(m)} \Rightarrow G$ obtains that $F_N^{(m)}(x_1, x_2, ..., x_n) \to G(x_1, x_2, ..., x_n)$ for all $(x_1, x_2, ..., x_n)$, and thus by (9.54) and (1):

$$C_G\left(G_1\left(x_1\right), ..., G_n\left(x_n\right)\right) = \lim_{m \to \infty} C_F^m\left(F_1\left(a_1^{(m)}x_1 + b_1^{(m)}\right), ..., F_n\left(a_n^{(m)}x_n + b_n^{(m)}\right)\right). \tag{2}$$

Let $(G_1\left(x_1\right), G_2\left(x_2\right), ..., G_n\left(x_n\right)) \in \prod_{j=1}^n Rng\left(G_j\right)$, the range of this vector transformation on $\mathbb{R}^n$, and note for completeness that depending on the γ_j-parameters:

$$(0, 1)^n \subset \prod_{j=1}^n Rng\left(G_j\right) \subset [0, 1]^n.$$

If it can be demonstrated that for all such points:

$$C_G\left(G_1\left(x_1\right),...,G_n\left(x_n\right)\right) = \lim_{m\to\infty} C_F^m\left(G_1^{1/m}\left(x_1\right),...,G_n^{1/m}\left(x_n\right)\right), \qquad (3)$$

then C_G is an extreme value copula and C_F is in the domain of attraction of C_G by Definition 7.28.

It is enough by (2) to show that as $m \to \infty$:

$$\left|C^{(m)}\left(G_1\left(x_1\right),..,G_n\left(x_n\right)\right) - C^{(m)}\left(F_1^m\left(a_1^{(m)}x_1 + b_1^{(m)}\right),..,F_n^m\left(a_n^{(m)}x_n + b_n^{(m)}\right)\right)\right| \to 0.$$

But from the Lipschitz continuity of copulas in (7.9):

$$\left|C^{(m)}\left(G_1\left(x_1\right),...,G_n\left(x_n\right)\right) - C^{(m)}\left(F_1^m\left(a_1^{(m)}x_1 + b_1^{(m)}\right),...,F_n^m\left(a_n^{(m)}x_n + b_n^{(m)}\right)\right)\right|$$

$$\leq \sum_{j=1}^n \left|G_j\left(x_j\right) - F_j^m\left(a_j^{(m)}x_j + b_j^{(m)}\right)\right|.$$

This final summation converges to 0 for all such $(x_1, x_2, ..., x_n)$ by the Fisher-Tippett-Gnedenko theorem, and the proof of (3) is complete.

∎

Corollary 9.53 (Multivariate FTG theorem; $A_j = 1$, $B_j = 0$ all j) *If sequences $\left\{\left(a_1^{(m)}, a_2^{(m)}, ..., a_n^{(m)}\right)\right\}_{m=1}^{\infty}$ with $a_j^{(m)} > 0$ for all j, m and $\left\{\left(b_1^{(m)}, b_2^{(m)}, ..., b_n^{(m)}\right)\right\}_{m=1}^{\infty}$ exist to produce (9.53), then there exist sequences so that $A_j = 1$ and $B_j = 0$ for all j. That is:*

$$G_j\left(x_j\right) = G_{\gamma_j}\left(x\right).$$

Proof. *This follows from Corollary 9.31.* ∎

9.6.2   The Extreme Value Distribution G

With the aid of the multivariate Fisher-Tippett-Gnedenko theorem, it is possible to derive additional details on the form of multivariate extreme value distribution G and the associated extreme value copula C_G. We begin by reformulating the weak convergence result in a way that more directly defines G in terms of F.

Recall that given a joint distribution function $F(x_1, x_2, ..., x_n)$ with marginal distributions $\{F_j(x_j)\}_{j=1}^n$, $F \in \mathcal{D}(G)$ denotes that F is in the domain of attraction of G. By definition then, G is a distribution function with nondegenerate marginal distributions $\{G_j(x_j)\}_{j=1}^n$, and there exist vector sequences $\left\{\left(a_1^{(m)}, a_2^{(m)}, ..., a_n^{(m)}\right)\right\}_{m=1}^{\infty}$ with $a_j^{(m)} > 0$ for

all j, m and $\left\{\left(b_1^{(m)}, b_2^{(m)}, ..., b_n^{(m)}\right)\right\}_{m=1}^{\infty}$ so that (9.52) is satisfied for all continuity points of $G(x_1, x_2, ..., x_n)$. By Proposition 9.52, this means for all $(x_1, x_2, ..., x_n)$ in the domain of G.

Proposition 9.54 (G from F when $F \in \mathcal{D}(G)$) *Given a joint distribution function $F(x_1, x_2, ..., x_n)$ with marginal distributions $\{F_j(x_j)\}_{j=1}^n$, assume that $F \in \mathcal{D}(G)$. Then there exist real constants $\{\gamma_j\}_{j=1}^n$, so that for all $x_1, x_2, ..., x_n > 0$:*

$$G(h_1\left(x_1\right), ..., h_n\left(x_n\right)) = \lim_{m\to\infty} F^m\left(U_1(mx_1), ..., U_n(mx_n)\right). \qquad (9.55)$$

Here:

$$U_j(x) \equiv F_j^*(1 - 1/x),$$

$$h_j(x_j) = \begin{cases} \left(x_j^{\gamma_j} - 1\right)/\gamma_j, & \gamma_j \neq 0, \\ \ln x_j, & \gamma_j = 0. \end{cases}$$

Proof. *As in the proof of Proposition 9.52, using a corollary to the Cramér-Wold theorem,* $F_N^{(m)} \Rightarrow G$ *implies that for all* j:

$$F_j^m\left(a_j^{(m)}x_j + b_j^{(m)}\right) \Rightarrow G_j(x_j).$$

By Corollary 9.53, Corollary 9.27, and (9.33) of Corollary 9.35, for each j *the sequence* $\left\{\left(a_j^{(m)}, b_j^{(m)}\right)\right\}_{m=1}^{\infty}$ *can be replaced by functions* $\{(a_j(t), b_j(t))\}$ *with* $b_j(t) = U_j(t)$ *so that for all* $x_j > 0$ *and* $h_j(x_j)$ *defined above:*

$$\frac{U_j(tx_j) - U_j(t)}{a_j(t)} \to h_j(x_j),$$

as $t \to \infty$.

Letting $t = m$ *for integer* $m \to \infty$, *this obtains:*

$$\frac{U_j(mx_j) - b_j^{(m)}}{a_j^{(m)}} \to h_j(x_j),$$

where $a_j^{(m)} \equiv a_j(m)$ *and* $b_j^{(m)} \equiv U_j(m) = F_j^*(1 - 1/m)$. *We now need a technical result which follows from weak convergence of* $F_N^{(m)} \Rightarrow G$, *and generalizes (9.52).*

Claim: If $\left(y_1^{(m)}, y_2^{(m)}, ..., y_n^{(m)}\right) \to (x_1, x_2, ..., x_n)$ *as* $m \to \infty$, *then:*

$$G(x_1, x_2, ..., x_n) = \lim_{m \to \infty} F^m\left(a_1^{(m)}y_1^{(m)} + b_1^{(m)}, a_2^{(m)}y_2^{(m)} + b_2^{(m)}, ..., a_n^{(m)}y_n^{(m)} + b_n^{(m)}\right).$$

Once proved, applying this result to $y_j^{(m)} \equiv \left(U_j(mx_j) - b_j^{(m)}\right)/a_j^{(m)}$ *will obtain (9.55) since* $y_j^{(m)} \to h_j(x_j)$ *for* $x_j > 0$.

By assumption, given $\epsilon > 0$ *there is an* M *so that* $\left|y_j^{(m)} - x_j\right| < \epsilon$ *for all* j *and all* $m \geq M$. *Since* $a_j^{(m)} > 0$ *and* F *is increasing, it follows that for* $m \geq M$:

$$F^m\left(a_1^{(m)}x_1^{(m)} + \left(b_1^{(m)} - a_1^{(m)}\epsilon\right), ..., a_n^{(m)}x_n^{(m)} + \left(b_n^{(m)} - a_n^{(m)}\epsilon\right)\right)$$
$$\leq F^m\left(a_1^{(m)}y_1^{(m)} + b_1^{(m)}, ..., a_n^{(m)}y_n^{(m)} + b_n^{(m)}\right)$$
$$\leq F^m\left(a_1^{(m)}x_1^{(m)} + \left(b_1^{(m)} + a_1^{(m)}\epsilon\right), ..., a_n^{(m)}x_n^{(m)} + \left(b_n^{(m)} + a_n^{(m)}\epsilon\right)\right).$$

Applying (9.23) with $a_j^{(m)\prime} = a_j^{(m)}$ *and* $b_j^{(m)\prime} = b_j^{(m)} \pm a_j^{(m)}\epsilon$, *it follows that:*

$$G(x_1 - \epsilon, ..., x_n - \epsilon)$$
$$\leq \lim_{m \to \infty} F^m\left(a_1^{(m)}y_1^{(m)} + b_1^{(m)}, ..., a_n^{(m)}y_n^{(m)} + b_n^{(m)}\right)$$
$$\leq G(x_1 + \epsilon, ..., x_n + \epsilon).$$

The claim is proved by continuity of G. ∎

Remark 9.55 (Continuous $\{F_j(x_j)\}_{j=1}^n$) *From (9.50):*

$$\left(X_1^{(m)}, X_2^{(m)}, ..., X_n^{(m)}\right) \equiv \left(\max_{k\leq m} X_{1k}, \max_{k\leq m} X_{2k}, ..., \max_{k\leq m} X_{nk}\right),$$

and so (9.55) can be restated with:

$$F^m\left(U_1(mx_1), ..., U_n(mx_n)\right)$$
$$= \Pr\left[\max_{k\leq m} X_{1k} \leq F_1^*(1 - 1/mx_1), ..., \max_{k\leq m} X_{nk} \leq F_n^*(1 - 1/mx_n)\right].$$

Recall that $\Pr$ *is defined in terms of* μ, *the probability measure on the space* $(\mathcal{S}, \mathcal{E}, \mu)$ *on which all* X_{jk} *are defined. Specifically, letting* $Y_j \equiv \max_{k\leq m} X_{jk}$:

$$\Pr\left[Y_1 \leq c_1, ..., Y_n \leq c_n\right] \equiv \mu\left[\bigcap_{j=1}^n Y_j^{-1}(-\infty, c_j]\right].$$

If the marginal distributions $\{F_j(x_j)\}_{j=1}^n$ *of* F *are continuous, so* $F_j\left(F_j^*(y)\right) = y$ *by Proposition 3.22, then:*

$$F^m\left(U_1(mx_1), ..., U_n(mx_n)\right)$$
$$= \Pr\left[\max_{k\leq m} F_1\left(X_{1k}\right) \leq 1 - 1/mx_1, ..., \max_{k\leq m} F_n\left(X_{nk}\right) \leq 1 - 1/mx_n\right]$$
$$= \Pr\left[\max_{k\leq m}\left(\frac{1}{1 - F_1\left(X_{1k}\right)}\right) \leq mx_1, ..., \max_{k\leq m}\left(\frac{1}{1 - F_n\left(X_{nk}\right)}\right) \leq mx_n\right].$$

In the case of continuous marginals, the result in (9.55) can be transformed to reveal a simpler connection between G and the initial distribution function F. We use much the same approach as that seen in the analysis preceding the proof of the Fisher-Tippett-Gnedenko theorem of Proposition 9.30.

Corollary 9.56 (G from F when $F \in \mathcal{D}(G)$) *If* $x_1, x_2, ..., x_n > 0$, *and:*

$$0 < G(h_1\left(x_1\right), ..., h_n\left(x_n\right)) < 1,$$

then with $h_j\left(x_j\right)$ *defined above:*

$$-\ln G(h_1\left(x_1\right), ..., h_n\left(x_n\right)) = \lim_{m\to\infty} m\left[1 - F\left(U_1(mx_1), ..., U_n(mx_n)\right)\right]. \tag{9.56}$$

Further, this limit can be expressed as $t \to \infty$ *for real* t.
If the marginals $\{F_j(x_j)\}_{j=1}^n$ *are continuous, the limit in (9.56) can be expressed:*

$$-\ln G(h_1\left(x_1\right), ..., h_n\left(x_n\right)) \tag{9.57}$$
$$= \lim_{m\to\infty} m\Pr\left[F_j\left(X_j\right) > 1 - 1/mx_j, \text{ for at least one } j\right].$$

Proof. *Given the restriction on* $G(h_1\left(x_1\right), h_2\left(x_2\right), ..., h_n\left(x_n\right))$, *take logarithms in (9.55) to produce:*

$$\ln G(h_1\left(x_1\right), ..., h_n\left(x_n\right)) = \lim_{m\to\infty} m\ln F\left(U_1(mx_1), ..., U_n(mx_n)\right).$$

This implies that $\ln F\left(U_1(mx_1), U_2(mx_2), ..., U_n(mx_n)\right) \to 0$ *and thus:*

$$F\left(U_1(mx_1), U_2(mx_2), ..., U_n(mx_n)\right) \to 1.$$

As $\ln x = x - 1 - O(x-1)^2$ *obtains that* $-\ln x/(1-x) \to 1$ *as* $x \to 1$, *the result in (9.56) follows.*

The limit in (9.56) with integers $m \to \infty$ can be expressed as a limit of real $t \to \infty$ since $U_j(tx_j)$ is monotonically increasing in t, and if $x'_j \geq x_j$ all j, then $F(x'_1, x'_2, ..., x'_n) \geq F(x_1, x_2, ..., x_n)$. Thus if $m \leq t \leq m+1$:

$$m\left[1 - F\left(U_1((m+1)x_1), ..., U_n((m+1)x_n)\right)\right]$$
$$\leq \quad t\left[1 - F\left(U_1(tx_1), ..., U_n(tx_n)\right)\right]$$
$$\leq \quad (m+1)\left[1 - F\left(U_1(mx_1), ..., U_n(mx_n)\right)\right],$$

and the outer expressions have the same limits.

When the marginals are continuous, $F_j\left(F_j^*(y)\right) = y$ by Proposition 3.22, so using the approach of Remark 9.55:

$$1 - F\left(U_1(mx_1), U_2(mx_2), ..., U_n(mx_n)\right)$$
$$= \quad \Pr\left[X_j > F_j^*(1 - 1/mx_j), \text{ for at least one } j\right]$$
$$= \quad \Pr\left[F_j(X_j) > 1 - 1/mx_j, \text{ for at least one } j\right].$$

■

Remark 9.57 (In general, (9.57) is an upper bound) *In the case of general marginals, meaning not necessarily continuous, the probability statement for $F_j(X_j)$ in (9.57) provides an upper bound for $-\ln G(h_1(x_1), h_2(x_2), ..., h_n(x_n))$.*

In detail for the general case:

$$1 - F\left(U_1(mx_1), ..., U_n(mx_n)\right)$$
$$= \quad \Pr\left[F_j(X_j) > F_j F_j^*(1 - 1/mx_j), \text{ for at least one } j\right]$$
$$\leq \quad \Pr\left[F_j(X_j) > 1 - 1/mx_j, \text{ for at least one } j\right],$$

since $F_j F_j^*(1 - 1/mx_j) \geq 1 - 1/mx_j$ by (3.10).

9.6.3 The Extreme Value Copula C_G

We now turn to an investigation into the form of the extreme value copula C_G associated with G. The multivariate Fisher-Tippett-Gnedenko theorem proved that if the distribution function F is in the **domain of attraction** of the distribution function G, denoted $F \in \mathcal{D}(G)$, then for all $(u_1, u_2, ..., u_n) \in \prod_{j=1}^n Rng(G_j)$:

$$C_G(u_1, u_2, ..., u_n) = \lim_{m \to \infty} C_F^m\left(u_1^{1/m}, u_2^{1/m}, ..., u_n^{1/m}\right).$$

Here C_F and C_G are copulas associated with F and G, respectively. Thus C_G is an extreme value copula by (7.24) of Definition 7.28.

Proposition 9.58 (C_G from C_F when $F \in \mathcal{D}(G)$) *If $F \in \mathcal{D}(G)$, then for $y_1, y_2, ..., y_n \geq 0$:*

$$-\ln C_G(e^{-y_1}, e^{-y_2}, ..., e^{-y_n}) \tag{9.58}$$
$$= \quad \lim_{m \to \infty} m\left[1 - C_F(1 - y_1/m, ..., 1 - y_n/m)\right].$$

Further, this limit can be expressed as $t \to \infty$ for real t.
Proof. *We repeat many of the steps of Corollary 9.56. If $0 < C_G(u_1, u_2, ..., u_n) \leq 1$, then taking logarithms of the above expression for $C_G(u_1, u_2, ..., u_n)$ obtains:*

$$\ln C_G(u_1, u_2, ..., u_n) = \lim_{m \to \infty} m \ln C_F\left(u_1^{1/m}, u_2^{1/m}, ..., u_n^{1/m}\right),$$

and so $C_F\left(u_1^{1/m}, u_2^{1/m}, ..., u_n^{1/m}\right) \to 1$ *as* $m \to \infty$. *Thus* $\ln C_F$ *can be replaced by* $C_F - 1$ *in this limit.*

For $u_j \in (0,1]$, *let* $u_j = e^{-y_j}$ *with* $y_j \geq 0$ *yielding:*

$$-\ln C_G(e^{-y_1}, ..., e^{-y_n}) = \lim_{m \to \infty} m \left[1 - C_F(e^{-y_1/m}, ..., e^{-y_n/m})\right]. \tag{1}$$

By (7.9):

$$\left| C_F(e^{-y_1/m}, ..., e^{-y_n/m}) - C_F(1 - y_1/m, ..., 1 - y_n/m) \right|$$
$$\leq \sum_{j=1}^{n} \left| (e^{-y_j/m} - (1 - y_j/m)) \right|.$$

Using a Taylor series approximation for the exponentials, this summation is $O(m^{-2})$, *and hence (9.58) follows from (1).*

That the limit with integers $m \to \infty$ *in (9.58) can be expressed as a limit of real* $t \to \infty$, *note that if* $m \leq t \leq m + 1$:

$$m\left[1 - C_F(1 - y_1/(m+1), ..., 1 - y_n/(m+1))\right]$$
$$\leq t\left[1 - C_F(1 - y_1 t, ..., 1 - y_n t)\right]$$
$$\leq (m+1)\left[1 - C_F(1 - y_1/m, ..., 1 - y_n/m)\right].$$

The bounding expressions have the same limit since $m/(m+1) \to 1$, *and the result follows.* ∎

Remark 9.59 (Continuous $\{F_j(x_j)\}_{j=1}^n$**)** *The copula limit in (9.58) can be expressed more directly in terms of "tail" probabilities of* F *when the marginal distributions* $\{F_j(x_j)\}_{j=1}^n$ *are continuous, and this then yields an alternative version of (9.57).*

By continuity of marginal distributions and Proposition 7.18:

$$1 - C_F(1 - y_1/m, ..., 1 - y_n/m) = 1 - F\left(F_1^*(1 - y_1/m), ..., F_n^*(1 - y_n/m)\right)$$
$$= \Pr\left[X_j > F_j^*(1 - y_j/m), \text{ for at least one } j\right]$$
$$= \Pr\left[F_j(X_j) > 1 - y_j/m, \text{ for at least one } j\right].$$

By this same result, since $\{G_j(x_j)\}_{j=1}^n$ *are continuous and strictly increasing on* $\{x_j | 0 < x_j < 1\}$:

$$-\ln C_G(e^{-y_1}, e^{-y_2}, ..., e^{-y_n}) = -\ln G\left(G_1^{-1}(e^{-y_1}), ..., G_n^{-1}(e^{-y_n})\right).$$

Thus (9.58) can be restated:

$$-\ln G\left(G_1^{-1}(e^{-y_1}), ..., G_n^{-1}(e^{-y_n})\right) \tag{9.59}$$
$$= \lim_{m \to \infty} m \Pr\left[F_j(X_j) > 1 - y_j/m \text{ for at least one } j\right].$$

When the marginals $\{F_j(x_j)\}_{j=1}^n$ *are continuous, the result in (9.59) and that above in (9.57) are equivalent formulations. Comparing the limits on the right, these expressions are equivalent if* $x_j = 1/y_j$ *with* $y_j \neq 0$, *and thus it must be the case that:*

$$h_j(1/y_j) = G_j^{-1}(e^{-y_j}).$$

This is verified by substitution, checking the cases $\gamma_j \neq 0$ *and* $\gamma_j = 0$, *respectively.*

The tail probability statement implied by Proposition 9.58 gives rise to the following definition. This notion was introduced in a 1992 thesis of **Xin Huang** and further developed in a 1998 joint paper with **Holger Drees.**

Definition 9.60 (Stable tail dependence function) *A joint distribution function* $F(x_1, x_2, ..., x_n)$ *is said to have a **stable tail dependence function** $l_F(y_1, y_2, ..., y_n)$, if the following limit exists for all $y_1, y_2, ..., y_n \geq 0$:*

$$l_F(y_1, y_2, ..., y_n) \equiv \lim_{t \to \infty} t\left[1 - C_F(1 - y_1/t, ..., 1 - y_n/t)\right] \tag{9.60}$$
$$= \lim_{t \to \infty} t \Pr\left[F_j(X_j) > 1 - y_j/t, \text{ for at least one } j\right].$$

Proposition 9.61 ($C_G \Longleftrightarrow l_F$ if $F \in \mathcal{D}(G)$) *If a distribution function $F(x_1, x_2, ..., x_n)$ is in the domain of attraction of an extreme value distribution function $G(x_1, x_2, ..., x_n)$, that is $F \in \mathcal{D}(G)$, then F has a stable tail dependence function:*

$$l_F(y_1, y_2, ..., y_n) \equiv -\ln C_G(e^{-y_1}, e^{-y_2}, ..., e^{-y_n}). \tag{9.61}$$

Equivalently, for $(u_1, u_2, ..., u_n) \in (0, 1]^n$:

$$C_G(u_1, u_2, ..., u_n) = \exp\left[-l_F(-\ln u_1, -\ln u_2, ..., -\ln u_n)\right]. \tag{9.62}$$

Proof. *If $F \in \mathcal{D}(G)$, then the limit in (9.60) exists by Proposition 9.58, and the equivalent formulas follow from (9.58).* ∎

Remark 9.62 *We make two observations on l_F:*

1. l_F **without** $F \in D(G)$: *If $F(x_1, x_2, ..., x_n)$ has a stable tail dependence function, then the first limit in (9.60) can be restated using by (7.9) as in the proof of Proposition 9.58:*

$$\lim_{m \to \infty} m\left[1 - C_F(1 - y_1/m, 1 - y_2/m, ..., 1 - y_n/m)\right]$$
$$= \lim_{m \to \infty} m\left[1 - C_F(e^{-y_1/m}, e^{-y_2/m}, ..., e^{-y_n/m})\right].$$

This implies that $1 - C_F(e^{-y_1/m}, e^{-y_2/m}, ..., e^{-y_n/m}) \to 0$ and $(1 - C_F)/\ln C_F \to -1$, and so:

$$-l_F(y_1, y_2, ..., y_n) = \lim_{m \to \infty} \ln C_F^m(e^{-y_1/m}, e^{-y_2/m}, ..., e^{-y_n/m}),$$

and equivalently:

$$\exp\left[-l_F(y_1, y_2, ..., y_n)\right] = \lim_{m \to \infty} C_F^m(e^{-y_1/m}, e^{-y_2/m}, ..., e^{-y_n/m}).$$

Substituting $u_j = e^{-y_j}$, this proves that for $(u_1, u_2, ..., u_n) \in (0, 1]^n$ that the following limit exists:

$$\lim_{m \to \infty} C_F^m\left(u_1^{1/m}, u_2^{1/m}, ..., u_n^{1/m}\right) \equiv L(u_1, u_2, ..., u_n). \tag{1}$$

By (7.23),

$$C^m(u_1^{1/m}, u_2^{1/m}, ..., u_n^{1/m}) = C^{(m)}(u_1, u_2, ..., u_n),$$

where $C^{(m)}(u_1, u_2, ..., u_n)$ denotes a copula associated with $F^{(m)}(x_1, x_2, ..., x_n)$, the distribution function of maximal random vectors.

Thus, if $F(x_1, x_2, ..., x_n)$ has a stable tail dependence function $l_F(y_1, y_2, ..., y_n)$, then (1) holds for all $(u_1, u_2, ..., u_n) \in (0, 1]^n$. Since $l_F(y_1', y_2', ..., y_n') \leq l_F(y_1, y_2, ..., y_n)$ when $y_j \leq y_j'$ for all j, this limit can be extended to $[0, 1]^n$. The function L defined in (1) and so extended can now be seen to be a copula. In fact $L(u_1, u_2, ..., u_n)$ is an extreme value copula by Proposition 7.32 since it is max-stable:

$$L^r\left(u_1^{1/r}, u_2^{1/r}, ..., u_n^{1/r}\right) = \lim_{m \to \infty} C_F^{mr}\left(u_1^{1/mr}, u_2^{1/mr}, ..., u_n^{1/mr}\right)$$
$$= L(u_1, u_2, ..., u_n).$$

However, it is not possible to assert that $F \in \mathcal{D}(G)$ for some G without an assumption on the marginals $\{F_j(x_j)\}_{j=1}^n$ of F, and in particular, that these marginals are in the domains of attraction of extreme value distributions $\{G_j(x_j)\}_{j=1}^n$.

2. ***Reparametrizations:*** *The representation of C_G in (9.62) with uniform marginals $\{u_j\}_{j=1}^n$ locates the **right tail events** in the upper corner of the compact hypercube $[0, 1]^n$. It is sometimes convenient in applications to reparametrize this copula in terms of other marginal distributions.*

For example:

(a) ***Fréchet Marginals:*** *For $(x_1, x_2, ..., x_n) \in (0, \infty)^n$:*

$$C_G(e^{-1/x_1}, e^{-1/x_2}, ..., e^{-1/x_n}) = \exp\left[-l_F(1/x_1, 1/x_2, ..., 1/x_n)\right].$$

(b) ***Gumbel Marginals:*** *For $(x_1, x_2, ..., x_n) \in \mathbb{R}^n$:*

$$C_G\left(\exp\left(-e^{-x_1}\right), ..., \exp\left(-e^{-x_n}\right)\right) = \exp\left[-l_F\left(e^{-x_1}, ..., e^{-x_n}\right)\right].$$

(c) ***Reversed-Weibull Marginals:*** *For $(x_1, x_2, ..., x_n) \in (-\infty, 0)^n$:*

$$C_G(e^{x_1}, e^{x_2}, ..., e^{x_n}) = \exp\left[-l_F(-x_1, -x_2, ..., -x_n)\right].$$

Example 9.63 (Stable tail dependence functions) *Recall the examples of extreme value copulas from Example 7.34. Applying (9.61) obtains the following stable tail dependence functions. We omit the Pickands characterization of the prior example, and investigate that below.*

1. ***Independence Copula:*** *If $C_G = C^I$:*

$$C^I(u_1, u_2, ..., u_n) = u_1 u_2 ... u_n,$$

and $F \in \mathcal{D}(I)$, then:

$$l_F(y_1, y_2, ..., y_n) = \sum_{j=1}^n y_j.$$

2. ***Comonotonicity Copula:*** *If $C_G = C^{Co}$:*

$$C^{Co}(u_1, u_2, ..., u_n) = \min_j u_j,$$

and $F \in \mathcal{D}(Co)$, then:

$$l_F(y_1, y_2, ..., y_n) = \max y_j.$$

3. **Gumbel-Hougaard Copula:** *If $C_G = C^G$:*

$$C^G(u_1, u_2, ..., u_n) = \exp\left[-\left(\sum_{j=1}^{n}(-\ln u_j)^\theta\right)^{1/\theta}\right], \quad \theta \geq 1,$$

and $F \in \mathcal{D}(G)$, then:

$$l_F(y_1, y_2, ..., y_n) = \left(\sum_{j=1}^{n} y_j^\theta\right)^{1/\theta}.$$

We next investigate a result that motivates the Pickands characterization introduced in (7.26) in Example 7.34.

Proposition 9.64 (Properties of $l_F(y_1, y_2, ..., y_n)$) *If a joint distribution function $F(x_1, x_2, ..., x_n)$ has a stable tail dependence function $l_F(y_1, y_2, ..., y_n)$, then:*

1. **Value on Basis Vectors of $\mathbb{R}^n$:** *Define $e_k = (y_1, y_2, ..., y_n)$ with $y_k = 1$, $y_j = 0$ for $j \neq k$, then:*
$$l_F(e_k) = 1.$$

2. **Bounds:**
$$\max_j y_j \leq l_F(y_1, y_2, ..., y_n) \leq \sum_{j=1}^{n} y_j.$$

3. **Homogeneity of degree 1:** *For real $\lambda > 0$:*
$$l_F(\lambda y_1, \lambda y_2, ..., \lambda y_n) = \lambda l_F(y_1, y_2, ..., y_n).$$

Further, if $F \in D(G)$ then:

4. **Convexity:** *Given $y = (y_1, y_2, ..., y_n)$ and $x = (x_1, x_2, ..., x_n)$, where $x_j, y_j \geq 0$ for all j, then for all $0 \leq t \leq 1$:*
$$l_F((1-t)x + ty) \leq (1-t)l_F(x) + tl_F(y).$$

Proof.

1. *The values of l on the basis vectors in $\mathbb{R}^n$ follow immediately from the fact that copulas have uniform marginals. For example, since $C_F(1 - 1/t, 1, ..., 1) = 1 - 1/t$, it follows from (9.60) that:*
$$l_F(e_1) \equiv \lim_{t \to \infty} t[1 - C_F(1 - 1/t, 1, ..., 1)] = 1.$$

2. *Using the Fréchet-Hoeffding bounds of Proposition 7.12 obtains:*
$$1 - \min_j(1 - y_j/t) \leq 1 - C_F(1 - y_1/t, ..., 1 - y_n/t)$$
$$\leq 1 - \max\left\{0, \sum_{j=1}^{n}(1 - y_j/t) - (n-1)\right\},$$

which simplifies to:
$$\max_j y_j/t \leq 1 - C_F(1 - y_1/t, ..., 1 - y_n/t) \leq \sum_{j=1}^{n} y_j/t.$$

The result now follows from (9.60).

3. *The substitution $t = \lambda s$ produces:*

$$l_F\left(\lambda y_1, \lambda y_2, ..., \lambda y_n\right) = \lim_{t \to \infty} t\left[1 - C_F(1 - \lambda y_1/t, ..., 1 - \lambda y_n/t)\right]$$
$$= \lim_{t \to \infty} \lambda s\left[1 - C_F(1 - y_1/s, ..., 1 - y_n/s)\right].$$

4. *For convexity assume $0 < t < 1$, since the result for $t = 0, 1$ needs no comment. Simplifying notation, it follows from (9.61) that the convexity inequality is equivalent to:*

$$C_G^{1-t}(e^{-x})C_G^t(e^{-y}) \leq C_G\left(e^{-[(1-t)x+ty]}\right),$$

where for notational simplicity $e^{-y} \equiv (e^{-y_1}, e^{-y_2}, ..., e^{-y_n})$ with $y_j \geq 0$, and similarly for the other expressions. Substituting $u = e^{-x}$, $v = e^{-y}$, again defined componentwise so $u, v \in (0, 1]^n$, convexity is equivalent to:

$$C_G^{1-t}(u)C_G^t(v) \leq C_G\left(u^{1-t}v^t\right).$$

Now substitute $a = u^{1-t}$, $b = v^t$, again defined componentwise so $a, b \in (0, 1]^n$, and this is equivalent to:

$$C_G^{1-t}(a^{1/(1-t)})C_G^t(b^{1/t}) \leq C_G\left(ab\right).$$

Since extreme value copulas are max-stable by Proposition 7.32, it follows from (7.25) that this is equivalent to:

$$C_G(a)C_G(b) \leq C_G\left(ab\right). \tag{1}$$

We now prove this inequality for $a, b \in (0, 1]^n$ by contradiction. Assume that for some such a, b that $C_G(a)C_G(b) > C_G\left(ab\right)$. Another application of (7.25) then yields

$$C_G^r(a^{1/r})C_G^r(b^{1/r}) > C_G^r\left([ab]^{1/r}\right),$$

and taking rth roots it follows that for all $r > 0$:

$$C_G(a^{1/r})C_G(b^{1/r}) > C_G\left([ab]^{1/r}\right).$$

Letting $r \to \infty$ this produces the contradiction $1 > 1$, since $c^{1/r} \to 1$ for all $c > 0$ and thus all $C_G(c^{1/r}) \to 1$.

Consequently (1) is proved for all $a, b \in (0, 1]^n$. Retracing substitutions, this proves convexity for all x, y with nonnegative components.

∎

Remark 9.65 ($l\left(y_1, y_2, ..., y_n\right) \nRightarrow$ **EV copula**) *If $l\left(y_1, y_2, ..., y_n\right)$ is a function that satisfies properties 1-4 of Proposition 9.64, and one defines a function C as in (9.62),*

$$C(u_1, u_2, ..., u_n) = \exp\left[-l\left(-\ln u_1, -\ln u_2, ..., -\ln u_n\right)\right],$$

it is natural to wonder if C will be an extreme value copula.

Certainly $C \geq 0$, and the marginal distributions of C are uniform. Also, by homogeneity, this function will always satisfy the criterion in (7.25) to be max-stable:

$$C^r\left(u_1^{1/r}, u_2^{1/r}, ..., u_n^{1/r}\right) = C(u_1, u_2, ..., u_n).$$

But there is one outstanding question.

Is C so defined necessarily a copula?

*It is known that for $n > 2$, C need not be a copula. See for example **Beirlant, J., Goegebeur, Y., Segers, J., and Teugels, J. (2004).***

The following corollary provides bounds for extreme value copulas which refine the lower **Fréchet-Hoeffding bound** in (7.8), and also derives the **Pickands characterization** noted in (7.26) and named for a 1981 result of **James Pickands.**

Corollary 9.66 (C^{EV} **bounds and the Pickands characterization**) *Let* C^{EV} *be an extreme value copula. Then:*

1. **Bounds:** *For all* $(u_1, u_2, ..., u_n) \in [0,1]^n$:

$$\prod_{j=1}^n u_j \leq C^{EV}(u_1, u_2, ..., u_n) \leq \min_j u_j. \tag{9.63}$$

Thus, extreme value copulas are bounded by the independence copula and comonotonicity copula, both of which are extreme value copulas by Example 7.34.

2. **Pickands characterization:** *As in (7.26), for* $(u_1, u_2, ..., u_n) \in [0,1]^n$ *and* $(u_1, u_2, ..., u_n) \neq (1, 1, ..., 1)$:

$$C^{EV}(u_1, u_2, ..., u_n)$$
$$= \exp\left[\left(\sum_{j=1}^n \ln u_j\right) A\left(\frac{\ln u_1}{\sum_{j=1}^n \ln u_j}, \frac{\ln u_2}{\sum_{j=1}^n \ln u_j}, \cdots, \frac{\ln u_n}{\sum_{j=1}^n \ln u_j}\right)\right].$$

*The **Pickands dependence function** $A(x_1, x_2, ..., x_n)$ is defined on $\left\{\sum_{j=1}^n x_j = 1\right\} \subset [0,1]^n$, is convex, satisfies $A(x_1, x_2, ..., x_n) = 1$ on the n basis vectors, and:*

$$1/n \leq \max_j x_j \leq A(x_1, x_2, ..., x_n) \leq 1.$$

Proof. *Since C^{EV} is an extreme value copula, it is max-stable and hence in its own domain of attraction by Proposition 7.32, so by (9.58):*

$$-\ln C^{EV}(e^{-y_1}, ..., e^{-y_n}) = \lim_{m \to \infty} m\left[1 - C^{EV}(1 - y_1/m, ..., 1 - y_n/m)\right].$$

Thus the stable tail dependence function $l(y_1, y_2, ..., y_n)$ exists by Definition 9.60, and:

$$l(y_1, y_2, ..., y_n) = -\ln C^{EV}(e^{-y_1}, ..., e^{-y_n}).$$

The bounds for C^{EV} are now a direct consequence of the bounds for $l(y_1, y_2, ..., y_n)$:

$$\max_j(-\ln y_j) \leq -\ln C^{EV}(u_1, u_2, ..., u_n) \leq -\sum_{j=1}^n \ln y_j.$$

Noting that $\max_j(-\ln y_j) = -\ln[\min_j y_j]$, the result follows by exponentiation.

Similarly, the representation noted above is a restatement of (9.62). By homogeneity of $l(y_1, y_2, ..., y_n)$ from Proposition 9.64:

$$C^{EV}(u_1, u_2, ..., u_n)$$
$$= \exp[-l(-\ln u_1, -\ln u_2, ..., -\ln u_n)]$$
$$= \exp\left[\left(\sum_{j=1}^n \ln u_j\right) l\left(\frac{\ln u_1}{\sum_{j=1}^n \ln u_j}, \frac{\ln u_2}{\sum_{j=1}^n \ln u_j}, \cdots, \frac{\ln u_n}{\sum_{j=1}^n \ln u_j}\right)\right].$$

The properties of $A(x_1, x_2, ..., x_n) = l(x_1, x_2, ..., x_n)$ follow from Proposition 9.64, other than the lower bound of $1/n$, which follows from the definition of the domain of A. ∎

Remark 9.67 (On the Pickands dependence function) *Based on an earlier 1977 characterization by **Laurens de Haan** and **Sidney Resnick**, **James Pickands** derived the above result in 1981 and also derived the following characterization of $l(y_1, y_2, ..., y_n)$ for $(y_1, y_2, ..., y_n) \in [0, \infty)^n$:*

$$l(y_1, y_2, ..., y_n) = \int_{\Delta_{n-1}} \max_j(y_j w_j) dH(w_1, w_2, ..., w_n), \qquad (9.64)$$

*where the simplex $\Delta_{n-1} \equiv \left\{ \sum_{j=1}^n w_j = 1 \right\} \subset [0,1]^n$, and H is a **Borel measure** on Δ_{n-1}, called the **spectral measure**, with the constraints that for all j:*

$$\int_{\Delta_{n-1}} w_j dH(w_1, w_2, ..., w_n) = 1.$$

From this characterization, the homogeneity and convexity of l can be confirmed, though the formal definition of this **Lebesgue-Stieltjes integral** requires the more advanced tools of Book V. It then follows that the **Pickands dependence function** $A(u_1, u_2, ..., u_n)$ can be defined as $l(y_1, y_2, ..., y_n)$ restricted to Δ_{n-1}.

When $n = 2$, the above representation of A characterizes all extreme value copulas. As noted in Example 7.35, if $A_0(t) \equiv A(1-t, t)$ is convex on $[0, 1]$ and satisfies the bounds:

$$\max(t, 1-t) \leq A_0(t) \leq 1,$$

then $C^{EV}(u, v)$ defined above is always an extreme value copula. In addition, $C^{EV}(u, v)$ can then be equivalently expressed for $(u, v) \in (0, 1]^2 - \{(1, 1)\}$ by:

$$C^{EV}(u, v) = \exp\left[(\ln uv) A_0\left(\frac{\ln v}{\ln(uv)} \right) \right] = (uv)^{A_0[\ln v / \ln(uv)]}.$$

When $n > 2$, convexity of A and satisfaction of the above bounds on A do not assure that $C^{EV}(u_1, u_2, ..., u_n)$ defined above is an extreme value copula.

References

I have listed below a number of textbook references for the mathematics and finance presented in this series of books. Many provide both theoretical and applied materials in their respective areas that are beyond those developed here, and would be worth pursuing by those interested in gaining a greater depth or breadth of knowledge in the given subjects. This list is by no means complete and is intended only as a guide to further study. In addition, a limited number of research papers will be identified in each book if they are referenced therein. A more complete guide to published papers can be found in the references below.

The reader will no doubt observe that the mathematics references are somewhat older than the finance references and upon web searching will find that several of the older texts in each category have been updated to newer editions, sometimes with additional authors. Since I own and use the editions below, I decided to present these editions rather than reference the newer editions which I have not reviewed. As many of these older texts are considered "classics," they are also likely to be found in university and other libraries.

That said, there are undoubtedly many very good new texts by both new and established authors with similar titles that are also worth investigating. One that I will, at the risk of immodesty, recommend for more introductory materials on mathematics, probability theory, and finance is the first entry in this list:

1. Reitano, Robert, R. *Introduction to Quantitative Finance: A Math Tool Kit.* Cambridge, MA: The MIT Press, 2010.

Topology, Measure, Integration, Linear Algebra

2. Doob, J. L. *Measure Theory.* New York, NY: Springer-Verlag, 1994.

3. Dugundji, James. *Topology.* Boston, MA: Allyn and Bacon, 1970.

4. Edwards, Jr., C. H. *Advanced Calculus of Several Variables.* New York, NY: Academic Press, 1973.

5. Gemignani, M. C. *Elementary Topology.* Reading, MA: Addison-Wesley Publishing, 1967.

6. Halmos, Paul R. *Measure Theory.* New York, NY: D. Van Nostrand, 1950.

7. Hewitt, Edwin, and Karl Stromberg. *Real and Abstract Analysis.* New York, NY: Springer-Verlag, 1965.

8. Royden, H. L. *Real Analysis,* 2nd Edition. New York, NY: The MacMillan Company, 1971.

9. Rudin, Walter. *Principals of Mathematical Analysis,* 3rd Edition. New York, NY: McGraw-Hill, 1976.

10. Rudin, Walter. *Real and Complex Analysis,* 2nd Edition. New York, NY: McGraw-Hill, 1974.

11. Shilov, G. E., and B. L. Gurevich. *Integral, Measure & Derivative: A Unified Approach.* New York, NY: Dover Publications, 1977.

12. Strang, Gilbert. *Introduction to Linear Algebra*, 4th Edition. Wellesley, MA: Cambridge Press, 2009.

Probability Theory & Stochastic Processes

13. Billingsley, Patrick. *Probability and Measure*, 3rd Edition. New York, NY: John Wiley & Sons, 1995.

14. Chung, K. L., and R. J. Williams. *Introduction to Stochastic Integration*. Boston, MA: Birkhäuser, 1983.

15. Davidson, James. *Stochastic Limit Theory*. New York, NY: Oxford University Press, 1997.

16. de Haan, Laurens, and Ana Ferreira. *Extreme Value Theory, An Introduction*. New York, NY: Springer Science, 2006.

17. Durrett, Richard. *Probability: Theory and Examples*, 2nd Edition. Belmont, CA: Wadsworth Publishing, 1996.

18. Durrett, Richard. *Stochastic Calculus, A Practical Introduction*. Boca Raton, FL: CRC Press, 1996.

19. Feller, William. *An Introduction to Probability Theory and Its Applications*, Volume I. New York, NY: John Wiley & Sons, 1968.

20. Feller, William. *An Introduction to Probability Theory and Its Applications*, Volume II, 2nd Edition. New York, NY: John Wiley & Sons, 1971.

21. Friedman, Avner. *Stochastic Differential Equations and Applications, Volume 1 and 2*. New York, NY: Academic Press, 1975.

22. Ikeda, Nobuyuki, and Shinzo Watanabe. *Stochastic Differential Equations and Diffusion Processes*. Tokyo, Japan: Kodansha Scientific, 1981.

23. Karatzas, Ioannis, and Steven E. Shreve. *Brownian Motion and Stochastic Calculus*. New York, NY: Springer-Verlag, 1988.

24. Kloeden, Peter E., and Eckhard Platen. *Numerical Solution of Stochastic Differential Equations*. New York, NY: Springer-Verlag, 1992.

25. Lowther, George, *Almost Sure, A Maths Blog on Stochastic Calculus*, https://almostsure.wordpress.com/stochastic-calculus/

26. Lukacs, Eugene. *Characteristic Functions*. New York, NY: Hafner Publishing, 1960.

27. Nelson, Roger B. *An Introduction to Copulas*, 2nd Edition. New York, NY: Springer Science, 2006.

28. Øksendal, Bernt. *Stochastic Differential Equations, An Introduction with Applications*, 5th Edition. New York, NY: Springer-Verlag, 1998.

29. Protter, Phillip. *Stochastic Integration and Differential Equations, A New Approach*. New York, NY: Springer-Verlag, 1992.

30. Revuz, Daniel, and Marc Yor. *Continuous Martingales and Brownian Motion*, 3rd Edition. New York, NY: Springer-Verlag, 1991.

31. Rogers, L. C. G., and D. Williams. *Diffusions, Markov Processes and Martingales*, Volume 1, Foundations, 2nd Edition. Cambridge, UK: Cambridge University Press, 2000.

32. Rogers, L. C. G., and D. Williams. *Diffusions, Markov Processes and Martingales*, Volume 2, Itô Calculus, 2nd Edition. Cambridge, UK: Cambridge University Press, 2000.

33. Sato, Ken-Iti. *Lévy Processes and Infinitely Divisible Distributions*. Cambridge University Press, Cambridge, UK, 1999.

34. Schilling, René L. and Lothar Partzsch. *Brownian Motion: An Introduction to Stochastic Processes,* 2nd Edition. Berlin/Boston: Walter de Gruyter GmbH, 2014.

35. Schuss, Zeev, *Theory and Applications of Stochastic Differential Equations.* New York, NY: John Wiley and Sons, 1980.

Finance Applications

36. Etheridge, Alison. *A Course in Financial Calculus.* Cambridge, UK: Cambridge University Press, 2002.

37. Embrechts, Paul, Claudia Klüppelberg, and Thomas Mikosch. *Modelling Extremal Events for Insurance and Finance.* New York, NY: Springer-Verlag, 1997.

38. Hunt, P. J., and J. E. Kennedy. *Financial Derivatives in Theory and Practice,* Revised Edition. Chichester, UK: John Wiley & Sons, 2004.

39. McLeish, Don L. *Monte Carlo Simulation and Finance.* New York, NY: John Wiley, 2005.

40. McNeil, Alexander J., Rüdiger Frey, and Paul Embrechts. *Quantitative Risk Management: Concepts, Techniques, and Tools.* Princeton, NJ.: Princeton University Press, 2005.

Research Papers for Book II

41. Balkema, A., and de Haan, L. Residual life time at great age, *Annals of Probability*, 2, 792–804 (1974).

42. Beirlant, J., Goegebeur, Y., Segers, J., Teugels, J.: *Statistics of Extremes: Theory and Applications.* Wiley Series in Probability and Statistics. John Wiley & Sons Ltd., Chichester (2004).

43. Clayton, D.G. A model for association in bivariate life tables and its application in epidemiological studies of familial dependency in chronic disease incidence. *Biometrika* 65, 141–151 (1978).

44. de Haan, L. and S. Resnick. Limit theory for multivariate sample extremes. Z. Wahrscheinlichkeitstheorie verw. *Gebiete*, 40:317–337 (1977).

45. Drees, H. & Huang, X. Best attainable rates of convergence for estimates of the stable tail dependence function, *J. Mult. Anal.*, 64, 25–47 (1998).

46. Fisher, R.A. and Tippett, L.H.C. Limiting forms of the frequency distribution of the largest and smallest member of a sample, *Proc. Camb. Phil. Soc.*, 24, 180-190 (1928).

47. Frank, M.J. On the simultaneous associativity of $F(x, y)$ and $x + y - F(x, y)$. *Aequationes Math.* 19 (2-3), 194–226 (1979).

48. Genest, C. and L.P. Rivest. A characterization of Gumbel's family of extreme value distributions. *Statist. Probab. Lett.*, 8:207–211 (1989).

49. Gnedenko, B. V. Sur la distribution limite du terme maximum d'une série aléatoire. *Ann. Math.*, 44, 423-453 (1943).

50. Gumbel, E.J. Distributions des valeurs extrêmes en plusieurs dimensions. *Publ. Inst. Statist. Univ. Paris* 9, 171–173 (1960).

51. Hougaard, P. A class of multivariate failure time distributions. *Biometrika* 73 (3), 671–678 (1986)

52. Huang, X. Statistics of bivariate extreme values, Ph. D. thesis, Tinbergen Institute Research Series (1992).

53. Kimberling, C. H. A probabilistic interpretation of complete monotonicity. *Aequationes Mathematicae*, 10, 152–164 (1974).

54. McNeil, Alexander J. and Johanna Nešlehová: Multivariate Archimedean copulas, d-monotone functions and l_1-norm symmetric distributions. *Ann. Statist.* 37 (5B), 3059–3097, (2009).

55. Pickands, J. Statistical inference using extreme order statistics, *Annals of Statistics*, 3, 119–131 (1975).

56. Pickands, J. Multivariate extreme value distributions. In: *Proceedings of the 43rd Session of the International Statistical Institute*, Vol. 2 (Buenos Aires, 1981), vol. 49, pp. 859–878, 894–902 (1981).

57. Rüschendorf, L. On the distributional transform, Sklar's Theorem, and the empirical copula process. *Journal of Statistical Planning and Inference* 139, 3921-3927 (2009).

58. Sibuya, M. Bivariate extreme statistics, *Ann. Inst. Statist. Math.* 11 195-210 (1960).

59. Sklar, A. Fonctions de répartition à n dimensions et leurs marges. *Publications de l'Institut Statistique de l'Université de Paris* 8, 229-231 (1959).

60. Von Mises, R. La distribution de la plus grande de n valeurs. [Reprinted (1954) in Selected Papers II 271-294]. *Amer. Math. Soc.*, Providence, RI (1936).

Index

stable tail dependence function, 243

Archemedian
    copulas, 150
    generator function, 150

Bayes, Thomas
    Bayes' theorem, 32
Bayesian statistics, 32
Bernoulli, Jacob
    Ars Conjectandi, 2
    Bernoulli's theorem, 100
Bernstein, Sergei Natanovich
    Bernstein's inequality, 102
    on monotone functions, 151
big-O, 196
binomial coefficient, 14
    binomial theorem, 14
binomial probability measures
    Bernoulli distribution, 13
Borel sigma algebra, 63
Borel, Émile
    Borel's theorem (strong law), 103
    Borel-Cantelli Theorem, 38
Borel-Cantelli lemma, 38
boundary
    of a set, 183

càdlàg (French)
    "continu à droite, limite à gauche", 126
Cantelli, Francesco
    Borel-Cantelli Theorem, 38
Carathéodory measurable
    with respect to $\mu_{\mathcal{A}}^{*}$, 127, 132
Carathéodory, Constantin
    Carathéodory Extension theorem, 5
Cardano, Gerolamo
    Liber de ludo aleae, 1
Cauchy, Augustin-Louis, 2
characteristic function
    of a set, 121, 185
    of a set A, 17
Clayton, David George

Clayton copula, 153
complement
    of a set, 3
complete
    metric space, 188
complete
    probability space, 3
conditional distribution function
    of a random vector, 67
conditional probability measure, 28
continuous almost everywhere (a.e.), 20
continuous from above, 131
continuous function
    right continuous, 126
continuous mapping theorem, 114
convergence almost everywhere
    with probability one, 107
convergence in probability, 103
convex hull
    of a set, 56
copula, 140
    empirical, 144
    parametric class, 145
    survival, 174
copulas
    Archimedian, 150, 169
    Clayton, 153, 170
    comonotonicity, 149, 157, 168, 244
    countermonotonicity, 149, 168
    explicit, implicit, 149
    extreme value, 154
    Frank, 154, 170
    Gaussian, 149, 168
    Gumbel, 153, 170
    Gumbel-Hougaard, 153, 157, 245
    independence, 149, 157, 168, 244
    max-stable, 156
    Pickands characterization, 157, 247
    Student T, 150, 169
cumulative distribution function (c.d.f.)
    of a random variable, 47
    of a random vector, 62, 63, 131
cylinder set

infinite dimensional product space, 80, 88, 93

de Moivre, Abraham
  De Moivre-Laplace Theorem, 198
  The Doctrine of Chances, 2
De Moivre-Laplace Theorem
  Central Limit Theorem, 198
degenerate
  distribution function, 217
degenerate random variable, 92
Delta method, 195
dense set, 139
density function, 68
discontinuity set, 192
discrete
  probability theory, 91
  random variable, 91
distribution function
  cumulative distribution function, 11, 16
distribution function (d.f.)
  conditional distribution function, 67
  left continuous inverse, 53
  marginal distribution function, 65
  of a random variable, 47, 127, 129
  of a random vector, 62, 63, 131
distributional transform, 160
Distributions
  beta, 208
  binomial, 13
  continuous uniform, 17
  discrete uniform, 13
  exponential, 18, 208
  Fréchet, 208
  generalized extreme value, 209
  generalized extreme value (GEV), 217
  generalized Pareto (GPD), 226
  Gumbel; double exponential, 208
  lognormal, 19
  multivariate extreme value, 236
  normal, 18
  Pareto, 208, 227
  Poisson, 14
  reverse-Weibull, 208
domain of attraction
  extreme value copula, 156
  extreme value theory, 217

endpoint continuity, 213
equivalence
  of probability spaces, 89

essential subset
  of a probability space, 89
event
  disjoint events, 3
  mutually exclusive events, 3
  null event, 3
exceedances
  over threshold, 226
explicit copulas, 149
exponential probability measure, 18
extreme value
  index, 217
extreme value
  distribution, 207, 209, 217, 236
extreme value index, 209

Fisher, Ronald
  Fisher-Tippett-Gnedenko theorem, 212
Fisher-Tippett theorem
  extreme value theory, 212
Fréchet, Maurice
  Fréchet class, 137, 140
  Fréchet distribution, 208
  Fréchet-Hoeffding bounds, 137
Frank, Maurice J.
  Frank copula, 154
Fubini, Guido
  Fubini's theorem, 18

Gauss, Carl Friedrich
  Gaussian probability measure, 18
generalized extreme value (GEV)
    distribution, 209, 217
generalized Pareto distribution (GPD), 226
Gnedenko, Boris
  Fisher-Tippett-Gnedenko theorem, 212
Gosset, William Sealy
  Student's T distribution, 150
greatest integer function, 184, 216
Gumbel, Emil Julius
  Gumbel copula, 153
  Gumbel distribution, 208

Hahn, Hans
  Hahn-Kolmogorov theorem, 5
Helly, Eduard
  Helly selection theorem, 185
Hoeffding, Wassily
  Fréchet-Hoeffding bounds, 137

i.i.d.

independent, identically dostributed,
    202
i.i.d.-X
    independent, identically distributed, 77
implicit copulas, 149
inclusion-exclusion formula, 132
independent
    events, 20
    random variables, 71
    random vectors, 71
independent events, 6
    in a probability space, 20
    pairwise, 20
indicator function
    of a set, 121
induced measure, 28
inverse
    of a function, 58

Karamata, Jovan
    Karamata's Representation theorem,
        228
Kolmogorov, Andrey
    *Foundations of the Theory of
        Probability*, 2
    Hahn-Kolmogorov theorem, 5
    Kolmogorov's zero-one law, 43, 118

L'Hôpital's rule, 153
l'Hôpital, Guillaume de
    l'Hôpital's rule, 153
Laplace, Pierre-Simon
    Bayes' theorem, 32
    De Moivre-Laplace Theorem, 198
    Laplace transform, 152
    Théorie analytique des probabilités, 2
large deviation theory, 202
Law of total probability, 29
Lebesgue, Henri
    Lebesgue integral, 2, 51
left inverse
    of a function, 58
left limit function, 161
left-continuous inverse
    of a distribution funciton, 49, 53, 147
limit
    left, right, 54
limit inferior
    of sets, 35
limit superior
    of sets, 35

Lipschitz, Rudolf
    Lipschitz continuity, 143
lognormal probability measure, 19
loss given default (LGD), 9, 51
loss model
    individual, 52

Mann, Henry
    Mann-Wald theorem, 194
mapping theorem, 114
mapping theorem on $\mathbb{R}$
    induced measures, 192
marginal distribution function
    of a random vector, 65
marginal survival function
    of a random vector, 174
measurable (Carathéodory)
    with respect to $\mu_{\mathcal{A}}^*$, 127, 132
measurable function sequence
    convergence almost everywhere, 110
    convergence in measure, 109
measure
    induced by a transformation, 192
    induced measures, 28
multivariate extreme value (MEV)
    distribution, 236

n-increasing, 131
normal probability measure, 18
normalize
    r.v. sequence, 207
normalized random variable, 198
normalizing sequence, 195, 207
normalizing sequences
    extreme value theory, 210

open ball
    in $\mathbb{R}^n$, 63
open rectangle
    in $\mathbb{R}^n$, 63
outer measure, 127

Pareto distribution, 227
Pareto, Vilfredo
    Pareto distribution, 208
peaks
    over threshold, 226
Pickands characterization, 247
Poisson, Siméon-Denis
    Poisson Limit theorem, 15
    Poisson probability measure, 14
polar coordinates, 19

probability function
    probability density function, 11, 16
probability measure
    conditional, 28
    on a general sample space, 2
probability space, 2
    complete, 3
    discrete, 92
    general model, 2
    on ℝ induced by X, 79, 88, 92, 99
product space
    infinite dimensional, 80, 88, 93
projection mapping, 80
Prokhorov, Yuri Vasilyevich
    Prokhorov's theorem, 188

quantile
    lower α-quantile, 163

Rüschendorf, Ludger
    distributional transform, 163
random sample, 77, 83, 154
random variable
    degenerate, 92
    discrete, 91
random variable (r.v.), 47
    independent, 71
random variable sequence
    convergence in distribution, 111, 181
    convergence in probability, 109
    convergence with probability 1, 110
random vector, 62, 63, 74, 131, 174
random vectors
    independent, 71
rank invariance
    of copulas, 143
recovery rate, 9, 51
rectangular distribution
    discrete , 17
rectangular probability measure
    discrete , 13
Riemann, Bernhard, 2
right inverse
    of a function, 58

sample
    of a random variable, 77
sample space, 2
    n-trial sample space, 6
separable
    metric space, 188

Sibuya, Masaaki
    tail independence, 168, 234
sigma algebra
    finer; coarser, 4
    generated by a random variable, 70
    generated by a random vector, 70
    in a sample space, 2
simulations
    stochastic, 78
Sklar, Abe
    Sklar's theorem, 140, 175
Skorokhod, A. V. (Anatoliy
            Volodymyrovych)
    Skorokhod's representation theorem,
            191
Slutsky , Evgeny "Eugen"
    Slutsky's theorem, 115
strong law
    of large numbers, 103
Student's T distribution, 150
survival function (s.f.)
    of a random variable, 174
    of a random vector, 174

tail dependence measure
    lower, upper, 165, 170
tail event
    of a distribution function, 201
    tail sigma algebra, 43, 118, 119
tight
    sequence of distribution functions, 187
    sequence of probability measures, 187
Tippett, L. H. C.
    Fisher-Tippett-Gnedenko theorem, 212
Tonelli, Leonida
    Tonelli's theorem, 18

uniform distribution
    discrete , 13
uniform probability measure
    continuous, 17

Von Mises, Richard
    von Mises' condition, 229

Wald, Abraham
    Mann-Wald theorem, 194
weak convergence
    of distribution functions, 181, 182
    of increasing functions, 182
    of probability measures, 181, 182

weak law
    of large numbers, 100
Weibull, Waloddi
    reverse-Weibull distribution, 208

Printed in the United States
by Baker & Taylor Publisher Services